W9-CNP-415

Synthesis and Characterization of Advanced Materials

ACS SYMPOSIUM SERIES **681**

Synthesis and Characterization of Advanced Materials

Michael A. Serio, EDITOR
Advanced Fuel Research, Inc.

Dieter M. Gruen, EDITOR
Argonne National Laboratory

Ripudaman Malhotra, EDITOR
SRI International

Developed from a symposium sponsored by the Materials Chemistry Secretariat at the 212th National Meeting of the American Chemical Society, Orlando, Florida, August 25–29, 1996

American Chemical Society, Washington, DC

Library of Congress Cataloging-in-Publication Data

Synthesis and characterization of advanced materials / Michael A. Serio, Dieter M. Gruen, Ripudaman Malhotra, editors.

p. cm.—(ACS symposium series 681)

"Developed from a symposium sponsored by the Division of Materials Chemistry Secretariat, at the 212th National Meeting of the American Chemical Society, Orlando, Fla., August 25–29, 1996."

Includes bibliographical references and indexes.

ISBN 0–8412–3540–6

1. Materials.

I. Serio, Michael A., 1956– . II. Gruen, Dieter M., III. Malhotra, Ripudaman IV. American Chemical Society. Division of Materials Chemistry Secretariat. V. American Chemical Society. Meeting (212th: 1996: Orlando, Fla.) VI. Series.

TA403.6.S88 1997
620.1'1—dc21

97–46577
CIP

This book is printed on acid-free, recycled paper.

PRINTED IN THE UNITED STATES OF AMERICA

Advisory Board

ACS Symposium Series

Foreword

THE ACS SYMPOSIUM SERIES was first published in 1974 to provide a mechanism for publishing symposia quickly in book form. The purpose of the series is to publish timely, comprehensive books developed from ACS sponsored symposia based on current scientific research. Occasionally, books are developed from symposia sponsored by other organizations when the topic is of keen interest to the chemistry audience.

Before agreeing to publish a book, the proposed table of contents is reviewed for appropriate and comprehensive coverage and for interest to the audience. Some papers may be excluded in order to better focus the book; others may be added to provide comprehensiveness. When appropriate, overview or introductory chapters are added. Drafts of chapters are peer-reviewed prior to final acceptance or rejection, and manuscripts are prepared in camera-ready format.

As a rule, only original research papers and original review papers are included in the volumes. Verbatim reproductions of previously published papers are not accepted.

ACS BOOKS DEPARTMENT

Contents

POLYMERIZATION SYNTHESIS AND CHARACTERIZATION

INDEXES

Preface

A WIDE VARIETY OF CHEMICAL APPROACHES can be brought to bear on the synthesis and characterization of new and technologically advanced materials. Advanced materials are usually defined as those that have superior properties, are well characterized, and have been obtained by careful control of the synthesis process. Examples of such materials include diamonds, fullerenes and other carbon materials, glasses, ceramics, semiconductors, polymers, and composites.

Materials science is an area of national priority from both a scientific and economic perspective. The development of new materials is the basis of many emerging high-technology industries (e.g., semiconductors and ceramics) as well as a key element in the transformation of traditional industries (e.g., automotive and plastics). In 1991, the American Chemical Society (ACS) recognized the importance of materials science and the central role of materials chemistry by establishing the Materials Chemistry Secretariat (MTLS). The MTLS provides a forum for 14 member ACS divisions to present work in the interdisciplinary field of materials chemistry.

This volume is based on an MTLS symposium titled "High-Temperature Synthesis of Materials", at the 212th National Meeting of the American Chemical Society in Orlando, Florida, August 25–29, 1996. The symposium included 55 papers that covered various aspects of materials synthesis and characterization, and it was organized into five major subtopics: thin films, diamond and group III nitrides, fullerenes and carbon materials, ceramics and catalysts, and polymeric materials. This volume contains 21 chapters that are representative of these major groupings. In order to best reflect the content of the chapters and to make the book more useful for those new to the field, this volume has been organized according to the types of synthesis processes rather than the type of material.

A plenary session was held and included overview talks by scientists from the U.S. Government, national laboratories, and universities on some key areas of research in materials chemistry. Four of these six plenary talks appear as chapters in this volume, including a chapter by Sir Harold W. Kroto, who recently shared the Nobel Prize in Chemistry for his contributions to the discovery of fullerenes.

This volume should be of interest to chemical scientists actively engaged in materials research as well as those considering entering this field who want to have an overview of the major types of materials synthesis processes and characterization methods. The book will also be useful to graduate and undergraduate chemistry students by giving them information on the roles they can play in the emerging high technology fields of materials chemistry.

The symposium was sponsored by the ACS Materials Chemistry Secretariat and was cosponsored by the ACS Committee on Science and the following ACS Divisions: Physical Chemistry, Colloid and Surface Chemistry, Inorganic Chemistry, Inc., Fuel Chemistry, and Polymer Chemistry, Inc.

Acknowledgments

We thank the many individuals and organizations who were responsible for the symposium and this volume. The ACS Committee on Science and the ACS Materials Chemistry Secretariat provided encouragement and financial support. Additional financial support was provided by several organizations interested in materials chemistry, namely Advanced Fuel Research, Inc.; BuckyUSA; Corning, Inc; the Electric Power Research Institute; the Materials and Electrochemical Research Corporation; NAMAR Scientific, Inc; On-Line Technologies, Inc; SRI International; and TDA Research.

The authors and reviewers of the chapters deserve special thanks for their contributions. In addition, we thank the staff at ACS Books, Anne Wilson, Vanessa Johnson-Evans, and Tracie Barnes, for their help and encouragement. Karin Dutton of Advanced Fuel Research, Inc., provided invaluable assistance in the preparation for the symposium and this volume.

MICHAEL A. SERIO
Advanced Fuel Research, Inc.
87 Church Street
East Hartford, CT 06108

DIETER M. GRUEN
Materials Science and Chemistry Divisions
Argonne National Laboratory
9700 South Cass Avenue
Argonne, IL 60439

RIPUDAMAN MALHOTRA
Organic Materials and Energy Department
SRI International
333 Ravenswood Avenue
Menlo Park, CA 94025

August 18, 1997

Overview

Chapter 1

Chemistry of Advanced Materials

Dieter M. Gruen

Materials Science and Chemistry Divisions, Argonne National Laboratory, 9700 South Cass Avenue, Argonne, IL 60439

With the end of the Cold War and the increasing globalization of the U.S. economy, R&D expenditures must be increasingly justified based on future economic benefits. In the case of materials science, this justification is not difficult to make, since the results of these efforts are often closely linked on a fairly short time scale to important new industrial processes and products. In large measure, this is due to the fact that the hallmark of the discipline is its dependence on a mixture of basic and applied research and on interdisciplinary borrowing. In the organization of this volume, subject areas of current technological interest were chosen to provide an understanding of the important roles of material scientists, in general, and materials chemists, in particular, in the synthesis and characterization of advanced materials.

The U.S. economy is experiencing fundamental structural change in the wake of the Cold War and the increasing globalization of economic activity. These changes are having far-reaching effects on the nation's science and technology enterprise. The complex issues associated with the role of technological innovation in economic growth and change need to be better understood and addressed. One of the factors complicating this task is that we can no longer justify large expenditures in scientific research in Cold War terms. From the point of view of the business community, these expenditures rarely -- if ever -- yield returns that reflect favorably in a corporation's near term earnings report. In the longer term, substantial profits may be made but, by that time, the scientific discovery that gave rise to the new business opportunity -- be it the transistor, the diode laser, carbon fibers, or nylon -- is lost sight of. The connection between the discovery and the profit derived therefrom has been lost and cannot be forged anew in a convincing way.

That is the dilemma in which the scientific community finds itself today. Materials scientists and materials chemists, in particular, understand very well the role they play in inventing new technologies from which future economic benefits will be derived. They also know very well that, without sufficient investments in skills, research and development, infrastructure, plants, and equipment, the whole edifice of our complex society will be undermined.

The scientific community can only hope that policy makers and society as a whole will come to recognize that support of the scientific enterprise is absolutely crucial to continued economic prosperity. Of course, scientists can help shape the course of events by communicating to others their deeply held convictions. The materials chemistry community can argue strongly for this point of view because the results of that scientific effort are often closely and demonstrably linked on a fairly short time scale to important new industrial processes and products. In large measure, this is due to the fact that the hallmark of the discipline is its dependence on a mixture of basic and applied research and on interdisciplinary borrowing. These characteristics -- coupled with intense personal motivation -- lead to high levels of scientific innovation.

Materials chemistry applies the insights of chemical thermodynamics, kinetics, and quantum mechanics to problems in material science. In the organization of this volume, it was felt that subject areas of great current technological interest should be chosen which illustrate the important contributions of materials chemists. The subjects covered include the synthesis and characterization of thin films, Group III nitrides, fullerenes and other carbon materials, ceramics, catalysts, and polymeric materials, covering a broad spectrum of interdisciplinary activities.

Rather than in bulk form, more and more materials are finding wide-ranging uses as films varying in thickness from a few nanometers to many microns. Quantum well, multi-layer and thin film composites with diamond-like hardness are just a few of the exciting "atomically engineered" materials that are transforming the approaches to modern materials science. Materials chemistry is fulfilling an important function by moving many of these areas closer to practicality through improvements in synthesis processes and in the performance characteristics of advanced materials.

For example, although the story of carbon is as old as mankind itself, it is not fully told, and new chapters -- sometimes new volumes -- are written from time to time. We appear to be in a period of a rapidly evolving carbon science right now. Not only is diamond film growth by chemical vapor deposition (CVD) an active area of research and development, but the discovery of fullerenes -- the third allotrope of carbon -- has opened up a whole new field of science with vast potential for both basic knowledge and applications. The paper that follows, authored by Professor Sir Harold W. Kroto, discusses the profound implications of the discovery of C_{60} on the way we think about the structure of graphite and other layered materials. Because an important branch of chemistry -- organic chemistry -- deals with carbon and its compounds, carbon as a material quite naturally arouses the interests of chemists, particularly of materials chemists.

There can be no doubt that the discovery of the fullerenes by Kroto, Smalley and Curl, and their collaborators at Rice University in 1985, was one of the great moments in carbon chemistry. The recipe given us by Krätschmer and Hoffman in 1990 for the synthesis of macro amounts of C_{60} has given rise not only to the new science of fullerene chemistry, but also to an intensive search for other new forms of carbon. Carbon nanotubes and carbon onions are examples of forms of carbon that owe their existence to the curiosity caused by C_{60}. The vast, and hitherto unknown, ways in which carbon atoms can arrange and rearrange themselves in space continues as a subject of intense interest in the scientific community, along with the possibilities for using these materials in novel applications.

One of the outstanding problems confronting the large-scale utilization of fullerenes is the cost associated with present-day methods of production based largely on carbon arc techniques. New -- and hopefully more cost-effective -- approaches to the synthesis of the fullerenes will have to be found. Those currently being studied range from the use of solar rather than electric energy to the use of premixed hydrocarbon/oxygen flames to the pyrolysis of hydrocarbon precursors and plasma-enhanced CVD methods. The wonder is that, once the fullerenes were shown to exist, they now show up everywhere, even in nature. The synthesis of the fascinating boron- and nitrogen-doped fullerenes, as well as of endohedral metal-doped fullerenes has been accomplished. C_{60} can also serve as a "combinatorial pincushion" for the efficient synthesis of new drugs.

Research on the Group III nitrides has seen a spectacular increase in interest in the last few years as a result of new uses that have been discovered for these materials. Interest has centered on cubic boron nitride for surface hardness, on A1N for its thermal conductivity, and on GaN as a wide bandgap semiconductor.

Another forefront area in materials science, to which materials chemistry is making important contributions, is research on oxides, nitrides, and carbides. Although known to inorganic and physical chemists for many decades, these materials are coming into their own in high-technology applications. Such applications depend critically on one's ability to tailor structure and properties in such a way as to optimize the interplay between form and function. A very important use of these materials is as catalysts. Special structural features of metal vanadomolybdates, for example, enhance their function as selective oxidative dehydrogenation catalysts. Transition metal carbides and nitride catalysts are of interest because of their resistance to poisoning. It is important to develop new methodologies for preparing catalysts with better activity, selectivity, and long term stability. Catalysis was one of the first technological areas that required close cooperation between materials chemists and other materials scientists.

A particularly challenging emerging area is that of nanoparticles and particularly the synthesis of nanoscale oxide powders. Many properties -- mechanical, electrical, and optical -- of nanoscale materials are very different from polycrystalline materials composed of micron-sized crystallites. The reason is that the surface free energy becomes important, even in determining phase stability. Several percent of the atoms making up the material reside at the grain boundaries with profound consequences for material properties. Again, as in the case of the fullerenes, in order to find applications, cost-effective methods of production must be found. The insights of materials chemists are therefore vitally important in devising new, more efficient methods of synthesis. These methods, in addition to being more efficient, should also enable one to have better control over the nanaocrystallinity, and hence over the properties of the material.

Nucleation and formation mechanisms, of course, form the basis for understanding the synthesis of any nanocrystallite, whether produced at relatively low temperatures from solutions or at higher temperatures, as in flames. Efficient solvents and process conditions have been developed for the synthesis of ceramic powders, such as alpha aluminum oxide, at sizes ranging from tens of microns to tens of nanometers. The synthesis of nano powders such as β"-alumina, SiO_2 and TiO_2 by flame spray, bulk pyrolysis, and combustion spray techniques has been accomplished by invoking new methodologies which draw heavily on the insights most familiar and available to materials chemists. All of these high temperature methodologies will profit from improved spectrometry diagnostic capabilities, which are now available.

A discussion of the chemistry of advanced materials would be incomplete without the inclusion of polymeric materials. This is a classic illustration of an area where materials chemists can make important contributions to materials science because of the preeminent role of chemistry in the polymerization step. In some cases, polymers are used as precursors for other types of advanced materials. The synthesis of aluminosilicates from alkoxide precurors is a case in point. To accomplish this task requires, first of all, the synthesis and characterization of the polymer alkoxide precursors. In most cases, a polymeric material is the end result, as with the synthesis of high temperature polymers (e.g., thermosets)

The interdisciplinary nature of materials science and the central role of chemistry has been clearly demonstrated in this volume by bringing together in a cohesive way a wide diversity of topical areas which are of current interest in materials science. It is hoped that volumes of this type will promote an understanding of the important roles of material scientists, in general, and materials chemists, in particular, in the synthesis and characterization of advanced materials.

Acknowledgments

The author gratefully acknowledges the support of this work by the U.S. Department of Energy, BES Materials Sciences, under contract number W-31-109-ENG-38.

Chapter 2

New Horizons in the Structure and Properties of Layered Materials

Harold W. Kroto

Physics and Environmental Science, School of Chemistry, University of Sussex, BN1 9QJ Brighton, United Kingdom

The discovery that C_{60} *(buckminsterfullerene)* self-assembles from a condensing chaotic plasma causes us to look back and recognize certain misconceptions that existed over the structure of graphite and other similarly layered materials. The discovery has not only fundamentally changed our understanding of synthetic carbon chemistry, but has also opened up whole new and exciting possibilities in materials science - in particular at nanoscale dimensions. Furthermore, this new perspective also rationalizes numerous properties of bulk graphite that have hitherto been known but unexplained.

In 1985, the proposal that the stable C_{60} cluster, detected during laser ablation of graphite, might be a closed spheroidal cage (*1*) was greeted, by some, with skepticism. After all, most chemists had been brought up to take for granted the "fact" that carbon had an innate propensity to form flat sheets of carbon atoms in hexagonal arrays stacked to form graphite sandwich-like micro-crystals. Such a traditional image had been propagated throughout the standard literature and textbooks. Indeed, few appear to have given this received wisdom a second thought.

If, however, the C_{60} cluster, which had been created spontaneously from the chaos of a hot carbon plasma as it cooled, was indeed a closed spheroidal cage as proposed (*1*), then it clearly indicated that the whole concept of the intrinsic flatness of graphite might need some re-evaluation. Perhaps, flatness was not an invariable property of carbon after all. It soon became apparent, after a little careful deliberation, that on the scale of a few tens to a few hundreds of carbon atoms, graphite-like sheets of hexagonally arrayed carbon atoms would possess significant instabilities (dangling bonds) at the edges which might be relieved by closure into a cage network. In the bulk, edge instabilities would be insignificant but, on a small scale, the need to satisfy the dangling bonds would probably be decisive and lead to energy-driven closure. [We shall not here deal with the important question of how bulk (mainly flat) structures might arise]. Euler's law applies to such structures and thus closure could only be achieved by such flat (hexagonal network) graphite sheets with the inclusion of non-hexagonal disclinations. In 1966, Jones suggested that 12 pentagonal disclinations would lead to closure resulting in graphene network balloons (*2*), and Osawa and Yoshida (*3,4*) described the possibility that C_{60} itself might be stable. The breakthrough in 1990 by Krätschmer, Lamb, Fostiropoulos and Huffman (*5*), who managed to produce C_{60} on a macroscopic scale, has, at a stroke, brushed away any lingering doubts about the structure of C_{60}. It also reinforced the need to recognize that the intrinsic behavior of extended carbon arrays was not well understood - at least on a microscopic scale.

Those who have looked at graphitic materials carefully will have found that the flat sheets of graphite, extending essentially infinitely, are actually rather rare. Perhaps "perfect", or near-perfect, *or* even anywhere-near-perfect macroscopic flat stacked graphite objects, which really are isolated single crystals, may not actually exist. Sheet-like material with relatively large domains of near-perfect graphite can sometimes be found in mines such the Ticonderoga Mine.

However, after the original fullerene proposal was made, Iijima's early transmission electron microscope (TEM) studies of concentric graphitic shell structures (*6*) became a focus of interest (*7*), as they appeared to bear some relation not only to the structure of C_{60} but also to the mode of its formation. Iijima's TEM studies were recognized as giant graphite quasi-crystals (*7,8*) and later work by Ugarte (*9*) shed further light on these objects. Essentially, each crystal consists of a set of concentric giant fullerene shells in an onion-like infrastructure and TEM simulations (*8*) confirmed this new picture.

However, one might ask "What about the traditional graphite sandwich single crystal?" It was mainly researchers who specialized in carbon materials who were aware of the complex nature of the problem and how difficult it really is to describe the many apparently different forms of graphite-related carbons that appear to exist. In general, almost all experiments involving graphite purporting to involve perfect graphene surfaces hare been carried out on HOPG (highly ordered pyrolytic graphite) - a rather flaky material which actually consists of myriads of aligned small graphite domains each of which consists of a stack of relatively flat graphene sheets. Ticonderoga graphite consists of relatively large domains but appears to be a rather rare material. *[Certainly this author has never seen material which could be described as a single crystal of graphite of the kind often depicted schematically in traditional textbooks.]* Graphite is usually made by thermolysis of various types of highly aromatic feedstocks, such as mesophase pitches, which often consist of large more-or-less planar aromatic molecules which have presumably lined up in more-or-less planar stratified order, prior to reaching the solidification stages in the production process.

In HOPG production, it is the existence of H, OH and other terminating (non-carbon contaminants) at the edges of the polycyclic aromatic precursor species which tends to ensure that relatively large areas of fairly planar graphite can form. Presumably, edge effects do not become important until very large planar regions have already developed and then the graphene sheets are so large and interplanar forces are so cumulatively strong that they have little impact on the gross structure. The major breakthrough was made by Smalley and co-workers [*10*] in which the laser vaporization techniques were developed which enabled small refractory clusters to be made and studied for the first time. In general, when such small aggregates of refractory atoms form, the reconstruction effects that govern the arrangements of atoms at surfaces become the dominant structure-controlling factors. Upon laser vaporization of graphite, pure carbon *molecules* form *in the gas phase*. In the case of graphene sheets and other small planar species, an *edge* reconstruction analogue of the well studied surface reconstruction effect appears to be all-important and may be a major factor in the resulting creation of C_{60}.

With hindsight it all appears to be really rather obvious, but the story only goes to show how easy it is for such a simple result to remain overlooked, even though the evidence appears to have been staring us in the face for decades. There are thus major implications for the creation of extended carbonaceous materials. The overall dynamics of such processes depends on a myriad of nanoscale restructuring events which take place under conditions where edge effects are ever-present and impact on the resulting structure. The growth of a graphite "crystal" is thus quite different from normal crystallization processes which occur when a melt solidifies and ions or atoms lock into place at a vacant surface site under the influence of the highly anisotropic potential that exists at a nucleation site. In the case of graphite, a carbon atom or a carbon aggregate forms a covalent bond at an edge by some chemical condensation reaction. The intermediate structure is subject to the whole range of competing covalent carbon-carbon bonding factors that occur locally at the moment of condensation. These might be very complicated and highly variable depending on the degree of unsaturation, the temperature of the process, the level of hydrogenation (in the case of thermolytic dehydrogenation of aromatic precursors) as well as (of course) the structure of the precursor, etc.

Finally, one should note that the fullerene cage concept is not just confined to carbon. Studies of boron nitride indicate that it too can form nanotubes [*11,12*]. Furthermore, most interestingly, Tenne and coworkers have shown that molybdenum and tungsten sulphides - long known to form layered materials also produce fullerene-related giant quasi-crystals [*13,14*].

Summary

The proposal that C_{60} forms spontaneously was greeted with some skepticism in 1985 when it was first made and it is difficult now to recreate the climate in which so many had so much difficulty in accepting that carbon would form so elegant and symmetric an object which was so obviously *non-flat*. In addition to C_{60}, there was fleeting evidence of smaller structures [*15*] as well as larger ones such as the giant fullerenes [*16*]. Molecular model investigations of the latter indicated that they would not be smooth round spheres like Buckminster Fuller's domes. The study yielded a crucial result: in large cage structures, the curvature is focused in the region near the 12 pentagons. In the case of structures in which the 12 pentagonal disclinations (necessitated by the Euler Closure Criterion) are symmetrically distributed among the hexagons, an icosahedral shape results. This general shape prediction finds elegant confirmation in the electron microscope images of nanotubes, in which the 12 pentagons are split so that they are located in 6-packs and, as a result, *elongated* quasi-icosahedral shapes are produced, i.e., the nanotube [*17-19*].

Thus we see that the picture of graphitic materials must be tempered by a clear understanding of the conditions which are obtained during formation. Perusal of the literature on such materials indicates that round carbon structures are really quite ubiquitous, but the intrinsic details of the infrastructure were poorly understood and the driving factors governing the dynamics of the growth misunderstood. The discovery of C_{60} has changed all this. In particular, the relatively efficient self-assembly of a closed all-carbon cage, as a chaotic carbon plasma constrained by an inert gas bath cools, was totally unexpected and causes us to re-assess our assumptions about the factors governing the growth of graphite, particularly in the absence of epitaxial factors. It now appears that, at least when we discuss a 60 atom cluster of carbon in a graphitic sheet arrangement, the most stable configuration is a closed one in which the energy associated with the 20 or so dangling bonds that reside at the edge of a flat sheet have been accommodated by closure. We still do not completely understand the details of the self-assembly of C_{60}, but the edge energy appears to be a decisive factor. These factors are intimately involved in all aspects of graphitic carbon formation, but only now are recognized as the major force governing the structure and dynamics.

Acknowledgments

I wish to acknowledge many helpful discussions with Tony Cheetham, Morinobu Endo, Lawrence Dunne, Jon Hare, Wen-Kuang Hsu, Douglas Reid, Mauricio Terrones, Daniel Ugarte, David Walton. I also wish to thank the Royal Society, Alfred Bader, Peter Doyle (Zeneca) and EPSRC for support.

Literature Cited

1) Kroto, H. W.; Heath, J. R.; O'Brien, S. C.; Curl, R. F.; Smalley, R. E. *Nature (London)* **1985**, *318*, 162-163.

2) Jones, D. E. H., *New Scientist (3rd Nov 1966)* p 245; *The Inventions of Daedalus*; Freeman: Oxford, 1982; pp. 118-119.

3) Osawa, E. *Kagaku (Kyoto)* **1970**, *25*, 854-863 (in Japanese); *Chem. Abstr.* **1971**, *74*, 75698v.
4) Yoshida, Z.; Osawa, E. *Aromaticity*; Kagakudojin: Kyoto, 1971, pp. 174-178, (in Japanese).
5) Krätschmer, W.; Lamb, L. D.; Fostiropoulos, K.; Huffman, D. R. *Nature (London)*, **1990**, *347*, 354-358.
6) Iijima, S. *J. Cryst. Growth*, **1980**, *5*, 675-683.
7) Kroto, H. W. and McKay, K. G. *Nature*, **1988**, *331*, 328-331.
8) McKay, K. G., Wales, S. W. and Kroto, H. W., J. Chem Soc. Farad Trans *in press*.
9) Ugarte, D., *Nature* **1992**, *358*, 707.
10) Dietz, T. G.; Duncan, M. A.; Powers, D. E.; Smalley, R. E. *J. Chem. Phys.* **1981**, *74*, 6511-6512.
11) Terrones, M.; Hsu, W. K.; Terrones, H.; Zhang, J. P.; Ramos, S.; Hare, J. P.; Castillo, R.; Cheetham, A. K.; Kroto, H. W.; Walton, D. R. M. *Chem. Phys. Letts*. **1996**, *259*, 568-573.
12) Chopra, N. G.; Luyken, R. J.; Cherrey, K.; Crespi, V. H.; Cohen, M. L.; Louie, S. G.; Zettl, A., *Science*, **1995**, *269*, 966.
13) Tenne, R., *Adv. Mat*. **1995,** *7*, 965-995.
14) Feldman, Y.; Wasserman, E.; Srolovitz, D. J.; Tenne, R. *Science*, **1995,** *267*, 222.
15) Kroto, H. W. *Nature (London)* **1987**, *329*, 529-531.
16) Kroto, H. W. *Chem. Brit.* **1990**, *26*, 40-45.
17) Iijima, S., *Nature*, **1991**, *354*, 56-58 (see also: Ball. P., *Nature*, **1991**, *354*, 18).
18) Ebbesen, T. W.; Ajayan, P. M. *Nature*, **1992**, *358*, 220-222.
19) Dresselhaus, M. S.; Dresselhaus, C.; Eklund, P. C. *Science of Fullerenes and Carbon Nanotubes,* Academic Press: London, 1996.

Chemical Vapor Deposition Synthesis and Surface Characterization

Chapter 3

Chemical Considerations Regarding the Vapor-Phase Epitaxy of Binary and Ternary III-Nitride Thin Films

Robert F. Davis, Michael D. Bremser, Ok-Hyun Nam, William G. Perry, Tsvetanka Zheleva, and K. Shawn Ailey

Department of Materials Science and Engineering, Box 7907, North Carolina State University, Raleigh, NC 27695–7907

Monocrystalline GaN(0001) films were grown via OMVPE at 950°C on AlN(0001) deposited at 1100°C on α(6H)-SiC(0001)$_{Si}$ substrates. $Al_xGa_{1-x}N$ films ($0 \leq x \leq 1$) were deposited at 1100°C directly on SiC. X-ray rocking curves for 1.4 μm GaN(0004) revealed FWHM values of 58 and 151 arcsec for materials simultaneously grown on the on-axis and off-axis SiC, respectively. Silicon donor-doping in highly resistive GaN and $Al_xGa_{1-x}N$ (for $x \leq 0.4$) was achieved for net carrier concentrations ranging from approximately 2×10^{17} cm^{-3} to 2×10^{19} ($Al_xGa_{1-x}N$) or to 1×10^{20} (GaN) cm^{-3}. Mg-doped, p-type GaN was achieved with $n_A - n_D \approx 3 \times 10^{17}$cm^{-3}, $\rho \approx 7$ Ω·cm and $\mu \approx 3$ cm^2/V·s.

The numerous potential and recently realized commerical applications of the III-N materials has prompted considerable research regarding their growth, characterization and device development. Gallium nitride (wurtzite structure), the most studied of these materials, has a room temperature band gap of 3.39 eV and forms continuous solid solutions with both AlN (6.28 eV) and InN (1.95 eV). As such, materials with engineered direct band gaps are feasible for optoelectronic devices tunable in wavelength from the visible (600 nm) to the deep UV (200 nm). The relatively strong atomic bonding and wide band gaps of these materials also points to their potential use in high-power and high-temperature microelectronic devices. Selected thin film alloys with engineered bandgaps and p-n junction, double heterostructure and quantum well blue and green light emitting diode (LED) structures (*1-7*) and blue laser diodes (*8*) containing these compounds and alloys have been produced and either are or soon will be commercially available. High electron mobility transistors (*9*), heterostructure field-effect transistors (*10*), metal-semiconductor-field-effect transistors (*10-13*) and surface acoustic devices (*14*) have also been reported. Concomitant with the realization and/or optimization of these devices is the need for improved film quality.

Bulk single crystal wafers of AlN and GaN are not commercially available (*15*); therefore, heteroepitaxial films must be grown. The principal method of deposition of these films is organometallic vapor phase epitaxy (OMVPE); however, gas source molecular beam epitaxy (GSMBE) is being increasingly employed. Sapphire(0001) is the most commonly used substrate, although its a-axis lattice parameter and coefficients of thermal expansion are significantly different from that of any of the nitrides. It was first observed by Yoshida *et al.* (*16,17*). that the electrical and luminescence properties of GaN films grown via reactive MBE improved markedly when an AlN "buffer layer" was initially deposited on the sapphire(0001) substrate. Amano *et al.* (*18,19*) and Akasaki *et al.* (*20*) were the first to use an AlN buffer layer on the same substrate for improving MOVPE grown GaN. By decreasing the growth temperature of this intermediate layer from 1000°C to 600°C, further improvements in surface morphology and the electrical and luminescence properties of the GaN were realized (*19,20*) by these investigators. The use of an AlN buffer layer has also been used in the present research; however, the substrate was 6H-SiC(0001)$_{Si}$ and the growth temperature was 1100°C.

Ternary $Al_xGa_{1-x}N$ solid solutions are ideally suited for UV emitters and photodetectors in the range of 365 nm (x = 0) to 200 nm (x = 1). A negative electron affinity (NEA) effect has been observed in AlN and $Al_xGa_{1-x}N(x \geq 0.5)$ (*21*). As such, these materials may prove applicable as cold-cathode emitters for flat panel displays. $Al_xGa_{1-x}N$ alloys have also been used for cladding layers for the InGaN-based LEDs (*22*). Conductive $Al_xGa_{1-x}N$ buffer layers which can be grown directly on the substrate and which do not compromise the resultant GaN film quality are sought and constitute a part of the research reported herein.

An important and, from an historical perspective, surprising result of the deposition of pure GaN on buffer layers or the deposition of $Al_xGa_{1-x}N$ directly on a given substrate is the achievement of insulating films which can be controllably doped. The incorporation and activation of the n-type dopants of Si and Ge are easily achieved during film growth of both materials; the activation of the p-type dopants of Mg, Zn or Cd are correspondingly difficult. The realization that sample heating and the concomitant dissociation of H-acceptor dopant complexes (*23*) allowed the achievement of controlled p-type doping via post growth annealing in N_2 paved the way for the development of the LEDs noted above. The doping of $Al_xGa_{1-x}N$ has proved more difficult. Only the investigations of Tanaka *et al.* (*24*) and those of the present investigators have resulted in n-type doping of the alloys; p-type doping of these materials has not been previously reported.

The recent and considerable progress accomplished in these areas in the intervening years has been reviewed in Refs. (*25-29*). In the present research, GaN and $Al_xGa_{1-x}N$ films were deposited either on monocrystalline high-temperature (HT) AlN buffer layers previously deposited on vicinal and on-axis 6H-SiC(0001) substrates (GaN) or directly on the SiC ($Al_xGa_{1-x}N$ alloys). The n- and p-type doping using Si and Mg dopants, respectively, has also been accomplished. Extensive chemical, optical, microstructural and electrical characterization of the binary and ternary films has also been conducted. The following sections present an overview of the precursors which have been used by various groups to deposit III-N films, describe the experimental procedures and results and provide a discussion and conclusions regarding the research of the present authors.

MOVPE Growth of III-N Materials—Brief Review of Precursors

To provide an enhanced chemical perspective for this paper, a brief review of the variety of precursors which have been employed for MOVPE growth of III-N films is presented because of their importance to the quality of the films which are addressed in the subsequent sections. Neumayer and Ekerdt (*30*) have co-authored an excellent, in-depth review of this effort to early 1995 from which some of the following information is taken. The most commonly employed precursors are trimethylalluminum (TMAl), trimethylgallium (TMG), trimethylindium (TMIn) and ammonia. The ethyl analogues, (TEAl) and (TEGa) are less commonly used. All of these trialkyl compounds are sufficiently volatile (though the lines carrying TEAl must be heated to prevent condensation) and reactive with ammonia to produce high quality films with very small concentrations of impurities. However these compounds react readily with water and oxygen. By contrast, ammonia has a poor cracking efficiency (*31*), forms adducts with the trialkyls with lowered volatility or homogeneous nucleation of the nitride in the gas phase in poorly controlled flow systems and is very corrosive. However, it is normally employed in substantial quantities, especially when In incorporation is desired in the growing film. It is to address these concerns that new precursors have been developed and used for the deposition of these compounds.

Sauls *et al.* (*32*) have reviewed the available literature concerning the successive displacement surface reactions occurring during the formation of III-N films from the alkyls and ammonia. These reactions are illustrated for the formation of AlN as follows; however, analogous reactions are expected for the formation of GaN (*33*) and InN. Additional details regarding the formation of AlN have been reported by Liu *et al.* (*34*).

$$(CH_3)_3Al + NH_3 = (CH_3)_3Al{\bullet}NH_3 \qquad \text{ammonia adduct} \quad (1)$$

$$(CH_3)_3Al{\bullet}NH_3 = CH_4 + [(CH_3)_2AlNH_2]_n \qquad \text{amide} \quad (2)$$

$$[(CH_3)_2AlNH_2]_n = CH_4 + [CH_3AlNH]_n \qquad \text{imide} \quad (3)$$

$$[CH_3AlNH]_n = CH_4 + AlN \qquad \text{nitride} \quad (4)$$

The largest concern in the use of the methyl alkyls is the introduction of C into the growing nitride film. Metal-containing triethly and triisopropyl alkyls have been used to reduce this problem. These sources pyrolyze below 250°C. The reduction in C levels has been attributed by Kuech *et al.* to the efficient β-hydride elimination process.

Nonpyrophoric, water and oxygen insensitive compounds with a reduced tendency to form adducts have been used to grow low oxygen III-V materials. Examples of such precursors are 1,3-[dimethylamino)-propyl]-1-gallacyclohexane, and 1,3-[dimethyl amino)-propyl]-1-alacyclo-hexane. The principal drawback to using these compounds is their low vapor pressure. Other compounds including trimethylamine alane and gallane are very reactive with ammonia in the gas phase and on an alumina surface resulting in trimethylamine, hydrogen and a $(H_xN_{metal}H_y)_n$ polymer. Thus, these precursors are more suitable for ultra high vacuum deposition techniques, e.g., MOMBE, than the higher pressure MOVPE.

Alternative sources to ammonia which have been employed to grow III-V nitride films include hydrazine (N_2H_4), NF_3, hydrogen azide (HN_3), isopropylamine ($^iPrNH_2$) and tert-butylamine ($^tBuNH_2$). Mackenzie *et al.* (*35*) obtained quasi-amorphous AlN films using isopropylamine and MOMBE. Essentially all the studies regarding the alkylamines for III-N deposition have shown that these are poor nitrogen sources in this application because of their poor decomposition efficiency which is related to the C–N bond. NF_3 has also not proven to be a good nitrogen source because of the ease of reaction in the gas phase and the incorporation of fluorides in the growing films, e.g., AlF_3 in AlN (*36*). Dissociation of N from hydrazine, hydrogen azide and 1,1-dimethylhydrazine can more easily occur than from ammonia, but the first two are so toxic and explosive that their use is limited for safety reasons, and the last precursor contributes C to the films.

The use of precursors which contain both the group III metal and N, e.g., triethylgallium monamine and selected amides have not been encouraging. The triethylgallium monamine was found to decompose to form an amide which further decomposed to form a material with a marked decreased in volatility (*37*). The use of the amides $[HAL(NR_2)_3]_2$ and $[HAL(NR_2)_2]_2$ {R = CH_3, C_2H_5}resulted in both significant C incorporation into the films and nonstoichiometry in terms of higher metal concentrations (*38*). Deposition studies with the dialkylaluminum azides (*39*) and diethylgallium azide (*40*) resulted in essentially stoichiometric, amorphous or highly oriented polycrystalline AlN and GaN films, but with C incorporation to ≈10%.

In summary, numerous precursors have been examined to find a low temperature route for the deposition of the group III nitrides. However, it is important to note that the surface diffusion rates of the product species of these reactions are low; thus elevated temperatures are necessary to achieve sufficient surface and growth kinetics to obtain monocrystalline thin films of these materials. Moreover, the films must be essentially free of C, as it may act as a deep level dopant and compensate the films, and semiconductor grade pure in terms of other elements. Only the triethyl- and to a lesser extent the trimethyl-species coupled with ammonia as the N source have allowed all of these materials parameters to be achieved. They are also the precursors which are receiving the most attention from the suppliers in terms of high purity and careful handling during transfilling of the user's containers. The following sections illustrate the use of these precursors for the growth of GaN and $Al_xGa_{1-x}N$ thin films.

Experimental Procedure

As-received vicinal (oriented 3°-4° off-axis toward the <11$\bar{2}$0>) and on-axis 6H-SiC(0001)$_{Si}$ wafers, the former containing an ≈1 μm n-type homoepitaxial layer, which had been thermally oxidized were cut into 7.1 mm squares. These substrates were dipped into a 10% HF solution for 10 minutes to remove the thermally grown oxide layer and blown dry with N_2 before being loaded onto a SiC-coated graphite susceptor contained in a cold-wall, vertical, pancake-style reactor. The continuously rotating susceptor was RF inductively heated to the AlN or $Al_xGa_{1-x}N$ deposition temperatures of 1100 and 1200°C (as measured optically on the susceptor) in 3 SLM of flowing H_2 diluent. Hydrogen was also used as the carrier gas for the various metalorganic reactants and dopants. Deposition of each AlN buffer layer or $Al_xGa_{1-x}N$ was initiated by flowing triethylaluminum (TEA), triethylgallium (TEG) and ammonia (NH_3) into the reactor at 23.6–32.8 μmol/min (total MO flow rate) and 1.5 SLM, respectively. The system pressure during growth was 45 Torr. Each AlN buffer layer was grown for 30 minutes resulting in a thickness of ≈100 nm. The TEA flow was subsequently terminated, the susceptor temperature decreased to 950°C and the system pressure increased to 90 Torr for the GaN growth. The flow rate of triethylgallium (TEG) was maintained at 24.8 μmol/min. The growth rate for GaN was ≈0.9 μm/hr. Silicon doped GaN and $Al_xGa_{1-x}N$ films were grown by additionally flowing SiH_4 (8.2 –12.4 ppm in a balance of N_2) at flow rates between ≈0.05 nmol/min and ≈15 nmol/min. Mg-doping of GaN was accomplished by introducing bis-cyclopentadienyl-magnesium (Cp_2Mg) at a flow rate of 0.2 μmol/min.

Scanning electron microscopy (SEM) was performed using a JEOL 6400FE operating at 5 kV and equipped with an Oxford Light Element Energy Dispersive X-ray (EDX) microanalyzer. Conventional and high resolution transmission electron microscopy (TEM) was performed on a Topcon EM-002B microscope operating at 200 kV. Double-crystal x-ray rocking curve (DCXRC) measurements were obtained using a Philips MR3 double-crystal diffractometer and Cu K_α. The photoluminescence (PL) and cathodoluminescence (CL) properties of the GaN and $Al_xGa_{1-x}N$ films, respectively, were determined at 8K using a 15 mW He-Cd laser (λ = 325 nm) or electron gun as the excitation source. Spectroscopic ellipsometry was performed using a rotating analyzer ellipsometer with a xenon arc lamp (1.5eV–5.75eV). The carrier concentrations and mobilities in the doped GaN and $Al_xGa_{1-x}N$ films were determined via Hall-effect measurements (Van der Pauw geometry) using a modified Keithley Model 80 equipped with a sensitive digital voltmeter (Keithley Model 182DMM). Thermally evaporated Al (GaN) and In ($Al_xGa_{1-x}N$) served as the ohmic contacts. Capacitance-voltage (CV) measurements were also conducted on the doped binary and ternary samples using a MDC Model CSM/2-VF6 equipped with a Hg probe.

Results and Discussion

Aluminum Nitride and Gallium Nitride. The surfaces of the 1100°C and 1200°C AlN buffer layers grown on the vicinal and on-axis 6H-SiC(0001)$_{Si}$ substrates had a smooth surface morphology. Reflection high energy electron diffraction studies indicated that these films were monocrystalline as-deposited. In contrast, RHEED results indicated that AlN deposited on 6H-SiC(0001)$_{Si}$ substrates in the range of 500°C-1050°C resulted in polycrystalline material.

Coalescence of GaN islands occurred on the monocrystalline HT-AlN buffer layers on vicinal and on-axis 6H-SiC(0001)$_{Si}$ substrates within the first several hundred angstroms of growth. The SEM image in Fig. 1 shows an intermediate stage of coalescence of ≈150Å thick GaN islands after one minute of growth. Growth subsequently occurred by a layer-by-layer mechanism. In contrast, for GaN film growth on low-temperature buffer layers on sapphire(0001) a similar growth scenario pertains but requires several thousand angstroms of deposition and crystallographic selection of the fastest growing plane before island coalescence and layer-by-layer growth result (*20,41*).

The GaN films deposited on the HT-AlN buffer layers on the on-axis 6H-SiC(0001)$_{Si}$ substrates had very smooth surfaces. A slightly mottled surface was observed for films deposited on vicinal 6H-SiC(0001)$_{Si}$ substrates, probably as a result of the higher density of steps coupled with the mismatch in the Si/C and Al/N bilayer stacking sequences at selected steps (*42*) on the growth surface of these substrate. For the vicinal and on-axis growth, there was no apparent differences in surface morphology between the GaN films deposited on the 1100°C and 1200°C AlN buffer layers.

The dislocation density within the first 0.5 μm of the GaN film on the vicinal 6H-SiC(0001)$_{Si}$ substrate was approximately 1×10^{9} cm^{-2}, as determined from initial plan view TEM analysis by counting the number of dislocations per unit area. This value is approximately an order of magnitude lower than that reported (*43*) for thicker GaN films deposited on sapphire(0001) substrates using low-temperature buffer layers. The dislocation density of the GaN film deposited on the vicinal 6H-SiC(0001)$_{Si}$ substrate decreased rapidly as a function of thickness. In contrast, the on-axis wafers had less step and terrace features; thus, the HT-AlN buffer layers on these substrates were of higher microstructural quality with smoother surfaces and fewer inversion domain boundaries. Consequently, the microstructural quality of the GaN films were better for on-axis growth as shown by the DCXRC data noted below.

DCXRC measurements taken on simultaneously deposited 1.4 μm GaN films on HT-AlN (1100°C) buffer layers revealed FWHM values to be 58 arc sec and 151 arc sec for deposition on the on-axis and off-axis 6H-SiC(0001)$_{Si}$ substrates, respectively. The FWHM values of the DCXRC values for the corresponding 100 nm AlN buffer layers were approximately 200 and 400 arc sec. These latter values were unchanged by increasing the growth temperature to 1200°C. A 2.7 μm GaN film deposited under identical conditions on a vicinal 6H-SiC(0001)$_{Si}$ substrate exhibited a FWHM value of 66 arc sec. The reduction in FWHM values is consistent with the decrease in the dislocation density as a function of thickness for GaN films grown on vicinal 6H-SiC(0001)$_{Si}$ substrates, as noted above.

The low-temperature (8K) PL spectra of the GaN films on both on-axis and vicinal 6H-SiC(0001)$_{Si}$ substrates showed strong near band-edge emission at 357.4 nm (3.47 eV). The FWHM values of these I_2 bound exciton peaks were 4 meV. The spectrum from the GaN film on the vicinal substrate revealed a very weak peak centered at ≈545 nm (2.2 eV), commonly associated with deep-levels (DL) in the band gap. A more intense 2.2 eV peak was observed in the GaN film grown on-axis.

Undoped high quality GaN films grown on HT-AlN buffer layers on both vicinal and on-axis 6H-SiC(0001)$_{Si}$ substrates were too resistive for Hall-effect measurements. Controlled n-type doping was achieved using SiH_4 for net carrier concentrations ranging from $\approx1\times10^{17}$ cm^{-3} to $\approx1\times10^{20}$ cm^{-3} in GaN films grown on vicinal 6H-SiC(0001)$_{Si}$ substrates. The net carrier concentrations and room temperature mobilities versus SiH_4 flow rate are plotted in Fig. 2. Films with a net carrier concentration of n_D-n_A = 2×10^{17}cm^{-3} had a room temperature Hall mobility of μ = 375 cm^2/V·s.

Mg-doped GaN films were deposited at 950°C on HT-AlN (1100°C) buffer layers on vicinal 6H-SiC(0001)$_{Si}$ substrates by introducing Cp_2Mg at a flow rate of

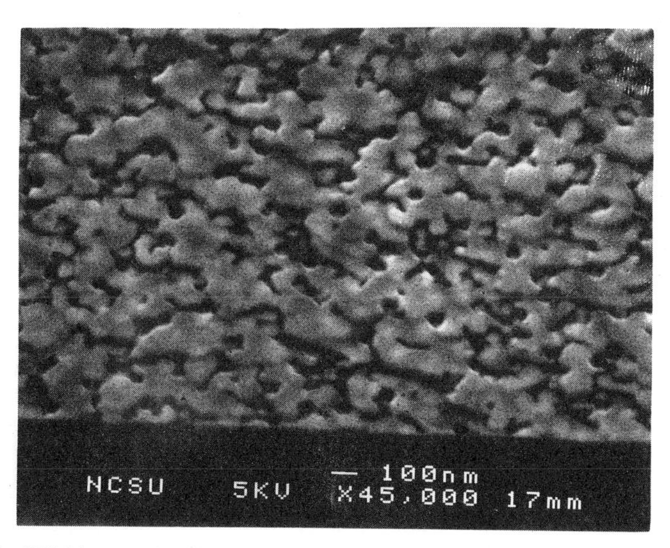

Figure 1. SEM image showing an intermediate stage of coalescence of ≈150Å thick GaN islands after 1 minute of growth on a monocrystalline AlN buffer layer deposited at 1100°C on a 6H-SiC(0001)$_{Si}$ substrate.

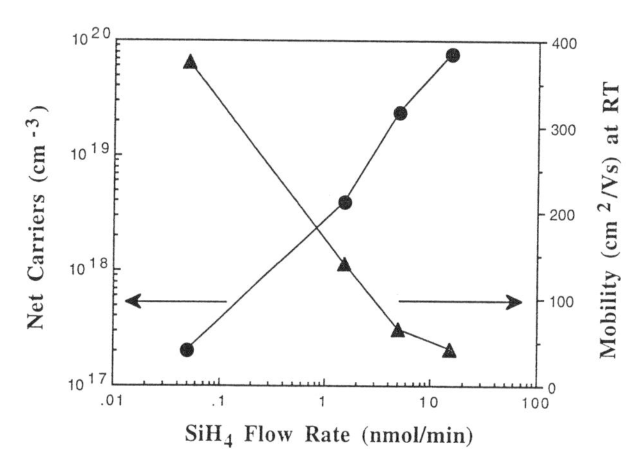

Figure 2. Net carrier concentration and room temperature mobilities in Si-doped, n-type GaN as a function of SiH_4 flow rate.

0.2 μmol/min. These samples were subsequently annealed at 700°C at 700 Torr in 3 SLM of N_2 for 20 minutes. These same samples were re-annealed at 900°C for 20 minutes under identical conditions. The PL (8K) spectral intensity of the blue emission was increased by the 700°C anneal and was dramatically decreased by the subsequent 900°C anneal. These results are similar to those reported by Nakamura *et al.*, (*44*). Hall-effect measurements made on the annealed samples revealed p-type GaN with a net hole carrier concentration of n_A-$n_D \approx 3\times10^{17}cm^{-3}$, a resistivity of $\rho \approx 7$ Ω·cm and a hole mobility of $\mu \approx 3$ $cm^2/V\cdot s$. Likewise, 4-point probe and Hg-probe C-V measurements verified p-type GaN.

$Al_xGa_{1-x}N$ Alloys. Thin films of $Al_xGa_{1-x}N$ ($0.05 < x < 0.96$) were deposited directly on both vicinal and on-axis 6H-SiC(0001) substrates. Films having $x < 0.5$ and deposited at 1100°C had smooth, featureless surfaces. Like GaN, smoother films were deposited on the on-axis substrates than on the vicinal ones. The surfaces of films with $x > 0.5$ were also very smooth but with occasional pits which increased in density with increasing Al composition as a result of the decreased surface mobility of the metal adatoms such that complete crystallographic coalescence of the two dimensional flat-top islands was inhibited (*43,45*). In order to eliminate this problem, deposition of the $Al_xGa_{1-x}N$ for $x > 0.5$ was conducted at 1150°C. The resulting films possessed smooth surfaces which were free of pits; however, the growth rate decreased by more than a factor of three. Therefore, deposition in the range of 1120-1130°C has been determined to be optimum for our system for these higher AlN concentrations.

The compositions of the films grown under different conditions were determined using EDX, AES and RBS. Standards of AlN and GaN grown in the same reactor under similar conditions were used for the EDX and AES analyses. Compositions were assigned to each film after carefully consideration of the errors (2 at.%) involved with each technique. The data from EDX and AES measurements showed excellent agreement. The RBS data did not agree well with the other two techniques due to small compositional variations through the thickness of the film. Simulation of the compositions determined by RBS was conducted only for the surface compositions. Analysis via EDX revealed that the $Al_xGa_{1-x}N$ grown on the on-axis SiC substrates tended to be 1-2 atomic percent more Al rich than those grown off-axis SiC. It is thought that the presence of steps on the growth surface promotes the adhesion of the gallium adatoms.

In Fig. 3, these compositions are compared with their respective CL near band-edge emission energies. Additionally, in Fig. 4, the CL emission peaks are compared with bandgap values obtained by scanning electron microscopy. Using a parabolic model, the following relationship describes the CL peak emission (I_2-line emission) (Eq. (1)) as a function aluminum mole fraction for $0 < x < 0.96$.

$$E_{I2}(x) = 3.50 + 0.64x + 1.78x^2 \quad (1)$$

Clearly, the model of the measurements shows a negative deviation from a linear fit. This is in agreement with earlier research by other investigators (*46-48*). However, these films are highly strained due to incomplete relaxation of the tensile stresses generated by the differences in the coefficients of thermal expansion. The effect of strain has yet to be quantified. It should be noted that other researchers have seen a linear relationship between composition and absorption edge for thick or relaxed films on sapphire (*49,50*). Comparison of the CL spectra of $Al_{0.12}Ga_{0.88}N$ grown on a pre-deposited 1000Å AlN buffer layer which is used for GaN deposition versus growth directly on 6H-SiC reveals a 75 meV red shift for the latter films. This indicates that the films deposited directly on SiC are in greater tension than the ones deposited using an AlN buffer layer. Furthermore, there is a noticeable decrease in the defect peak centered around 3.2 eV. The reason for the reduction in this peak is unclear at this time and is under investigation.

The low temperature (4.2K) CL of undoped $Al_xGa_{1-x}N$ films containing up to 0.96 mole fraction of aluminum exhibited near band-edge emission which has been

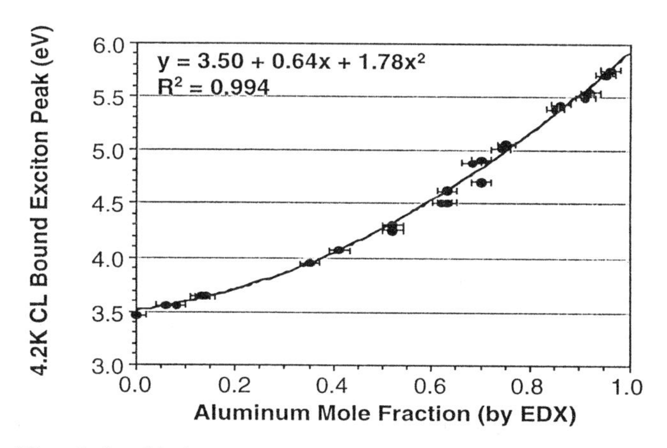

Figure 3. The relationship between aluminum mole faction and 4.2K CL near band-edge emission from AlGaN thin films deposited directly on vicinal and on-axis 6H-SiC (0001) substrates.

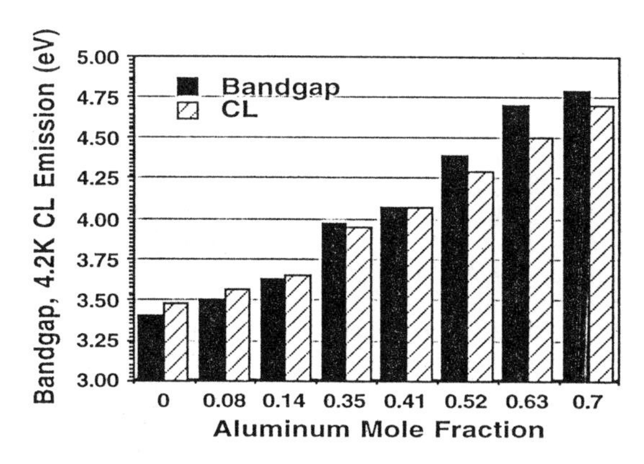

Figure 4. Comparsion of 4.2K CL near band-edge emission and bandgap as determined by spectroscopic ellipsometry at room temperature.

attributed to an exciton bound to a neutral donor (I_2-line emission) (*51*). For increasing concentrations of aluminum, this emission became gradually weaker. No emission was observed for pure AlN other than a broad peak centered around 3.1 eV which has been previously attributed to oxygen (*52*). The most narrow near band-edge CL FWHM was 31 meV for $Al_{0.05}Ga_{0.95}N$. This value increased with aluminum concentration to a maximum of approximately 100meV for $Al_{0.5}Ga_{0.5}N$. For higher aluminum compositions, this value did not increase. The broadening of the exciton features is likely due to both exciton scattering in the alloy, as well as small variations in alloy compositions in the film. Additionally, two strong defect peaks were present at energies less than the band gap. Both of these features have been previously attributed to donor-acceptor transitions. The first peak is centered around 2.2 eV for GaN and is commonly associated with deep levels in the bandgap (*53*). The second defect peak is centered around 3.27 eV which is attributed to a transistion between a shallow donor and a shallow acceptor (*54*) and its position also changes sublinearly with increasing aluminum concentration. This sublinear shift is attributed to the donor state moving deeper into the gap. For $Al_{0.05}Ga_{0.95}N$, the donor to acceptor pair (DAP) and its two LO-phonon replicas typically seen in GaN are still resolved, but for higher aluminum compositions, these peaks become one broad peak. Furthermore, for films deposited at 1150°C, this peak is no longer seen in the CL spectra. Increasing the growth temperature to eliminate this defect peak would indicate that it is impurity related. Attempts to determine this impurity are in progress.

Initial studies of the initial growth of $Al_xGa_{1-x}N$ directly on SiC give some insight into the strain relief mechanisms near the $Al_xGa_{1-x}N$/SiC interface. Despite the somewhat close lattice match between GaN and 6H-SiC(0001) (3.5%), previous research in our laboratories has shown that GaN epitaxy undergoes three dimensional island growth, since the critical thickness of GaN on 6H-SiC is <10Å. By contrast, AlN films, with a critical thickness of 45Å and a lattice mismatch of 0.9%, grow via coalescence of flat-top islands. The surface gradually roughens with increasing thickness (*45*). At a thickness of 1000Å, AFM measurements reveal an RMS roughness of 32Å when deposited on vicinal SiC substrates.

In contrast, $Al_xGa_{1-x}N$ films deposited at 1050°C undergo a considerably more complex growth mechanism. Figure 5 shows the results of five minutes of deposition with a gas phase mixture which results in a $Al_{0.2}Ga_{0.8}N$ thin film. It clearly shows two distinct regions of growth which are referred to as the "islands" and the "valleys". The compositions of the islands and the valleys were determined using field emission-AES (Fig. 6). Two important findings are evident from this graph. Firstly, this data clearly shows that the interface region is aluminum rich and that the bulk film composition is achieved at approximately 100Å of film thickness. Secondly, the interface of the valley regions is significantly more aluminum rich than the islands. The valley regions appear to contain approximately 0.65 mole fraction of aluminum at the interface while the islands contains only 0.40 mole fraction of aluminum at the interface. Assuming Vegard's Law, this would imply an interfacial mismatch of only 1.9% for the valleys versus 2.5% for the islands. It should be noted that the mismatch of the islands is the same as that between AlN and GaN and, moreover, their morphology is similar to that observed for the first few hundred angstroms of GaN deposited on a high temperature AlN buffer layer (*45*).

The process of coalescence results in smooth $Al_xGa_{1-x}N$ films after a few thousands angstroms of growth. It should be noted that profiling took place in 100Å increments. Therefore, the possibility exists that an extremely thin AlN or very aluminum rich $Al_xGa_{1-x}N$ layer occurs within the first few angstroms of the interface; thereby providing a graded buffer layer structure of less than 100Å in thickness. More noteable, however, is the fact that low temperature CL revealed near band-edge emission at 3.53 eV. This indicates that considerable tensile strain is present in this film since thicker films of the same composition have emission located at 3.83 eV. In summary, initial studies indicate that lattice strain is partial relieved at the interface during growth

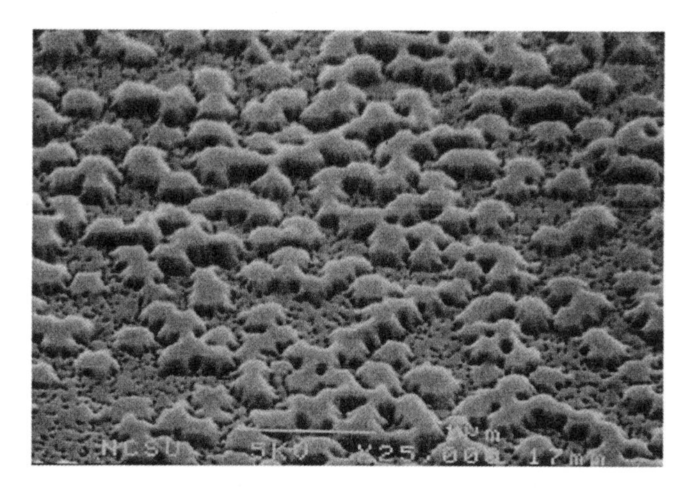

Figure 5. Scanning electron micrograph after 5 minutes of growth of $Al_{0.2}Ga_{0.8}N$ at 1050°C directly on 6H-SiC.

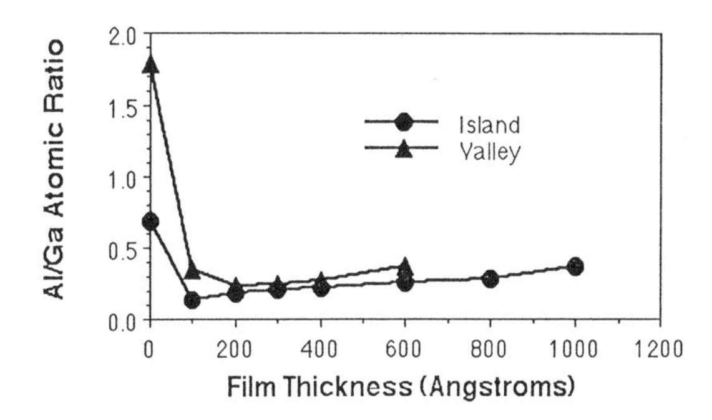

Figure 6. Field emission-Auger electron spectroscopy data from the various areas of the sample shown in Figure 5.

by the formation of an aluminum rich buffer layers which result in an island morphology. Subsequently, these islands coalescence into a smooth, coherent film containing tensial strain generated during cooling due to the differences in the thermal expansion coefficents as well as numerous threading dislocations.

The TEM results from a 1.8 micron thick $Al_{0.13}Ga_{0.87}N$ film deposited on an on-axis substrate revealed a microstructure dominated by threading dislocations, but free of low angle grain boundaries. Inspection of this and other micrographs reveals a progressive reduction in dislocation density as one moves away from the $Al_xGa_{1-x}N$/SiC interface. This is additionally supported by a narrowing of the full width at half maximum (FWHM) of the DCXRC of the $Al_{0.13}Ga_{0.87}N$ (0002) peak from 315 arcsec to 186 arcsec as the thickness of film increased from 0.9 microns to 1.8 microns. For films of similar thicknesses and compositions, films grown on vicinal SiC substrates exhibited higher FWHMs of the DCXRC of the AlGaN(0002). Furthermore, the growth rate on vicinal substrates was slightly higher due to increased density of steps on the surface; however, these steps probably act as formation sites for inversion domain boundaries (*55*).

Several structures have been grown using GaN and $Al_xGa_{1-x}N$. A TEM micrograph of an $Al_{0.2}Ga_{0.8}N$/GaN superlattice with periods of various thicknesses is shown in Fig. 7. A schematic of the structure is shown in Fig. 7a. The superlattice structure was deposited on 0.6 microns GaN which was deposited on a 1000Å, 1100°C AlN buffer layer. Each superlattice period was repeated 5 times and the structure was capped with 0.2 microns of GaN. Figures 7b and 7c show coherent interfaces and the high quality of the superlattice structure. Observation of the structure in plan-view TEM did not indicate a reduction in dislocation density below that normally observed in single layer GaN films on a buffer layer.

Undoped, high quality $Al_{0.05}Ga_{0.95}N$ films grown directly on vicinal 6H-SiC(0001) exhibited residual, ionized donor concentrations of 1×10^{18} cm^{-3}. The ionized donor concentration decreased rapidly with increasing Al content and was $<1\times10^{17} cm^{-3}$ for $Al_{0.12}Ga_{0.88}N$ and $<1\times10^{16}$ cm^{-3} for $Al_{0.35}Ga_{0.65}N$, as determined by CV measurements. The origin of these donors is under investigation, since concentrations of $<1\times10^{15}$ cm^{-3} have been measured for GaN films grown on AlN buffer layers in the same reactor. Moreover, layers of undoped AlN having N_D-N_A of 8×10^{15} cm^{-3} has also been deposited. However, the controlled introduction of SiH_4 allowed the reproducible achievement of ionized donor concentrations within the range of 2×10^{17} cm^{-3} to 2×10^{19} cm^{-3} in $Al_xGa_{1-x}N$ films for $0.12 \leq x \leq 0.52$. For $x>0.52$, additions of silicon resulted in films too resistive for CV measurements. The growth of p-type $Al_xGa_{1-x}N$ films for $x \leq 0.13$ via the introduction of Mg has been successful.

Conclusions

Organometallic vapor phase epitaxy has been used to grow monocrystalline GaN(0001) thin films at 950°C on high-temperature on 100 nm thick, monocrystalline AlN(0001) buffer layers previously deposited at 1100°C on α(6H)-SiC$(0001)_{Si}$ substrates. $Al_xGa_{1-x}N$ films ($0\leq x\leq1$) were grown directly on these substrates at 1100°C. All films possessed a smooth surface morphology and were free of low-angle grain boundaries and associated oriented domain microstructures. Cross-sectional TEM of $Al_{0.13}Ga_{0.87}N$ revealed a microstructure similar to that of GaN grown on a high-temperature AlN buffer layer. Analysis via EDX, AES and RBS were used to determine the compositions which were paired with their respective CL near bandedge emission energies. A negative bowing parameter was determined. The CL emission energies were similar to the bandgap values obtained by SE. Field effect AES of the initial growth of $Al_{0.2}Ga_{0.8}N$ revealed an aluminum rich layer near the interface. $Al_{0.2}Ga_{0.8}N$/GaN superlattices with coherent interfaces were fabricated. The PL spectra of the GaN films deposited on both vicinal and on-axis substrates revealed strong bound exciton emission with a FWHM value of 4 meV. The spectra of these films on the vicinal substrates were shifted to a

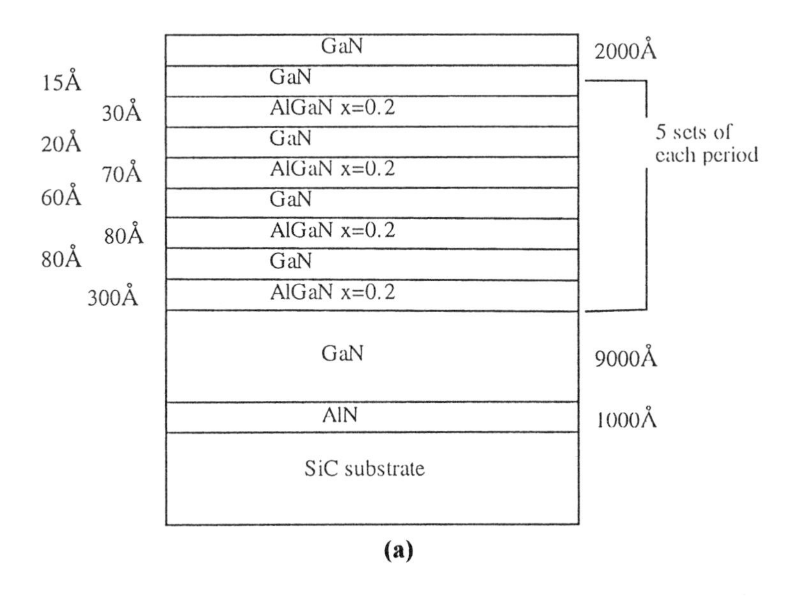

(a)

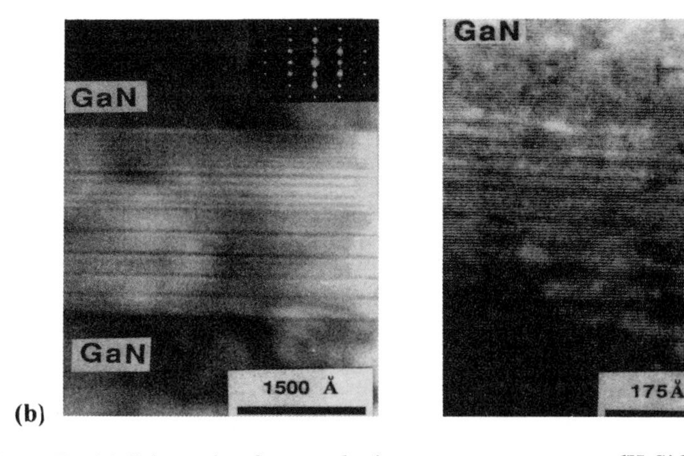

(b) (c)

Figure 7. (a) Schematic of a superlattice structure grown on 6H-SiC. (b) cross-section transmission electron micrograph showing superlattice region of the structure. (c) High resolution, cross-sectional TEM micrograph showing the 15Å GaN/30Å $Al_xGa_{1-x}N$.

lower energy, indicative of films containing residual tensile stresses. A peak believed to be associated with free excitonic emission was also observed in each on-axis spectrum. Cathodoluminescence of solutions with $x<0.5$ exhibited strong near band edge emission with a FWHM as low as 31 meV. The band gaps were determined via spectral ellipsometry. Controlled n-type Si-doping in GaN and $Al_xGa_{1-x}N$ (for $x \leq 0.4$) was achieved for net carrier concentrations ranging from approximately 2×10^{17} cm^{-3} to 2×10^{19} ($Al_xGa_{1-x}N$) or to 1×10^{20} (GaN) cm^{-3}. Mg-doped, p-type GaN was achieved with $n_A - n_D \approx 3\times10^{17} cm^{-3}$, $\rho \approx 7\ \Omega\cdot cm$ and $\mu \approx 3\ cm^2/V\cdot s$.

Acknowledgements

This work was supported by the Office of Naval Research on Contracts N00014-92-J-1720 and N00014-92-J-1477. Appreciation is expressed to Cree Research for the SiC wafers and for their assistance in performing the C-V measurements, N. Doyle of the Westinghouse Science and Technology Center for the DCXRC measurements and to Prof. K. Hiramatsu of Nagoya University, T. W. Weeks, Jr., S. Tanaka and A. D. Batchelor of NCSU for many helpful discussions.

Literature Cited

1. Amano, H.; Kito, M.; Hiramatsu, K. Akasaki, I. *Jpn. J. Appl. Phys.* **1989**, *28*, pp L2112-L2114.
2. Nakamura, S.; Mukai, T.; Senoh, M. *Jpn. J. Appl. Phys.* **1991**, *30*, pp L1998-L1990.
3. Nakamura, S.; Senoh, M.; Mukai, T. *Appl. Phys. Lett.* **1993**, *62*, pp 2390-2392; *Jpn. J. Appl. Phys.* **1993**, *32*, pp L8-L10.
4. Goldenberg, B; Zook, J. D.; Ulmer, R. *J. Appl. Phys. Lett.* **1993**, *64*, pp 381-383.
5. Khan, M. A.; Chen, Q.; Skogman, R. A.; Kuzina, J. N. *Appl Phys. Lett.* **1995**, *66*, pp 2046-2048.
6. Nakamura, S. J. Vac. Sci. Technol. A 1995, A13, pp 705-711.
7. Nakamura, S.; Senoh, M.; Iwasa, N.; Nagahama, S. I. *Appl. Phys. Lett.* **1995**, *67*, pp 1868-1870.
8. Nakamura, S. *Jpn. J. Appl. Phys.* **1996**, *35*, pp L74-L76.
9. Khan, M. A.; Bhattari, A.; Kuznia, J. N.; Olson, D. T. *Appl. Phys. Lett.* **1993**, *63*, pp 1214-1216.
10. Khan, M. A.; A.; Kuznia, J. N.; Olson, D. T.; Schaff, W. J.; Burm, J. W.; Shur, M. S. *Appl. Phys. Lett.* **1994**, *65*, pp 1121-1233.
11. Khan, M. A.; Kuznia, J. N.; Bhattari, A.; Olson, D. T. *Appl. Phys. Lett.* **1993**, *62,* pp 1786-1788.
12. Binari, S. C.; Rowland, L. B.; Kruppa, W.; Kelner, G.: Doverspike, K.; Gaskill, D. K. *Electron. Lett.* **1994**, *30*, pp 1248-1251.
13. Shin, M. W.; Trew, R. J. *Electron Lett.* **1995**, *31*, pp 498-501.
14. Okano, H.; Tanaka, N.; Takahashi, Y.; Tanaka, T.; Shibata, K.; Nakano, S. *Appl. Phys. Lett.* **1994**, *64*, pp 166-168.
15. Davis, R. F. *Physica B* **1993**, *185*, pp 1-11.
16. Yoshida, S.; Misawa, S.; S. Gonda, *Appl. Phys. Lett..* **1983**, *42*, pp 427-429.
17. Yoshida, S.; Misawa, S.; S. Gonda, *J. Vac. Sci. & Technol. B* **1983**, *1*, pp 250-256.
18. Amano, H.; Sawaki, N.; Akasaki, I.; Toyoda, Y. *Appl. Phys. Lett.* **1986**, *48*, pp 353-355.
19. Amano, H.; Akasaki, I.; Hiramatsu, K.; Koide, N.; Sawaki, N. *Thin Solid Films* **1988**, *163*, pp 415-421.
20. Akasaki, I.; Amano, H.; Koide, N.; Hiramatsu, K.; Sawaki, N. *J. Cryst. Growth* **1989**, *98*, pp 209-214.

21. Benjamin, M.C.; Wang, C.; Davis, R. F.; R.J. Nemanich, *Appl. Phys. Lett.* **1994**, *64*, pp 3288-3290.
22. Amano, H.; Tanaka, T.; Kunii,Y.; Kato, K; Kim,S.T.; Akasaki, I. *Appl. Phys. Lett.* **1994**, *64*, pp 1377-1399.
23. Nakamura, S.; Senoh, M.; Mukai, T. *Jpn. J. Appl. Phys.* **1992**, *31*, pp 1258-1266.
24. T. Tanaka, A. Watanabe, H. Amano, Y. Kobayashi, I. Akasaki, S. Yamazaki and M. Koide, *Appl. Phys. Lett.* **1994**, *65*, pp 593-595.
25. R. F. Davis, *Proc. IEEE* 1991, *79*, pp 702-710.
26. S. Strite and H. Morkoc, *J. Vac. Sci. Technol.* **1992**, *B10*, pp 1237-1243.
27. J. H. Edgar, *J. Mater. Res.* **1992**, *7*, pp 235-241.
28. M. Henini, *Microelectron. J.* **1992**, *23*, pp 500-505.
29. Morkoç H; Strite, S.; Gao, G. B.; Lin, M. E.; Sverdlov, B.; Burns, M. *J. Appl Phys.* **1994**, *76*, pp 1363-1371.
30. Neumayer, D. A.; Ekerdt, J. G., *Chem. Mater.* **1996**, *8*, pp 9-25.
31. Liu, S. S.; Stevenson, D. A., *J. Electrochem. Soc.* **1978**, *125*, pp 1161-1166.
32. Sauls, F. C., Interrante, L. V. *Coord. Chem. Rev.* **1993**, *128*, pp 193-204.
33. Bertolet, D. C.; Rogers, Jr., J. W., *J. Phys. Chem.* **1991**, *95*, pp 4453-4462.
34. Liu, H; Bertolet, D. C.; Rogers, J. W., *Jr. Surf. Sci.* **1994**, *320*, pp 145-152.
35. Mackenzie, J. D.; Abernathy, C. R.; Pearton, S. J.; Krishnamoorthy, V; Gharatan, S.; Jones, K. S.; Wilson, R. *Appl. Phys. Lett.* **1995**, *67*, pp 253-255.
36. Edgar, J. H.; Yu, Z. J.; Ahmed, A. U.; Rys, A., *Thin Solid Films* **1990,** *189*, pp L11-L13.
37. Andrews, J. E.; Littlejohn, M. A.; *J. Electrochem. Soc.* **1975**, *122*, pp 1273-1280.
38. Takahashi, Y.; Yamashita, K.; Motojima, S.; Sugiyama, K. *Surf. Sci.* **1979**, *86*, pp 238-244.
39. Boyd, D. C.; Haasch, R. T.; Mantell, D. R.; Schulze, R. K.; Evans, J. F. Gladfelter, W. L. *Chem. Mater.* **1989**, *1*, pp 119-126.
40. Kouvetakis, J.; Beach, D. B. *Chem. Mater.* **1989**, *1*, pp 476-483.
41. Hiramatsu, K.; Itoh, S.; Amano, H.; Akaski, I.; Kuwano, N.; Shiraishi, T. Oki, K.; *J. Crystal Growth* **1991**, *115*, pp 628-635.
42. Tanaka, S.; Kern, R. S.; R. F. Davis; *Appl. Phys. Lett.* **1995**, *66*, pp 37-39.
43. Qian, W.; Skowronski, M.; De Graef, M.; Doverspike, K.; Rowland, L.B.; Gaskill, D.K.; *Appl. Phys. Lett.* **1995**, *66*, pp 1252-1255.
44. Nakamura, S.; Iwasa, N.; Senoh, M.; Mukai, T.; *Jpn. J. Appl. Phys.* **1992**, *31*, pp 1258-1266.
45. Weeks, Jr., T.W.; Bremser, M.D.; Ailey, K.S.; Carlson, E.P.; Perry, W.G.; Davis, R.F.; *Appl. Phys. Lett.*, **1995**, *67*, pp 401-403.
46. Weeks, T.W.; Bremser, M.D.; Ailey, K.S.; Perry, W.G.; Carlson, E.P.; Piner, E.L.; El-Masry, N.A.; Davis, R.F.; *J. Mat. Res.* **1996**, *11*, pp 1081-1087.
47. Yosida, S.; Misawa, S.; Gonda, S.; *J. Appl. Phys.*, **1982**, *53*, pp 6844-6850.
48. Koide, Y.; Itoh, H.; Khan, M.R.H.; Hiramatsu, K.; Sawaki, N.; Akasaki, I., *J. Appl. Phys.* **1987**, *61*, pp 4540-4546.
49. Khan, M.R.H.; Koide, Y.; Itoh, H.; Sawaki, N.; Akasaki, I.; *Solid State Commun.* **1986**, *60*, pp 509-511.
50. Wickenden, D.K., Bargeron, C.B.; Bryden, W.A.; Miragliotta, J.; Kistenmacher, T.J.; *Appl. Phys. Lett.* **1994**, *65*, pp 2024-2026.
51. Khan, M.A.; Skogman, R.A.; Schulze, R.G.; Gershenzon, M.; *Appl. Phys. Lett.* **1983**, *43*, pp 492-494.
52. Dingle, R.; Sell, D.D.; Stokowski, S.E.; Ilegems, M.; *Phys. Rev. B* **1971**, *4*, pp 1211-1218.
53. Fischer, A.; *Physik. Verhandl.* **1957**, *7*, pp 204-209.
54. Wolff, G.A.; Mellichamp, J.W.; *Phys. Rev.* **1959**, *114*, pp 1262-1267.
55. Adams, I.; Aucoin, T.R.; Wolff, G.A. *J. Electrochem. Soc.* **1962**, *109*, pp 1050-1056.

Chapter 4

Autocompensated Surface Structure of GaN Film on Sapphire

M. M. Sung[1], J. Ahn[1], V. Bykov[1], D. D. Koleske[2], A. E. Wickenden[2], and J. W. Rabalais[1,3]

[1]Department of Chemistry, University of Houston, Houston, TX 77204–5641
[2]Naval Research Laboratory, Code 6861, 4555 Overlook Avenue, SW, Washington, DC 20375–5347

The surface composition and structure of a GaN film on sapphire has been determined through the use of time-of-flight scattering and recoiling spectrometry (TOF-SARS), classical ion trajectory simulations, low energy electron diffraction (LEED), and thermal decomposition mass spectrometry (MS). Elastic recoil detection (ERD) was used to determine the bulk hydrogen concentration. The totality of this data leads to the conclusions that the (1x1) surface is not reconstructed, that it is terminated in a N layer, and that Ga comprises the 2nd-layer. Hydrogen atoms are bound to 3/4 of the N atoms in the outerlayer and protrude outward from the surface, facilitating autocompensation of the otherwise unstable (1x1) structure.

The group III nitrides (*1,3*), specifically GaN, have a large cohesive energy compared to group III phosphides and arsinides. This arises from the high electronegativity of nitrogen and the partial ionic character of the Ga-N bonds (*4*). The increased cohesive bond energy of the nitrides results in the wide band-gap emission which is observed in GaN based diodes (*5,6*) and lasers (*7*). The nitrides are also chemically stable, making them attractive for high power and high temperature device applications (*1-3*). To the best of our knowledge, there are no experimental surface structure measurements on any surface of GaN to date. Understanding the GaN surface structure will provide a starting point for determining how precursor molecules decompose leading to the growth of GaN. Also, the success of GaN electronic devices is dependent on the atomic level control of the elemental compositions and structures involved. One simple question is to what extent does the GaN {0001} surface reconstruct or relax, and if so what are the possible surface structures. In this paper the surface and bulk composition, termination layer, and structure of a GaN{0001}-(1x1) film grown on a sapphire wafer are investigated. The techniques of time-of-flight scattering and recoiling spectrometry (TOF-SARS),

[3]Corresponding author

classical ion trajectory simulations, low energy electron diffraction (LEED), and thermal decomposition mass spectrometry (MS) were used in the surface analysis. The bulk hydrogen concentration was determined by elastic recoil detection (ERD). The combination of these techniques allows characterization of the elemental composition in the outermost two atomic layers, the element which constitutes the surface termination layer, the surface symmetry, and possible reconstructions or relaxations. In addition, TOF-SARS has high sensitivity to surface hydrogen and allows one to probe the involvement of hydrogen in the surface structure.

Experimental Methods

GaN Sample. GaN was grown on the c plane of polished sapphire using the NRL facilities. This consisted of a vertical, inductively heated, water-cooled quartz OMVPE reactor at reduced pressure (57 torr) using H_2 as the carrier gas (*8*). After annealing in H_2, the wafer was cooled to ~450 °C and a 200 Å nucleation layer of AlN was grown using 1.5 mmole/min triethylaluminum and a 2.5 slm flow of NH_3. The substrate was then heated to the growth temperature of 1040 °C and GaN was grown using 53 mmole/min of trimethlygallium under an NH_3 flow of 2.25 slm. The films were uniformly doped using Si_2H_6 at a flow rate of 0.22 sccm (*9*). After growth of a 2.7 μm thick GaN film, the substrate was cooled in the NH_3 flow at a rate of 50 °C per minute. The n-type films investigated in this study had an electron concentration of 1.2×10^{17} cm^{-3} and a mobility of 325 cm^2/V/s.

The ~1x1 cm^2 samples were cleaned in the UHV chamber used for TOF-SARS at the University of Houston. The samples were mounted on the TOF-SARS sample holder with the sapphire substrate in contact with a Ta plate and held together by small Ta strips over the sample edges. Annealing was achieved by radiative heating and electron bombardment from a tungsten filament mounted behind the Ta plate. Temperatures were measured by means of a pyrometer and a thermocouple which was attached to the sample; the pyrometer readings were calibrated by the thermocouple. The absolute temperature measurements have a maximum uncertainty of ± 30 °C. The clean surface was prepared by cycles of sputtering (1 keV N_2^+ ions, 0.5 μA/cm^2, 10 min) and annealing (10 min) using a dynamic N_2 backfill. A faint (1x1) LEED pattern became discernible at ~815°C and evolved into a sharp hexagonal (1x1) pattern at ~920°C. Upon continued heating to 1000°C, a diffuse background and satellite spots around the hexagonal spots were observed, indicating possible surface faceting. Sample cleanliness was established by the absence of carbon and oxygen recoil peaks in the TOF-SARS spectra and the presence of a sharp (1x1) LEED pattern. Hydrogen was always present on the surface when the sharp (1x1) pattern was observed. The TOF-SARS measurements were carried out on three different GaN samples which were all grown and cleaned as described above. All three samples exhibited similar behavior and yielded similar results.

Analysis Techniques. The low energy electron diffraction (LEED) patterns were obtained with Princeton Research Instruments, Inc. reverse view optics. Mass spectra (MS) were acquired with a Leybold-Inficon, Inc. quadrupole residual gas analyzer. The technique of elastic recoil detection (ERD) with MeV ions, which is capable of both bulk and near surface analysis, was used to determine the hydrogen concentration in GaN.

The time-of-flight scattering and recoiling spectrometry (TOF-SARS) technique was used for surface composition analysis and atomic structure characterization. Details of the TOF-SARS technique have been described elsewhere (*10*). Briefly, a pulsed noble gas ion beam irradiates the sample surface in a UHV chamber and the scattered and recoiled ions plus fast neutrals are measured by TOF techniques. The primary 4 keV beam employed herein was Ne^+ for scattering from Ga atoms and Ar^+ for recoiling of H, N, and Ga atoms. The ion pulse width was ~50 ns, the pulse repetition rate was 30 kHz, and average beam current was 0.5 nA/cm^2. The angular notation is defined as follows: α = beam incident angle to the surface, δ = crystal azimuthal angle, θ = scattering angle, ϕ = recoiling angle, and β = scattering or recoiling exit angle from the surface. A schematic drawing of the unreconstructed GaN$\{000\bar{1}\}$ surface is shown in Figure 1; the <1000> azimuth is defined as $\delta = 0°$.

Classical Ion Trajectory Simulations. Classical ion trajectory simulations were carried out by means of the three-dimensional scattering and recoiling imaging code (SARIC) developed in this laboratory. SARIC is based on the binary collision approximation, uses the ZBL universal potential to describe the interactions between atoms, and includes both out-of-plane and multiple scattering. Details of the simulation have been published elsewhere (*11*).

Results

Identification of the 1st-layer Species and Hydrogen Analysis. Elemental analysis was obtained by matching the observed TOF peaks to those predicted by the binary collision approximation (*12*). The 1st-layer elemental species was determined by using grazing incidence (α=6°) TOF-SARS at random azimuthal angles and a scattering-recoiling angle of $\theta=\phi=40°$. Using random azimuthal angles and a low incident angle avoids the anisotropic effects of scattering along the principal low-index azimuths and provides scattering intensities which are from the 1st-atomic layer and exposed 2nd-atomic layers.

A typical TOF spectrum from a GaN$\{000\bar{1}\}$-(1x1) surface taken along a random azimuthal direction (not aligned along a high symmetry azimuth) with a grazing incident angle α is shown in Figure 2. Under these conditions the observed scattering features are almost exclusively from the 1st- and 2nd-atomic layers. The spectrum exhibits peaks due to scattering of Ar from Ga atoms and recoiling of H, N, and Ga atoms. The TOF's and corresponding energies of these scattered and recoiled atoms are consistent with the binary collision approximation. The scattering angle used (θ=40°) is above the critical angle for Ar single scattering from N atoms (θ_c=20.5°); this results in a negligible contribution of the N atoms to the scattering peak intensity. The relative elemental concentrations in the surface layers were obtained from spectra similar to that of Figure 2 which were collected at five different random azimuthal angles. The relative H, N, and Ga recoil intensities were

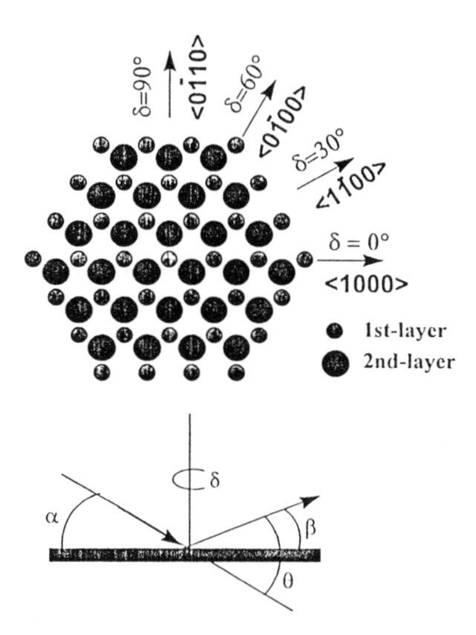

Figure 1. (top) Plan view of the ideal bulk-terminated GaN{0001} surface illustrating the azimuthal angle δ assignments. Another domain is obtained by 180° rotation of this surface about the surface normal. Open circles - 1st-layer N atoms; large solid circles - 2nd-layer Ga atoms. (bottom) Illustration of the angular notation used in TOF-SARS.

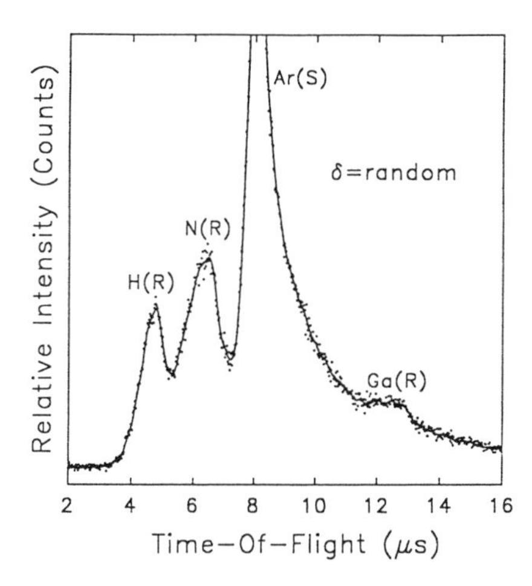

Figure 2. TOF-SARS spectrum of 4 keV Ar^+ scattering from a GaN{0001}-(1x1) surface with the ion beam aligned along a random azimuthal direction. Incident angle $\alpha = 6°$; Scattering angle $\theta = 40°$.

obtained from the average of these five spectra, from which concentrations were calculated after background subtraction and correction for the different recoiling cross sections. The cross sections were calculated in the binary collision approximation using the Moliere approximation to the potential function. The calculated cross sections and relative atomic concentrations are listed in Table I. The Ga recoiling peak rather than the scattering peak was used for the concentration determination because comparison between only scattering or only recoiling peaks provides more accuracy than comparison between both scattering and recoiling peaks. Relative concentrations obtained in this manner have an uncertainty on the order of 30%, primarily due to shadowing and blocking effects which are not corrected along the random azimuths.

Table I. Calculated cross sections (σ) for 4 keV Ar^+ scattering from Ga atoms and recoiling of H, N, and Ga atoms, relative atomic concentrations from experimental intensities along random azimuths, and ASEA values

Scattering (S) and Recoiling (R) Cross Sections (σ) [$Å^2$]			
$\sigma_{Ga}(S)$	$\sigma_H(R)$	$\sigma_N(R)$	$\sigma_{Ga}(R)$
0.21	0.13	0.071	0.099
Relative Surface Atomic Concentrations			
[N] = 1	[H] = 0.71		[Ga] = 0.091
Experimental* ASEA Values Along the 0° <1000> and 30° <1$\bar{1}$00> Azimuths			
$(I_{30°}/I_{0°})_N = 1.0$	$(I_{30°}/I_{0°})_H = 1.1$		$(I_{30°}/I_{0°})_{Ga} = 0.59$
Simulated* ASEA Values Along the 0° <1000> and 30° <1$\bar{1}$00> Azimuths			
$(I_{30°}/I_{0°})_N = 1.0$	$(I_{30°}/I_{0°})_H = 0.8$		$(I_{30°}/I_{0°})_{Ga} = 0.17$

[a]Ratios of recoiling peaks for N and H and scattering peaks for Ga were used in calculating both the experimental and simulated ASEA values.

The high ratio [N]/[Ga] = 6.3 from Table I provides compelling evidence that the surface is terminated in a layer of N atoms and that the Ga atoms occupy the second layer. Termination in a nitrogen layer may result from the method of cleaning, *i.e.* sputtering in N_2^+. The intense H recoil peak was present under all conditions that gave rise to the clear (1x1) LEED pattern, indicating that the H concentration on the (1x1) surface is comparable to the N concentration.

ERD analysis was performed with a 7.9 MeV C^{+4} beam at a grazing incident angle of 13° and exit angle of 15° using a GaN sample on which no cleaning attempt was made. A surface hydrogen recoil peak as well as a continuous background due to hydrogen in the bulk were observed as shown in Figure 3. The absolute hydrogen concentrations were obtained by comparison to a standard of amorphous silicon

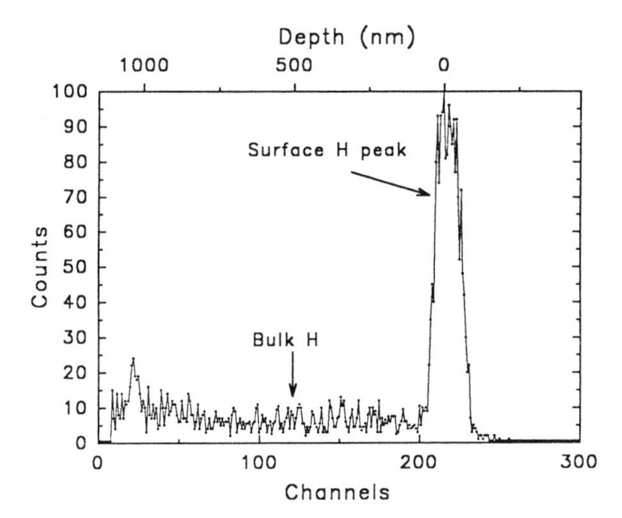

Figure 3. Elastic recoil detection (ERD) spectrum of H atoms from GaN{0001}-(1x1) using 7.8 MeV C^{+4} ions at a grazing incidence angle of 13° and exit angle of 15°.

which was implanted with $1x10^{17}$ H atoms/cm^2. The observed surface hydrogen peak corresponds to an areal density of $1.5x10^{16}$ H atoms/cm^2 and the continuous background corresponds to a bulk density of $4.0x10^{19}$ H atoms/cm^3. For an ideal GaN crystal, the N or Ga atom surface density is $1.1x10^{15}$ atoms/cm^2 and the bulk density is $4.4x10^{22}$ atoms/cm^3. The apparent abnormally high surface hydrogen concentration obtained from ERD results from the emergence of the high energy elastically recoiled H atoms from a depth of 5 - 10 atomic layers and from the surface atmospheric contamination.

Surface Periodicities of the H, N, and Ga Atoms. The surface periodicities (*13*) of the H, N, and Ga atoms were determined by monitoring I_{Ga} for Ne scattering from Ga atoms and I_H and I_N for Ar recoiling of H and N atoms as a function of crystal azimuthal angle δ. TOF spectra similar to that of Figure 2 were obtained and the intensities of the various peaks were plotted as a function of δ, resulting in the plots of Figure 4. Maxima and minima are observed as a function of δ. The minima are coincident with low-index azimuths where the surface atoms are inside of the shadowing or blocking cones cast by their aligned, closely spaced nearest neighbors, resulting in low intensities. As δ is scanned, the atoms move out of the shadow cones along the intermediate δ directions where the interatomic spacings between the atoms are long, resulting in an increase in intensity. The widths of the minima are related to the interatomic spacings along that particular direction. Wide, deep minima are expected from short interatomic spacings because of the larger degree of rotation about δ required for atoms to emerge from neighboring shadows. Schematic diagrams of the GaN surface illustrating the scattering and blocking directions are presented along with the δ-scans of Figure 4.

The I_N δ-scan was performed in the shadowing mode, *i.e.* a grazing incident angle of $\alpha = 10°$ was used. The data exhibit minima every 30°, with the minima labeled 0° and 60° being deeper than those at 30° and 90°. The 0° and 60° directions correspond to the <1000> and $<0\bar{1}00>$ azimuths, respectively, where the N-N interatomic spacings are shortest and shadowing of 1st-layer N atoms by neighboring 1st-layer N atoms is most efficient. The 30° and 90° minima correspond to the $<1\bar{1}00>$ and $<0\bar{1}10>$ azimuths, respectively, where the N-N interatomic spacings are longer and shadowing is less efficient. As noted earlier, shorter interatomic spacings result in deeper and wider minima due to the larger azimuthal rotations required for atoms to move out of the shadow cones of their aligned nearest neighbors.

The I_{Ga} δ-scan (Figure 4) was also performed in the shadowing mode with an incident angle of $\alpha = 12°$. The data also exhibit minima every 30°, but in contrast to the I_N minima, the minima labeled 30° and 90° are deeper than those at 0° and 60°. Shadowing and blocking of 2nd-layer Ga by 1st-layer N is most efficient along the 30° and 90° directions due to the aligned overlayer N atoms. Along the 0° and 60° directions, the overlayer N atoms are not aligned with the Ga atoms, resulting in less severe shadowing and blocking. Although the surface has only three fold symmetry, the I_{Ga} δ-scan (Figure 4) is indicative of the six fold symmetry of the surface. As noted above, the degree of shadowing and blocking along the 30° and 90° azimuths is not identical due to the relationship of the 1st- and 2nd-layer atoms. The minima

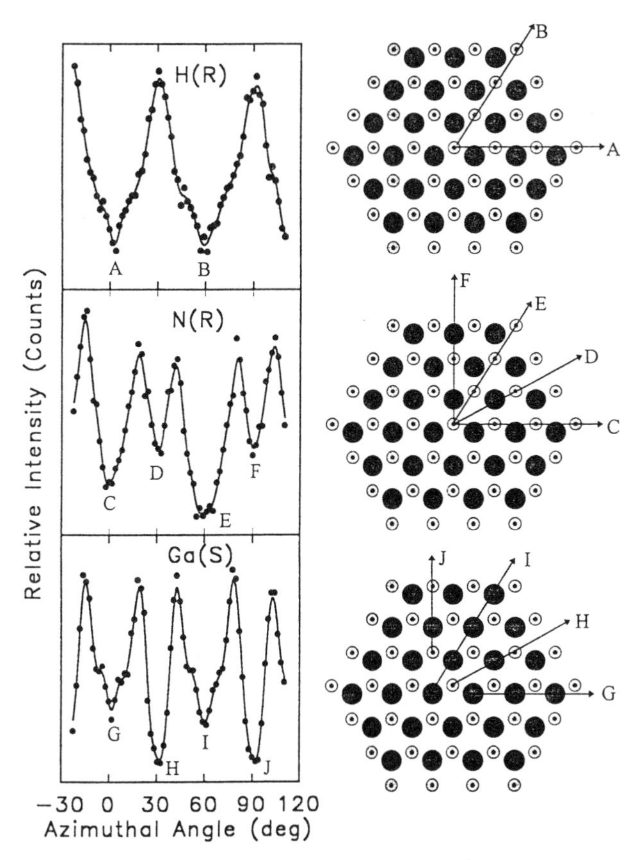

Figure 4. Experimental azimuthal angle δ-scans of Ne^+ scattering from Ga (I_{Ga}) along with H (I_H) and N (I_N) recoiling from Ar^+ on the GaN{0001}-(1x1) surface. The projectile was 4 keV Ne^+ and Ar^+, and the conditions were α = 20°, θ = 30° for H, α = 10°, θ = 40° for N, and α = 12°, θ = 90° for Ga.

observed at these two angles have identical widths and depths, suggesting that the surface exists in two different domains; if there would be only one domain, different widths and depths for these minima would be expected.

The I_H δ-scan (Figure 4) was performed in the blocking mode, *i.e.* a small exit angle $\beta = 10°$ was used. The data exhibit 60° periodicity with deep minima along the 0° and 60° directions, unlike the other two elements. The H atoms are too light to cause significant shadowing of the incoming Ar projectiles. However, H recoiling atoms can be blocked from the 1st-layer N atoms. The radii of the shadowing and blocking cones for H atom collisions with other H atoms is only <0.3 Å compared to > 1Å for H atom collisions with N and Ga atoms (*14*). By using a small exit angle β, recoiled H atoms can be blocked by nearest neighbor N atoms. The observed periodicity arises if the H atoms are bound to the outerlayer N atoms and protruding outward from the surface plane. The short H-N interatomic spacings along 0° and 60° result in overlayer recoiled H atoms being blocked by their 1st-layer N neighbors. Along the 30° and 90° directions, the H-N interatomic spacings are longer and the recoiling H trajectories are outside of the N blocking cones (Figure 4). The model in Figure 1 is but one of the several possible models consistent with the I_H δ-scan.

Ion trajectory simulations were carried out for the δ-scans of Figure 4. Good agreement of the the simulated and experimental data could only be obtained by combining two domains which were rotated by 60° from each other. The results are shown in Figure 5. Combining the two domains results in symmetrical δ-scans which are a good reproduction of the experimental scans.

Thermal Decomposition of GaN. The LEED pattern, TOF-SARS spectra, and mass spectra (MS) of the residual gases were monitored every 50° while heating a GaN crystal. Beginning at low temperature with an uncleaned sample, the LEED image was completely diffuse and the TOF-SARS spectrum was dominated by H, C, O, and N recoil peaks. At a temperature of ~700°, the recoil impurities in the TOF-SARS spectrum began to decrease until the spectrum was eventually similar to that of Figure 2. The diffuse LEED image evolved into a (1x1) pattern at ~800°C; also at this temperature the MS exhibited sharp increases in the N_2^+, NH_2^+, and H_2^+ signals. The absence of an NH^+ signal may be due to its relative instability and the absence of direct line of sight between the sample surface and the MS. The N_2^+ and NH_2^+ signals increased by a factor of $\sim 10^2$ between 800° and 900°, after which their intensities reached a plateau level which persisted to about 1,050°C. The H_2^+ signal increased monotonically by a factor of ~10 over this temperature range. The large amount of hydrogen evolved indicates that the hydrogen is not only on the surface but distributed throughout the bulk crystal. The (1x1) LEED pattern began to get increasingly more diffuse and the N and H recoil peaks in the TOF-SARS spectrum decreased as the temperature increased above 950°C. The sapphire substrate cracked at ~1000° and was not usable at higher temperatures.

Discussion

The bulk-terminated polar surfaces of compound semiconductor surfaces are typically unstable due to the partial electron occupancies of their dangling bonds

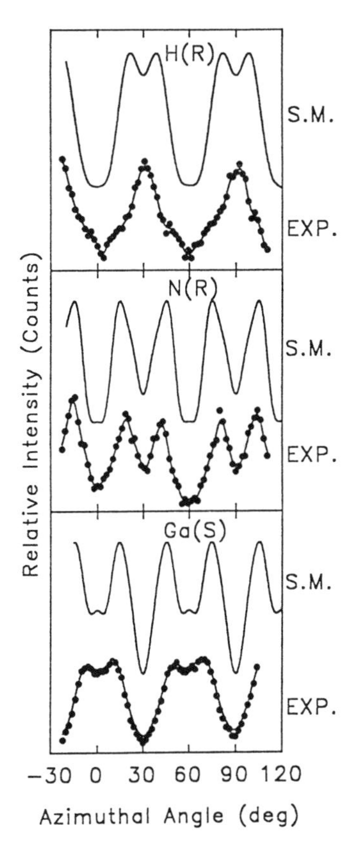

Figure 5. Trajectory simulations of the azimuthal angle δ-scan of Ne^+ scattering from Ga (I_{Ga}) along with H (I_H) and N (I_N) recoiling from Ar^+ projectiles on a GaN{0001}-(1x1) surface using the same conditions as listed in Fig. 4. Two different domains were used in the simulations. S.M. is the simulation and EXP. is the experimental data from Fig. 4.

(*15*). These surfaces achieve stability and semiconducting behavior by reconstructing in a manner in which the dangling bonds become completely filled or empty by electron transfer from the more electropositive to the more electronegative elements. Such a situation results in a surface in which all dangling bonds are either completely filled or empty, *i.e.* it is autocompensated (*13,16*). The clean unreconstructed GaN$\{000\bar{1}\}$-(1x1) surface is not autocompensated; the N dangling bonds protruding from the surface each have an electron deficiency of ¾e. Stability of this surface can be achieved either by reconstruction or by bonding of these N dangling bonds to some other atom; H atoms can fulfill this requirement. Indeed, under all conditions where a (1x1) LEED pattern was observed, H coverage of the N atoms was observed by TOF-SARS. Only ¾ of the N atoms in the outerlayer can be bound to H atoms in order to satisfy the autocompensation rule. The N dangling bonds are electron deficient by ¾e and H atoms can supply 1e. Therefore, for every three N-H bonds that are formed, there will be ¾e remaining which can be transfered to another N dangling bond. This results in only three of four N dangling bonds of the unit cell being passivated by H atoms; the other N dangling bond is stabilized by electron transfer from the three N-H bonds. There are several ways to distribute H atoms on a N-terminated GaN surface. The model in Figure 6 is one of the possible structures.

The nature of the termination layer, *i.e.* either N or Ga, is expected to be dependent on polarity matching considerations between the substrate and film. Recent *ab initio* total energy calculations by Capaz, et al. (*17*) have shown that the lowest energy interfaces are expected to be those on which cations(anions) of the substrate bind to anions(cations) of the film. Hence, the final polarity of the film is strongly dependent on the polarity of the substrate. Therefore, the nature of the atomic termination layers on sapphire surfaces is important to the understanding of the effects of substrate structure on the epitaxial relationship and overlayer lattice structure of thin film deposition. However, the exact termination elements of the $\{0001\}$ and $\{11\bar{2}0\}$ faces of sapphire are still unknown. The $\{0001\}$ and $\{11\bar{2}0\}$ surfaces of sapphire can be terminated with ether O-rich or Al-rich top surface monolayers. The packing sequence of the GaN wurtzite structure is (...ABAB...), where the layers are not equally spaced; the vertical spacing along the hexagonal <0001> direction is 0.63 Å within a bilayer and 1.95 Å between bilayers. Since the simplest stable unit of GaN is a single bilayer consisting of a layer of Ga and N atoms, if growth commences with a Ga layer, it must terminate with a N layer. The only other experimental work known to us on identifying polarity matching is for GaN grown on SiC by Sasaki and Matsuoka (*19*). This work addresses the problem indirectly by measuring the XPS chemical shift of the Ga atoms near the surface resulting from bonding to impurity oxygen atoms. Since the oxygen diffuses into the GaN to a depth of a few atomic layers and XPS samples a similar depth, the measurement is not specific to the outermost atomic layer. Nevertheless, the conclusion was that GaN layers on $\{0001\}_C$ and $\{0001\}_{Si}$ SiC substrates are terminated with Ga and N, respectively, in agreement with the theoretical prediction (*17*).

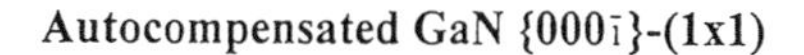

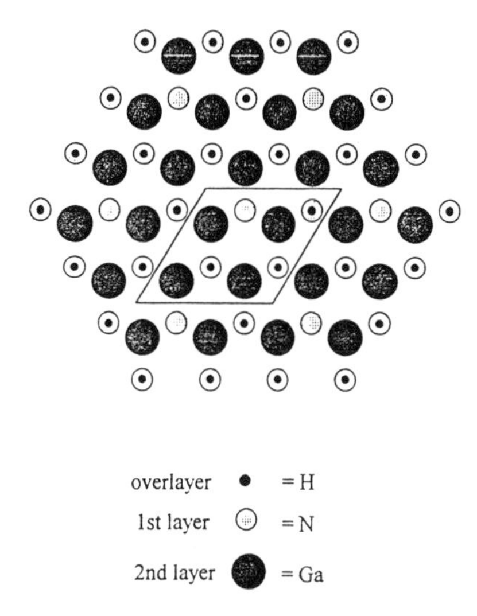

Figure 6. Plan view of the ideal bulk-terminated GaN{0001} surface illustrating the surface (1x1) unit cell and a proposed model for the surface H atom sites. Open circles - 1st-layer N atoms; large solid circles - 2nd-layer Ga atoms; small solid circles - overlayer H atoms.

Conclusions

The most significant results on GaN$\{000\bar{1}\}$-(1x1) derived from the LEED, TOF-SARS, MS, ERD, and simulation data can be summarized as follows.

(I) The experimental and simulated azimuthal scans show that the (1x1) surface is bulk terminated with no reconstruction.

(ii) The surface elemental analysis, ASEA values, and experimental and simulated azimuthal scans consistently show the following:

- (a) The surface is terminated in a N layer with Ga comprising the second layer, *i.e.* the $\{000\bar{1}\}$ plane.
- (b) Hydrogen atoms are bound to 3/4 of the N atoms in the outermost layer and protrude outward from the surface.

(iii) Thermal decomposition of the structure commences at ~850°C with the evolution of N_2, NH_2, and H_2. The presence of NH_x species provides strong evidence for N-H bonds on the surface. The high temperature required for evolution of the hydrogen indicates that it is as strongly bound as the Ga and N lattice atoms.

(iv) The large amount of H_2 evolved is consistent with hydrogen being distributed in the bulk of the crystal as well as on its surface.

(v) The azimuthal angle scans are consistent with the existence of two domains on the surface.

Acknowledgments

This material is based on work supported by the National Science Foundation under Grant No. CHE-9321899 and the R. A. Welch Foundation under Grant No. E656. We thank Z. Zhang and W. K. Chu of the Texas Center for Superconductivity for ERD analysis of the GaN.

References

[1] Davis, R.F. Proc. IEEE **1991**, *79*, 702.
[2] Strite, S.; Morkoc, H. *J. Vac. Sci. Technol. B* **1992**, *10*, 1237.
[3] Mohammad, S. N.; Salvador, A. A.; Morkoc, H. Proc. IEEE **1995**, *83*, 1306.
[4] Jaffe, J. E.; Pandey, R.; Zapol, *P. Phys. Rev. B* **1996**, *53*, 1.
[5] Nakamura, S.; Senoh, M.; Mukai, T. *Appl. Phys. Lett.* **1994**, *64*, 1687.
[6] Morkoc, H.; Mohammad, S. N. *Science* **1995**, *267*, 51.
[7] Nakamura, S.; Senoh, M.; Nagahama, S.; Iwasa, N.; Yamada, T.; Matsushita, T.; Kiyoku, H.; Sugimoto, Y. Jpn. *J. Appl. Phys.* **1996**, *35*, L74.
[8] Rowland, L. B.; Doverspike, K.; Giordana, A.; Fatemi, M.; Gaskill, D. K.; Skowronski, M.; Freitas, Jr., J. A. In *Silicon Carbide and Related Materials*; Spencer, M. G.; Devaty, R. P.; Edmond, J. A.; Khan, M. A.; Kaplan, R.; Rahman, M., Ed.; Inst. Phys.: Bristol, 1994, pp. 429.
[9] Wickenden, A. E.; Rowland, L. B.; Doverspike, K.; Gaskill, D. K.; Freitas, Jr., J. A.; Simons, D. S.; Chi. P. H. *J. Electron. Mat.* **1995**, *24*, 1547.
[10] Rabalais, J. W. *Science* **1990**, *250*, 521; Grizzi, O.; Shi, M.; Bu, H.; Rabalais, J. W. *Rev. Sci. Instrum.* **1990**, *61*, 740.
[11] Sung, M. M.; Bykov, V.; Al-Bayati, A.; Kim, C.; Todorov, S. S.; Rabalais, J. W. *Scanning Microsc.*, **1995**, *9*, 321.
[12] Parilis, E. S.; Kishinevsky, L. M.; Turaev, N. Yu.; Baklitzky, B. E.; Umarov, F. F.; Verleger, V. Kh.; Nizhnaya, S. L.; Bitensky, I. S. *Atomic Collisions on Solid Surfaces*; North-Holland, New York, 1993.
[13] Sung, M. M.; Kim, C.; Bu, H.; Karpuzov, D. S.; Rabalais, J. W. Surface *Science*, **1995**, *322*, 116.
[14] Shi, M.; Grizzi, O.; Bu, H.; Rabalais, J. W.; Rye, R. R.; Nordlander, *P. Phys. Rev. B* **1989**, *40*, 10163.
[15] Duke, C. B. *Scanning Microsc.* **1994**, *8*, 753.
[16] Pashley, M. D. *Phys. Rev. B* 1989, *40*, 10481.
[17] Capaz, R. B.; Lim, H.; Joannapoulos, J. D. *Phys. Rev. B* **1995**, *51*, 17755.
[18] Zhu, X. -Y.; Wolf, M.; Huett, T.; White, J. M. *J. Chem. Phys.* **1992**, *97*, 5856.
[19] Sasaki, T.; Matsuoka, T. *J. Appl. Phys.* **1988**, *64*, 4531.

Chapter 5

In Situ Monitoring by Mass Spectrometry of Laser Ablation Plumes Used in Thin Film Deposition

John W. Hastie, Albert J. Paul, David W. Bonnell, and Peter K. Schenck

Materials Science and Engineering Laboratory, National Institute of Standards and Technology, A215–223, Gaithersburg, MD 20899–0001

Molecular beam mass spectrometry has been used for real-time, *in situ* determination of species identities and velocity distributions of post-expansion laser ablation plumes. As an example of the measurement approach, the Al_2O_3 system is considered. The principal species found in Al_2O_3 plumes include Al, O, AlO, and Al_2O, in addition to Al^+. Several distinct populations of velocity distribution are often found for each species. For Al atoms, one distribution is centered around 1×10^5 cm s^{-1}, and another around 1.2×10^6 cm s^{-1}. The latter distribution agrees with the results of a plume expansion model. A new approach to the determination of species angular distributions is also described, and preliminary results are given for the C and Ag systems. The distributions observed for Ag species follow those expected from a supersonic expansion of the plume, whereas the results for Ag^+, C_3, and $C_3{}^+$ are somewhat anomalous and appear to be influenced by the presence of laser-plume interactions.

The technique of pulsed laser deposition (PLD) has been found to have unique advantages as a new approach to thin film deposition [1]. However, efforts to control, model, and optimize the process require a much improved characterization of the intermediate species. Here, we present the results of an *in situ* species measurement approach, based on molecular beam sampling mass spectrometry (MBMS) of fully-expanded laser ablation/vaporization plumes. Attention is also given to the essential features required of a MBMS system for representative sampling of laser ablation plumes. In order to verify and supplement the MBMS results, complementary investigations using emission spectroscopy and real-time imaging, coupled with gasdynamic, thermodynamic, and gas-kinetic models, have also been carried out, as discussed elsewhere [2-4].

Plumes generated by high-power pulsed (10 ns to 30 ns) laser vaporization or ablation of refractory materials can attain ultra-high temperatures (typically 5,000 K to 30,000 K) and relatively high pressures of condensible species ($> 10^5$ Pa ≡ 1 bar), *e.g.*, see [2]. Because of the difficulties of direct insertion of a molecular beam sampling probe into such an environment, the sampling approach used was to allow the plume to fully expand to collisionless flow before sampling. This approach is also more amenable to gasdynamic modeling [3, 4]. Although the plume conditions do not entirely preclude the use of an intrusive molecular beam sampling probe, modeling the intermediate expansion with the probe geometry included would appear to require a molecular dynamics approach. This discussion deals with measurements of a fully expanded beam, a necessary prerequisite to future consideration of probe extraction geometries.

Plume expansion into a high vacuum ($\leq 10^{-10}$ bar), maintained in a high conductance system, is sufficiently rapid that the initial plume species information can be retained and representative sampling with MBMS can be achieved [5]. In addition to plume species identities and abundances, MBMS analysis also provides beam time-of-flight information, yielding velocity distributions and gas temperatures. Such data form the basis of gasdynamic models of plume formation and dissipation [3,4] and are essential for future development of film growth models.

Significant progress in defining the general nature of laser vaporization /ablation plumes, used for thin film deposition, has been made in recent years [1,2]. Our current understanding of the main plume features is summarized by the idealized schematic of Figure 1. Note the use of extreme gradients to encompass time, distance, temperature, pressure, and velocity scales. As shown, the plume initiates at the laser focus spot on the target surface and moves away along an axis normal to the target (*i.e.* to the right in Figure 1). After a few ns time delay from the onset of the laser pulse, sufficient vapor density develops for collisional local equilibrium (labeled "loc. eq." in Figure 1) to become established, followed by a supersonic adiabatic expansion process. This expansion process is isentropic and rapidly leads to a free-flight, collision-free condition. In the present work, species are sampled from this free-flight region for MBMS analysis (note orifice shown at the extreme right in Figure 1).

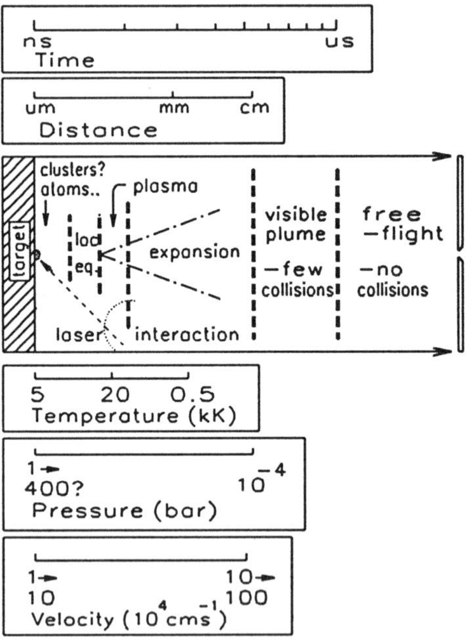

Figure 1 —Laser vaporization/ablation schema. Scales shown are logarithmic and are representative of typical laser ablation conditions present during pulsed laser deposition of thin films.

APPARATUS

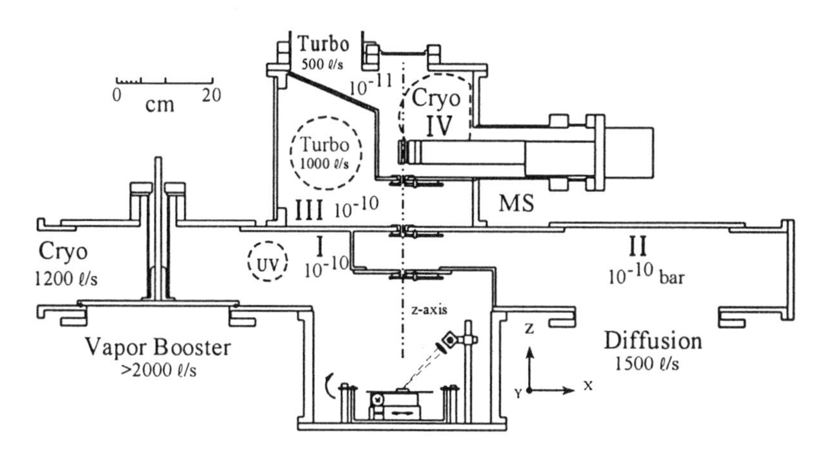

Figure 2 — Laser vaporization/ablation molecular beam mass spectrometer. Region I provides high conductance gas transport (exceeds 4000 ℓ/s before pumps) of the PLD-produced plume using either of two pumping systems. See text for discussion of the sample stage shown at the bottom center of Region I. The Y direction is the laser input direction (perpendicular to page). Regions II and III provide beam collimation and differential pumping of the expanded beam. Region IV is the MS region. An internal cryopump minimizes back scattering of the molecular beam.

The basic features of the original MBMS system have been described in detail elsewhere [5]. We present here only a description of recent modifications. Figure 2 shows schematically, but to scale, the main features of the MBMS apparatus. This system has very high conductance differential pumping to allow for rapid free-jet expansion and representative sampling of the expanded vapor. To accommodate the high velocities of beams formed from laser plumes, together with the pumping and optical access requirements, the flight distance between the plume and the mass spectrometer ion source was made relatively long (L=47.9±0.2 cm). The initial flight distance (Region I, target surface to the Region I-II interface in Figure 2) is sufficiently long (about 22 cm) to allow the plume to expand fully to molecular flow, particularly given that the conductance in Region I exceeds 2×10^6 cm^3 s^{-1}. Figure 2 shows that the differential pumping and collimation adds a distance of only 21 cm to the overall beam path length (*cf.* Regions II and III). The center of the MS ion source is about 5 cm above the Region III-IV differential orifice. Each of the adjustable differential interface orifices were set to 0.48 cm aperture, aligned with the entrance aperture of the MS, and provided a fully collimated molecular beam. As will be shown, a beam distance of 47.9 cm allows for sufficiently long flight times to permit observation of velocity distributions with good time resolution, over the range 10^4 to > 10^6 cm s^{-1}, which is characteristic of post-expansion plumes used for PLD.

In order to increase the area of film uniformity in PLD, particularly with respect to composition and thickness, it is desirable to monitor *in situ* and to understand the origin of the spatial distributions of plume species. For this purpose, we designed a kinematic target stage, based on commercially available linear translators in X-Y orientation. These translators are mounted on a rotatable (about the X-axis) platform whose orientation is controlled externally, using a vacuum feedthrough drive. This stage is shown schematically in Figure 2, mounted in Region I of the MBMS vacuum system. The stage rotates the target surface about the X axis *in situ*. This motion thus controls the tilt to the target normal axis (Z) relative to the fixed MBMS sampling axis (z). The rotatable platform operates like a simple swing and the target surface location is adjusted with jacking screws (not shown) to coincide with the pivot axis (X), using a laser alignment system through the pivot axis. The same system also aligns the Y axis of the surface to the plane defined by the two translation stages.

The geometrical relationships between the laser beam, plume, target, and MBMS sampling axis are shown, to scale, in more detail in Figure 3. The plume dimensions shown are based on images [4] obtained at early times in the expansion of ionized Al (Al^+) produced by 248 nm KrF excimer laser interaction (~30 ns pulse) with an Al_2O_3 target at a moderately high fluence. Note the significant interaction volume between the laser and plume during the laser pulse duration time (~30 ns in this case).

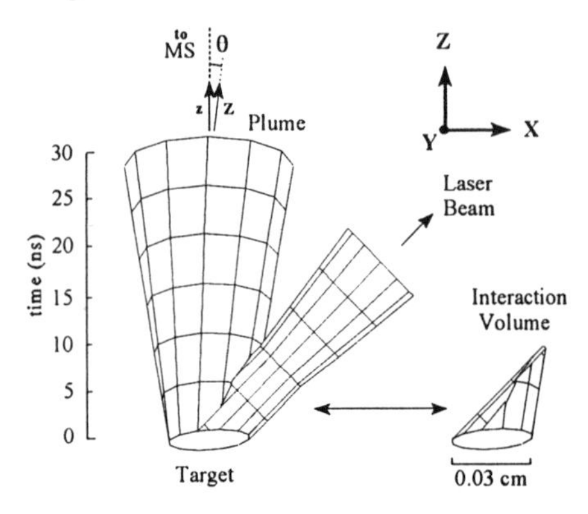

Figure 3 —Laser-plume-target interaction geometry, 20° rotation, viewed from -Y, at 30 ns after the laser pulse begins.

The ablating laser beam is focused at the target surface using an external telescope, with the last reflection being made through a fused silica prism. The long axis of the approximately elliptical footprint of the laser is coincident with the platform's rotating axis. Thus, as the stage is pivoted, the plume remains perpendicular to the target surface (Z-axis), presenting different angular views to the sampling axis (z) of the mass spectrometer (MS). The X- and Y-axes are referenced to the target surface, and the Z-axis (target normal and plume major axis) is coincident with the molecular beam axis (z) when the plume angle, θ, is 0°. The plume angle, defined in Figure 3 as the angular separation between the plume major axis (Z) and the molecular beam sampling axis (z), is the same as the stage angle.

For these studies, the laser beam was incident typically at a fixed azimuth angle of 45°, but the depth of focus was sufficiently long that the entire footprint could be considered to be uniform in the ablating surface. It was considered impractical to pivot the laser beam optics to allow the vertical plane containing the laser beam to remain perpendicular to the target. In the present arrangement, the elliptical laser footprint rotates by approximately the plume angle about the axis normal to the target surface. In the absence of laser-plume interactions, an integration of the area change due to the change in the laser footprint with plume angle indicates only cosine falloff in laser fluence at small angles. This geometric effect on fluence should be swamped by the observed $\cos^n\theta$ ($n \geq 4$) decrease in plume intensity that results from gasdynamic effects.

RESULTS AND DISCUSSION

Results from representative investigations are presented here to demonstrate application of the MBMS technique to the characterization of laser plume species. It is noteworthy that no evidence of cluster species, present in the plume or formed in the expansion cooling process, was found in these or earlier [2,5] studies. For the most part, the species observed are those we predict from their thermochemical stabilities for conditions near the atmospheric boiling points of the target materials [2,5-7].

Velocity Distributions. In a typical experiment, the mass spectral intensity (I) of each species is monitored as a function of time-of-arrival (TOA, t) at the detector, relative to the termination of the laser pulse. Conversion of time to velocity (v) is made using the molecular beam flight distance, L and

$$v = \frac{L}{t}\ ;\quad I'(t) = v\,I(t)\ ;\quad I'(v) = I'(t)\,\frac{t^2}{L}\ , \tag{1}$$

where I'(t) is the observed intensity ($I(t)$) corrected for the velocity (v) discrimination associated with the density detection nature of the electron impact ion source, as shown in equation 1 (*see also* [2]). $I'(v)$ is the intensity (velocity distribution) in velocity space equivalent to the molecular beam flux, $I'(t)$, obtained as shown from the observed number density, $I(t)$. Figure 4 shows an example TOA plot for Al produced by 1064 nm Nd/YAG laser interaction (~10 ns pulse) with an Al_2O_3 target at a moderate fluence.

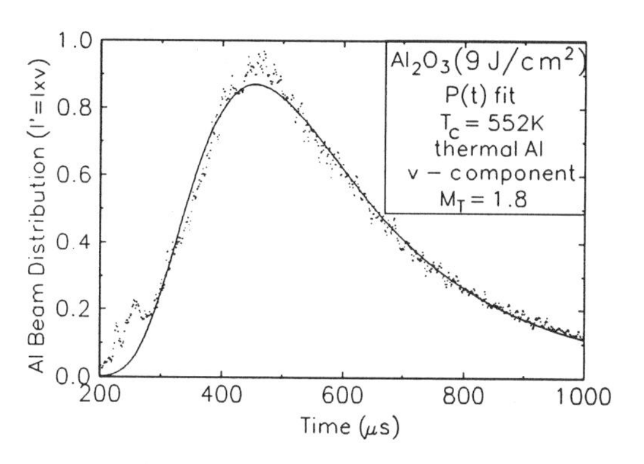

Figure 4 —Time-of-arrival for Al (Al_2O_3); λ = 1064 nm, fluence = 9 J cm^{-2}. Solid curve is a distribution fit ($\propto P(t)$) based on equations 1 and 2 (see text) giving values of T_c = 552±10 K and u = 0±400 cm s^{-1}. Measurement uncertainties are: I'±0.02 based on counting statistics at peak; t±5 μs.

An appropriate velocity distribution $\{P(v)\}$ approximation in this case [3] is that of a full-range Maxwellian distribution referenced to a center-of-mass net flow or drift velocity (u):

$$P(v_z) \propto v_z^3 \exp[-(m/2kT_c)(v_z-u)^2]. \quad \textbf{(2)}$$

The z axis is the molecular beam axis (as in Figures 2 and 3); T_c is the local terminal gas temperature at the onset of collisionless flow; m is molecular mass and k is the Boltzmann constant. In practice, as the experimental data are obtained (initially) in time rather than velocity space, an analogous time-dependent expression $P(t_z)$ may also be used to fit the TOA profiles (as in Figure 4). These nonlinear fits provide values of T_c and u. Similar results are obtained from the slope of a linear fit of the dependence of the time corresponding to the maximum intensity TOA values on the species molecular weights (see Figure 5). The general derivation of the relationship between derived slope and T_c may be found in [8].

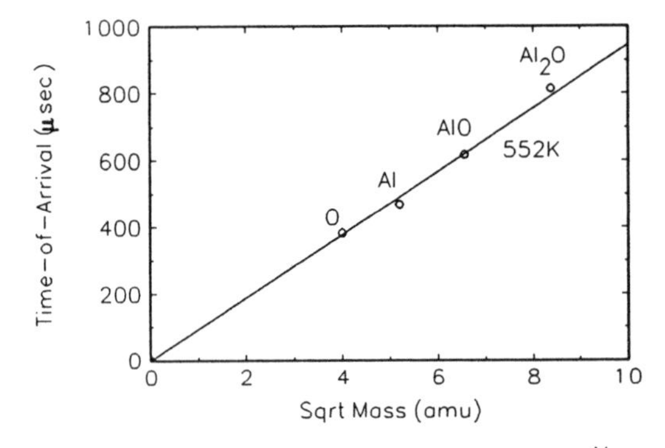

Figure 5 —Peak TOA dependence on (mass)$^{1/2}$ for species detected as singly charged positive ions produced by electron impact in the mass spectrometer; target = Al_2O_3; λ=1064 nm; Same conditions as for Fig. 4. Fitted slope gives T_c=550±50 K. Measurement uncertainties are: t_{max}≤±20 μs; MS-observed masses ±0.1 umu.

As shown in Figures 4 and 5, the low value of T_c (=552 K) is evidence of considerable expansion cooling. From the magnitude of the mass spectral intensities, we have estimated that the initial pre-expansion pressure is of the order of one bar for the laser conditions used [2]. For Al_2O_3, a surface temperature of 4500 K would be needed to generate a pressure of this magnitude under thermodynamic equilibrium conditions (*see* [2]). Conversion of the TOA profiles to a velocity scale shows that the data of Figure 4 correspond to relatively low velocities (~ 10^4 cm s^{-1} to 10^5 cm s^{-1} range), characteristic of a relatively low terminal Mach number (M_T ~ 1.8).

The small early-time peaks shown in Figure 4, and discussed in more detail below, are not noise, but are believed to be the result of separate processes involving laser-plume interactions. These features grow as the laser fluence is increased. Figure 6 provides an analysis of these fast and slow peaks in terms of velocity distributions. The faster velocity distribution becomes relatively more significant with increasing fluence and has a very high T_c (~31,000 K). These effects have been attributed to laser-plasma interactions [2-4]. The plasma gasdynamic model fit, shown in Figure 6, agrees well with the observed high-velocity distribution. The theoretical basis of the model is discussed in detail else where [2,3]. These results also complement those obtained by optical probes (not discussed here, see [3]).

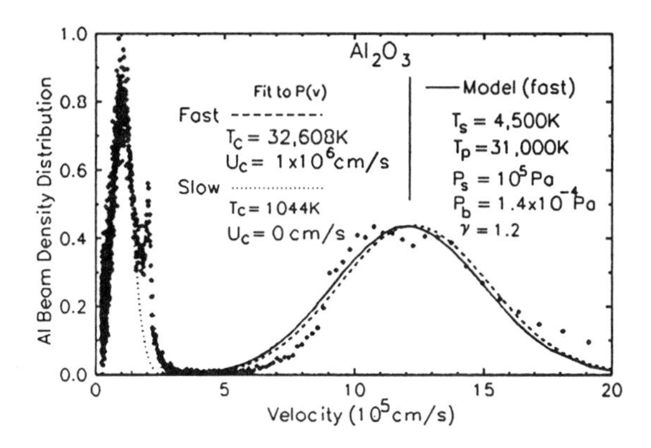

Figure 6 —Velocity distribution (peak normalized) for Al (Al_2O_3) at ~15 J cm^{-2}; λ=1064 nm. Solid curve is based on a plasma gasdynamic model [3,4] using the parameters shown in the figure, where γ is an effective heat capacity ratio used to model the expansion process; subscript *s* refers to target surface condition, *p* the plasma, and b the molecular beam. Measurement uncertainties are: v±1000 cm s^{-1} at 10^5 cm s^{-1} based on 5 μs channel time resolution at 47.9 cm; density distribution <±0.05 arb. units.

Sampling Fidelity. A key concern with any MBMS measurement approach is the reliability of the sampling process. In contrast with our earlier sampling of atmospheric pressure flames [9] and transpiration vapors [10,11], a sampling probe was not inserted into the active high pressure core of the plume. Instead, the sampling orifice was located intentionally well downstream of the shock front (Mach disc) location and in a region of collisionless flow. The target-to-orifice distance was 22 cm, compared with the calculated (from measured Mach numbers) distance to collisionless flow in the range of 0.05 cm to 5 cm. Also, the pumping speed and system conductances were high enough that the calculated Mach disc location was always downstream of the transition to collisionless flow, thus presenting a very weak shock front that should have a negligible effect on the expansion. Agreement between experimental and calculated expansion temperatures indicates that the sampling process was indeed located at a fully expanded jet location, and that no gasdynamic or other thermal perturbation had occurred.

There is strong evidence that local thermal equilibration of neutral species occurs in these laser plumes. It can be shown (*e.g.*, from equation 2) that the TOA peak should vary with species (mass)$^{1/2}$, provided the local thermal equilibration condition is maintained. In general, we have found such a dependence (Figure 5). Data analysis using this type of relationship also provides an additional benefit in assuring the reliability of the mass spectral species assignments. If electron impact fragmentation interference occurs, leading to a fragment ion, or both a fragment and a parent ion, at the same mass, it would be evident as a time displacement from the straight line in Figure 5, since the arrival times are for the pre-ionization condition of each species. As we have indicated elsewhere, free-jet MBMS can lead to very different fragmentation patterns than those found conventionally [6,12]. With a few notable exceptions [12], the expansion cooling process can be very beneficial in reducing electron impact fragmentation of parent molecular ions [6]. This is the case in Figure 5, where the excellent linear correlation of the assigned peaks indicates an absence of significant fragmentation during the electron impact ionization step in detection of the neutral species..

A more stringent, and more difficult to establish, test of sampling fidelity is provided by examination of possible shifts in chemical equilibria or kinetics. That is, are the relative concentrations of the sampled species modified by the sampling and analysis process? We have shown that, for systems that are demonstrably in thermodynamic equilibrium, sampling perturbations are negligible [10,11]. In laser-generated plumes, however, there is no *a priori* assurance that the pre-expansion plume is chemically equilibrated. We have shown that for moderate laser fluences, sufficient to attain a near-atmospheric boiling condition at the target surface, the pre-expansion species are frequently present in local thermodynamic equilibrium [2]. The MBMS species distributions are also found to agree well with thermodynamic equilibrium calculations, *e.g.* for C [6], BN [13], HfO_2 [7], $BaTiO_3$ [2,3] and Al_2O_3 [2]. For the Al_2O_3 case, we have also carried out detailed gas-kinetic calculations, showing that relatively small changes in species concentrations are possible but only during the first 20 ns of plume expansion [2].

Conversion of Final (Expansion) to Initial State Properties. The MBMS measurements yield information on terminal velocities (v_T) and temperatures (T_T). If we assume that the formation of an expanding vapor plume from a focused hot spot is gasdynamically similar to expansion through an orifice, then various standard gasdynamic relationships apply. The initial pre-expansion temperature of the gas, T_o, may then be obtained from:

$$T_T/T_o = \left[1 + \frac{\gamma - 1}{2} M_T^2\right]^{-1} \quad . \tag{3}$$

Typical values fall in the range 2500 K to 8000 K (for gas heat capacity ratios in the range, $\gamma = 1.4$ to 1.2). Pressure ratios may be obtained from:

$$T_T/T_o = \left[P_T/P_o\right]^{\frac{\gamma-1}{\gamma}} \quad . \tag{4}$$

Typical values of P_T/P_o in our system fall in the range 10^{-4} to 10^{-5}, or lower. When combined with P_T values determined from calibration of the mass spectral intensities, values of P_o typically in the range of 1 bar to 10 bar are obtained [2]. Direct and indirect measurement of surface temperatures also indicate target temperature and pressure conditions near the boiling point at moderate fluence conditions ($\lesssim$ 1 J cm^{-2} to 10 J cm^{-2}). Based on equation 3, we should then expect to find a linear correlation between the thermodynamically derived target boiling temperature ($T_{bpt} \sim T_o$) and the final, or beam, temperature (T_b) if T_o is strongly correlated with the surface temperature. Such a correlation has been found and was reported in [2], thus lending support to the analogy of expansion through an orifice.

Plume Angular Distribution. Velocity and angular distributions are expected to influence sticking coefficients, and hence film composition [14]. These distributions, in turn, are expected to be strongly influenced by the laser impact spot size. Angular distributions of visible emission plume components have been determined by us earlier, using OMA spectroscopy and CCD image analysis [4]. These results have been used to develop hydrodynamic models of angular distributions [4,8]. The focus of the MS-determined angular distributions is on the ground state (non-emitting) species, which are major plume components.

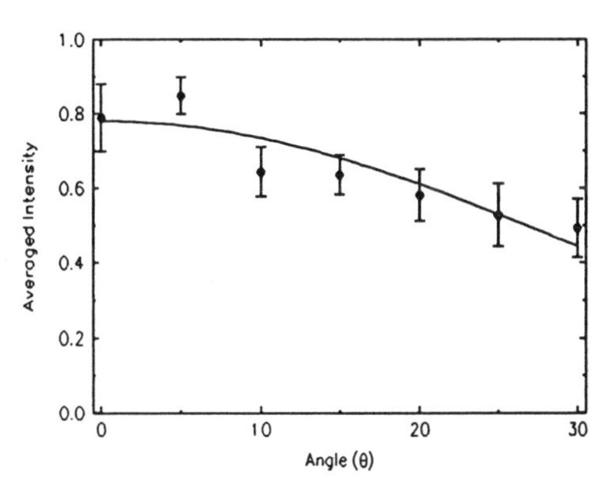

Figure 7 — Intensity (time-integrated, arbitrary units) as a function of plume angle for Ag; target=Ag; λ=1064 nm, moderate (< 10 J/cm^2) fluence. Intensities correspond to the areas under the curves of Fig 8(a), corrected for cos θ footprint effect (see text). Data (●) have been fitted (solid curve) to $\cos^{5.2\pm0.6}\theta$. Intensity uncertainties are based on single standard deviations for five replicate runs; angular uncertainty is <±1°.

When the plume major axis (Z) is oriented away from the molecular beam sampling axis (z), by tilting the kinematic target stage, the normal effect is a monotonic decrease in the mass spectral signal intensity (*e.g.* Figure 7). Generally, this decrease follows a $\cos^n\theta$ distribution law, expected for a supersonic expansion jet, where n is strongly dependent on gasdynamic parameters such as Mach number and pressure expansion ratio, and where the plume expansion is dominated by neutral species. In our studies, values of n from ~ 4 to 20 have been observed.

Figure 8 shows an example of the angular dependence of Time of Arrival (TOA) curves for Ag and Ag^+ plume species. Figure 8(a) is representative of the neutral and ion data observed using electron impact ionization of the plume molecular beam to analyze the neutrals. By turning off the ionizing electron beam, just the ionic species in the expanded plume molecular beam are seen (Figure 8(b)). For relatively fast ionic species, favored under higher laser fluence conditions, examples of anomalous angle-dependence behavior are found. Figure 8(b), representing the ionic species portion of Figure 8(a), shows such behavior. Note that the peak intensity for the faster ion signal occurs off-axis which, in our studies, appears to be typical when conditions favor laser-plume interaction. We believe that this off-axis behavior is consistent with the spatial asymmetry noted earlier in the plume imaging studies [4]. The plume wall, depicted in Figure 3 and based on optical images of the early expanding plume, is oriented at about 10° to 15° off vertical. This orientation, together with the MS angle-resolved data, suggests that, during the expansion process, ions formed within the high density conical volume are strongly quenched. On the other hand, ions formed in the lower density regions, resulting in the ion intensity peaking off-axis, appear able to escape with high velocity distributions distinct from those of the main high density region.

The non-monotonic form of the intensity vs. angle data in Figure 8(b) is suggestive of multiple $\cos^n\theta$ processes, with differing n-values, occurring at different times and with different angular dependence in the early expansion process. The fact that the individual TOA curves (*e.g. see* Figure 8) can all be fit well using multiple Maxwellian peaks with different peak velocities supports this conclusion. For fast ions, n should have a high value (>4), and similar effects should obtain for plume components affected by laser-plume interaction. Other intriguing observations, such as are shown in Figure 9, have also been made. The clear separation of slow and fast components is strongly indicative of multiple plume processes, separated either in space or time, or both. The complex angle dependencies shown for the fast neutral and charged species are attributed to laser-plume interactions, which are more significant for ions and can lead to complex geometry effects in the plume. Such interactions have been visualized using an Intensified Charge Coupled Device (ICCD)-based camera system [4].

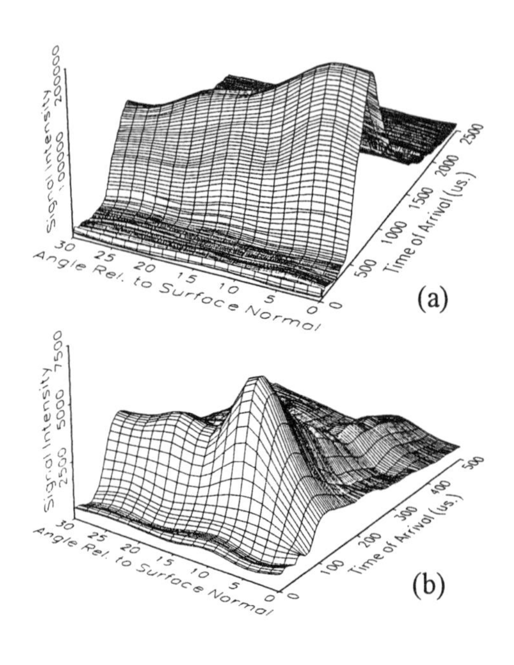

Figure 8 — (a) MBMS plume angle-dependence of the Ag atom (and small amount of Ag^+); (b) Ag^+ ion alone (MS ion source off). Conditions as for Figure 7; Signal Intensity in arbitrary units, relative uncertainties are similar to Figures 4 and 7; Angle (θ) in degrees $\pm1°$; Time of Arrival in μs ±5 μs.

In interpreting the angle-dependence results, the possible influence of geometrical effects requires consideration. Geometric computer-aided design representations have been used to estimate the extent of these effects (*e.g.*, as in Figure 3). Two notable effects are apparent as the stage and plume rotate away from the MS beam axis (z): (1) as the platform rotates, the interaction volume (separately depicted as the intersection of the plume and laser beam volumes in Figure 3) rotates around the target normal in the direction of the MS beam axis; (2) at about 10° rotation, the wall representing the "edge" of the visible, high-velocity emission region lines up with the MS sampling axis. Both effects suggest only minimal geometric effects on the plume volume element selected for MS sampling. The two mentioned laser-plume interaction effects should slightly enhance the observed off-axis intensity, probably somewhere between 10° and 20°, as observed (Figures 8 and 9). None of the geometric effects *per se*, however, account for the observed strong decrease in "fast" plume ion species observed inside the 10° sampling cone (*e.g.* as in Figure 7(b)). We suggest the existence of a preferential on-axis (θ~0°) loss mechanism for fast ion species. For example, the higher density of majority neutral plume species (Ag atoms in the case of Figure 7) on-axis should enhance the ion-electron-neutral recombination reactions. Work towards quantifying these effects and further testing the geometric models is in progress. Use of a lower laser fluence can be expected to lead to better behaved angular distributions for ions, as has been found for $YBa_2Cu_3O_x$ using Faraday cup detection, although the power of $\cos^n\theta$ (n=6) was somewhat lower than expected (n~10) for a fully developed supersonic expansion [15].

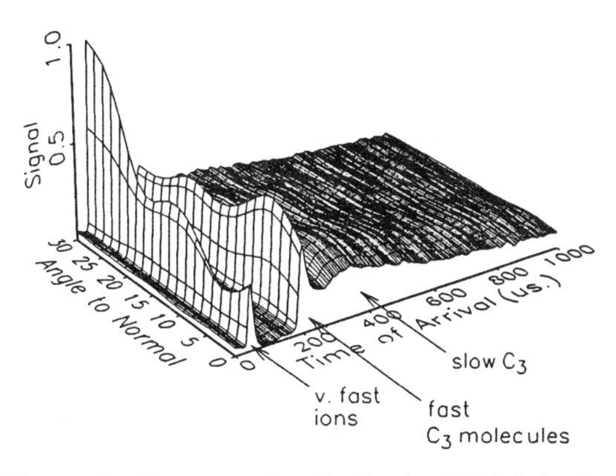

Figure 9 —Plume angular distribution TOA behavior for C_3 from graphite target; λ=1064 nm; fluence is high (≥10 J/cm²). Signal axis is normalized mass spectral intensity data, in arbitrary units, uncertainties are similar to Figures 4 and 7; Angle (θ) in degrees ±1°; Time of Arrival in μs ±5 μs.

It is pertinent to note the observations of Champeaux *et al* [16], concerning a highly directional ion distribution and its correlation with a modified film stoichiometry. Hence, we may expect to find a correlation between an observed plume asymmetry, with respect to the highly energetic ions, and a spatial non-uniformity in film properties.

CONCLUSIONS

The application of free-jet MBMS to laser plumes is of considerable value. Identities and concentrations of all significant species can be obtained. Velocity distributions and temperatures, combined with gasdynamic and plasma models, yield information on the pre-expansion plume properties. The species identity, flux and kinetic energy information obtainable is directly applicable to growth models of thin film formation. Information on spatial and temporal plume evolution can indicate optimal target, laser, and substrate positioning. Determination of species relationships with process variables (fluence, wavelength, pulse time, pressure, *etc.*) also allows for selection of optimum process conditions. Finally, MBMS results complement and enhance the development of more accessible optical probes for *in situ*, real-time process monitoring and control.

ACKNOWLEDGMENT

We wish to acknowledge the invaluable assistance of Mr. A.B. Sessoms in the construction and maintenance of the apparatus and in target preparation.

REFERENCES

1. *Pulsed Laser Deposition of Thin Films*; Chrisey, D.B.; Hubler, G.H.; John Wiley and Sons, Inc: New York, NY, 1994.
2. Hastie, J.W.; Paul, A.J.; Yeheskel, J.; Bonnell, D.W.; Schenck, P.K. *High Temp. Matls. Sci.* **1995**, *33*, 135-169.
3. Hastie, J.W.; Bonnell, D.W.; Paul, A.J.; Schenck, P.K. "Gasdynamics and Chemistry in the Pulsed Laser Deposition of Oxide Dielectric Thin Films" In *Gas-Phase and Surface Chemistry in Electronic Materials Processing*; Mountziaris, T.J., Paz-Pujalt, G.R., Smith, F.T.J., Westmorland, P.R., Eds.; MRS Symp. Series, Vol. 334; Materials Research Society: Pittsburgh, PA, 1994; 305-316.
4. Schenck, P.K.; Hastie, J.W.; Paul, A.J.; Bonnell, D.W. *Opt. Eng.* **1996**, *35*, 3199-3205.
5. Hastie, J.W.; Bonnell, D.W.; Schenck, P.K. *Molecular Basis for Laser-Induced Vaporization of Refractory Materials*; NBSIR 84-2983; National Technical Information Service: Washington, DC., 1984.
6. Hastie, J.W.; Bonnell, D.W.; Schenck, P.K. *High Temp. High Press.* **1988**, *20*, 73-89.
7. Bonnell, D.W.; Schenck, P.K.; Hastie, J.W.; Joseph, M. "Ultra-High Temperature Laser Vaporization Mass Spectrometry of SiC and HfO_2" In *5th Intl. Symposium on High Temperature Materials Chemistry*; Johnson, W.B., Rapp, R.A., Eds.; ECS Symp. Vol. PV90-18; Electrochemical Society: Pennington, NJ, 1990; 156-165.
8. Paul, A.J.; Schenck, P.K.; Bonnell, D.W.; Hastie, J.W. "*In Situ* Monitoring and Model Simulations of $BaTiO_3$ Pulsed Laser Thin Film Deposition" In *Film Synthesis and Growth Using Energetic Beams*; Atwater, H.A., Dickinson, J.T., Lowndes, D.H., Polman, A., Eds.; MRS Symp. Series Vol. 388; Materials Research Society: Pittsburgh, PA, 1996, 45-50.
9. Hastie, J.W. *Combust. and Flame*, **1973**, *21*, 187-194.
10. Hastie, J.W.; Bonnell, D.W. "Transpiration Mass Spectrometry: A New Thermochemical Tool" In *Thermochemistry and Its Applications to Chemical and Biochemical Systems*; Ribeiro da Silva, M. A. V.; Reidel Publishing: Boston, MA, 1984; 183-233.
11. Hastie, J.W.; Hager, J.P. "Vapor Transport in Materials and Process Chemistry" In *Proc. The Elliott Symposium on Chemical Process Metallurgy*, Iron and Steel Soc, Inc.: Warrendale, PA, 1990; 301-324.
12. Bonnell, D.W.; Hastie, J.W.; Zmbov, K.; *High Temp. High Press.* **1988**, *20*, 251-262.
13. Hastie, J.W.; Bonnell, D.W.; Schenck, P.K., *High Temp. Sci.* **1988**, *25*, 117-142.
14. Kools, J.C.S.; Riet, E. Van de.; Dieleman, J.; *Appl. Surf. Sci.* **1993**, *69*, 133-139.
15. Tyrrell, G.C.; York, T.; Cherief, N.; Givord, D.; Lawler J.; Lunney, J.G.; Buckley, M.; Boyd, I.W. *Microelectronic Eng.*, **1994**, *25*, 247-252.
16. Champeaux, C.; Damiani, D.; Girault, C.; Marchet, P.P.; Mercurio J.P.; Catherinot, A. "Plasma Formation from Laser-Target Interaction; Basic Phenomena and Applications to Superconducting Thin Film Deposition" In *Laser Ablation of Electronic Materials, Basic Mechanisms and Applications*; Elsevier: Amsterdam, Netherlands, 1992; 141-165.

Chapter 6

Fullerenes and Polymers Produced by the Chemical Vapor Deposition Method

Steve Kleckley[1], Hao Wang[1,4], Isaiah Oladeji[1], Lee Chow[1,5], Terry K. Daly[2], Peter R. Buseck[2], Touradj Solouki[3], and Alan Marshall[3]

[1]Department of Physics, University of Central Florida, Orlando, FL 32816–2385
[2]Departments of Geology, and Chemistry/Biochemistry, Arizona State University, Tempe, AZ 85287
[3]National High Magnetic Field Laboratory and Department of Chemistry, Florida State University, Tallahassee, FL 32310

Plasma-enhanced chemical vapor deposition has been used for the deposition of diamond and diamond-like thin films during the last decade with great success. Here we present experimental evidence that a chemical vapor deposition technique can also be used to synthesize other forms of carbon including fullerenes and hydrocarbon polymers. The mechanism of diamond nucleation and fullerene formation appear closely related.

Since the discovery that fullerenes form in supersonic molecular beams (*1*), many techniques (*2-9*) have been used to produce fullerenes. Out of these many techniques, carbon-arc and combustion methods have been used for commercial production of fullerenes. In the carbon-arc method, yields of 5% to 8% are common, and yields as high as 40% have been reported (*10*). It has been a puzzle for some time how a molecule as complex as fullerene can be formed so readily. Various formation mechanisms have been proposed (*11-14*). In particular the isolated pentagon rule (IPR) appears to explain many of the experimental observations.

Here we present observations of fullerene formation using hot-filament CVD and microwave-enhanced CVD methods. The yields of fullerene-containing soot are low, even though the yields of fullerene from the soot are quite reasonable. Our goals in this research are: (a) to determine the relationship between fullerene formation and diamond nucleation, and (b) to shed light on fullerene formation mechanisms.

Experimental

Hot filament CVD A schematic diagram of the apparatus (*15, 16*) used for this study is shown in Fig. 1. The chamber is constructed of stainless steel with a diameter of 35

[4]Current address: Shanghai Institute of Optics and Fine Mechanics, Academia Sinica, P.O. Box 800–216, Z01800 Shanghai, Peoples' Republic of China

[5]Corresponding author

cm and a height of 20 cm. The filament is typically straight 0.75 mm diameter tungsten wire approximately 8 cm in length. The filament hangs vertically, with the upper terminal fixed and the lower terminal attached to braided copper wire to avoid any stress on the filament. With this arrangement, the filament lifetime is improved. A stainless steel substrate holder is used for the diamond thin film deposition. The substrate holder is fixed to a micrometer, and the distance between the substrate holder and the tungsten filament can be controlled to less than 0.1 mm. The temperature of the substrate is controlled by the distance between the substrate and the filament and by the filament current. Typically filament currents are between 50 to 60 A, and filament temperatures are 2,000 °C to 2,200 °C, measured with a pyrometer. The typical substrate temperatures are between 950 °C and 1000 °C for the growth of CVD diamond thin films.

The feed gases are 99.8% pure CH_4 and 99.999% pure hydrogen. The flows are controlled by MKS mass flow controllers at 0.4 sccm for CH_4 and 99.6 sccm for H_2. The pressure of the chamber is controlled by the flow rate and the pumping speed through a valve. Typical chamber pressures are 30 to 100 torr. During CVD diamond deposition, we observed soot at the back of the substrate holder where the temperature was low.

Microwave enhanced CVD. Our microwave CVD apparatus is similar to the one used by Chang *et al.* (*17*). A 100 W 2.45 GHz generator was used as the excitation source, which was connected to an Evenson-type cavity. A 12.7 mm diameter 200 mm long quartz tube passing through the cavity was used as our reaction chamber. A particle trap for the soot was placed down-stream from the plasma. Another liquid nitrogen trap was placed between the pump and the chamber for some of the experiments. A schematic diagram is shown in Fig. 2. The feed gases we used include C_2H_2, H_2, and Ar. Gas ratios ranged from 5:5:1 to 2:2:1 for Ar, H_2, and C_2H_2. Typical pressures were between 1 and 10 torr. During these experiments, we did not put the substrate holder inside the microwave plasma, and no attempt was made to deposit diamond films. During the process, we found that at $P > 25$ torr, a conducting film was building up inside the quartz tube, which limited the deposition process to only a few minutes. At lower pressure we found that the plasma region was much more extended, and it resembled a glow discharge plasma. A yellowish film was deposited on the inside wall of the quartz tube at $P < 10$ torr. This film is non-conducting so it has little effect on the plasma characteristics. However, after being exposed to plasma for 30 minutes the film color changed to dark brown.

Mass Spectrometry. Time-of-flight mass spectrometry of the hot-filament CVD soot samples were carried out at Arizona State University. Details of the mass spectrometry used there have been described (*18*). The laser desorption FT-ICR mass spectra of the MW-CVD soot samples were acquired at the National High Magnetic Field Laboratory with an FTMS-2000 Fourier transform ion cyclotron resonance mass spectrometer equipped with a 3 tesla superconducting magnet, dual cubic Penning traps, and an Odyssey data system. Laser desorption/ionization was performed with a Nd:YAG laser operated at a wavelength of 355 nm with a pulse width of 7 ns. The laser beam was focused onto the probe tip by a 2:1 telescope. The laser power density is estimated to be 2×10^7 W/cm^2 at the probe tip. The laser beam spot size is approximately 400x600

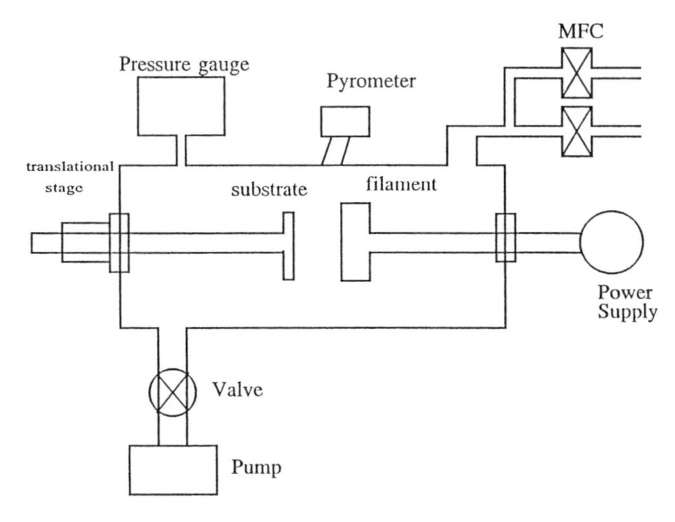

Figure 1. Schematic diagram of the hot filament CVD chamber.

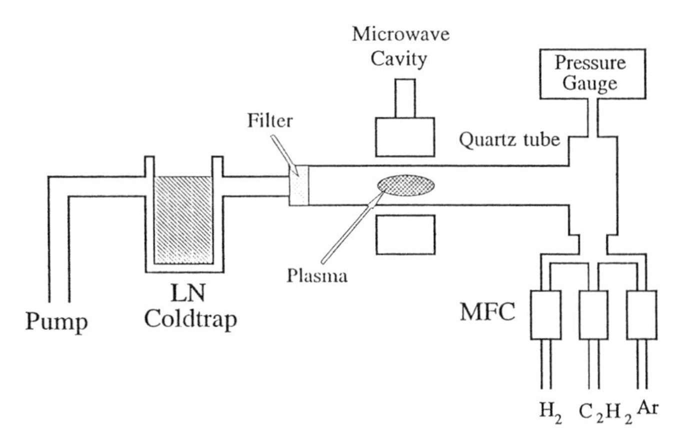

Figure 2. Schematic diagram of the microwave enhanced CVD chamber.

μm^2 (*19*). The soot sample solutions were applied to a thin stainless steel plate and air dried. The sample probe tip was inserted into the vacuum chamber and mass analyzed. All laser desorption FT-ICR mass spectra were acquired at a background pressure of $2x10^{-8}$ torr. After laser desorption/ionization, the ion x-axis translational energy was minimized by use of gated deceleration. After 60 - 150 μs to allow for ion transfer into the trap, the source trap and conductance limit plates were each restored to +2V. The trapped ions were excited by dipolar frequency sweep excitation. Fourier transformation of the resulting discrete time-domain signal, without zero-filling and Hamming apodization, followed by magnitude calculation and frequency-to-mass conversion yields an LD FT-ICR mass spectrum.

Although under high laser power density (>>10^7 W/cm^2), the laser desorption process may alter the sample and create fullerene species, careful adjustment of the laser power and experimental parameters can eliminate fullerene production (*20*). At $2x10^7$ W/cm^2 we did not observe any laser produced fullerene from graphite, and LD FT-ICR mass spectra of pure C_{60} and C_{70} samples at this laser power contained only C_{60} and C_{70} ions, respectively.

Results

Hot filament CVD. After the soot samples were collected, they were dissolved in benzene or carbon tetrachloride solvents and filtered. The solutions were then used for the mass analysis or UV-vis measurements. Some raw soot was analyzed directly by mass spectrometry. A time-of-flight mass spectrum of the soot solution is shown in Fig. 3; both C_{60} and C_{70} peaks are present. For pressure variations between 30 to100 torr, the samples also showed the fullerene peaks (*18*). In an attempt to increase the yield of soot, we increased the methane concentration to 2% and lowered the filament temperature to about 1700°C. Under these conditions, no diamond was formed and we were able to increase the soot yield by a factor of 6 to about 100 mg per run. However, no fullerene formed, only hydrocarbon clusters.

Microwave enhanced CVD. We collected samples from (a) inside the quartz tube, (b) at the fiber-glass trap, and (c) at the liquid nitrogen trap. The sample from inside the quartz tube consists mainly of polymer-like yellowish film. The film usually can be peeled or scratched off from the quartz tube. The film does not readily dissolve in most solvents. An absorption spectrum of the film on a quartz substrate is shown in Fig. 4. The spectrum shows high transmission above 300 nm and high absorption below 300 nm, which gives the film a predominant yellowish color. The film is insulating. A slightly burnt film gave an absorption edge at longer wavelength, as shown in the same figure. Exposure of the film to the microwave plasma for more than 30 minutes, especially at higher pressure, results in a hard, black, reflective, conducting, presumably graphitic film. It is highly resistant to scratching and adheres strongly to the quartz tube, possibly due to the formation of an intermediate SiC layer. X-ray diffraction measurement does not show any diffraction peaks, indicating that the film has amorphous structure or perhaps random cross-linked polymer structure.

The materials collected in the fiber-wool trap, on the other hand, are mostly soot particles. When the fiber-wool was washed in benzene or toluene, the solvents

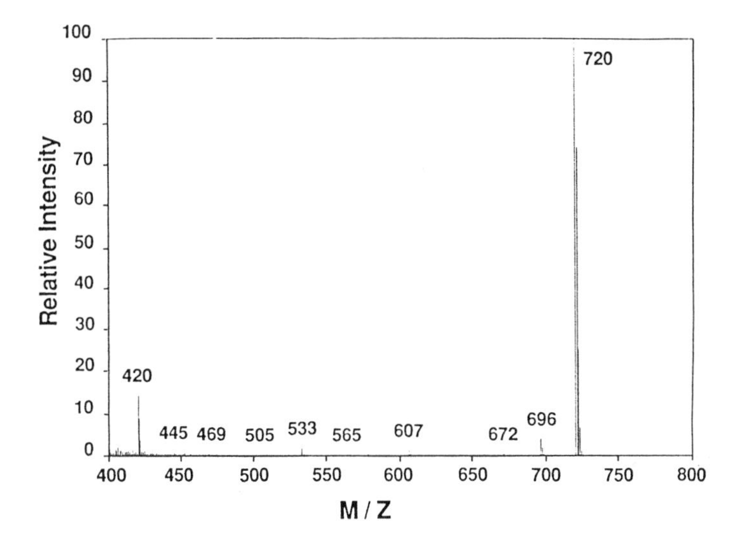

Figure 3. Time-of-flight mass spectrum of a soot sample collected in the hot filament CVD chamber under diamond-forming conditions. (Reproduced with permission from L. Chow, H. Wang, S. Kleckley et al, "Fullerene formation during production of chemical vapor deposited diamond", *Applied Physics Letters*, 66, 430 (1995). Copyright 1995 American Institute of Physics).

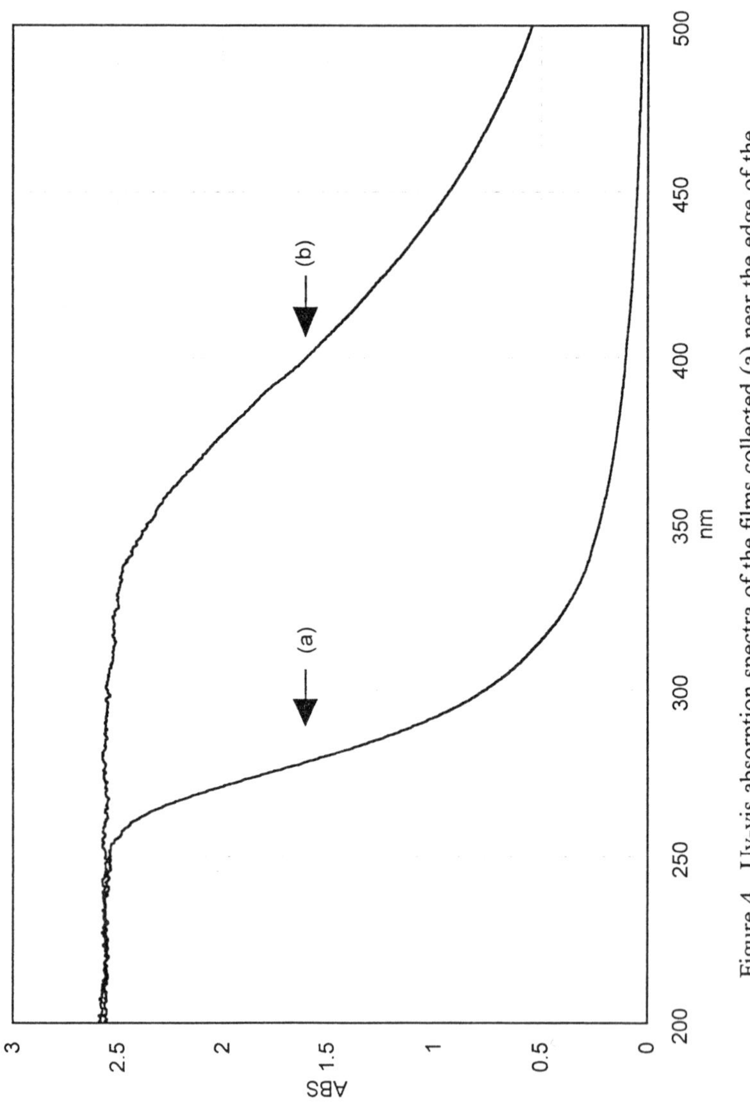

Figure 4. Uv-vis absorption spectra of the films collected (a) near the edge of the plasma, (b) inside the plasma produced by the microwave enhanced CVD method.

changed from colorless to a range of colors from light yellow to dark brown. The UV-vis spectra indicated some fullerenes and other hydrocarbons. However, in most cases, the hydrocarbon peaks obscure the identification of fullerene. When the solute is analyzed using LD FT-ICR mass spectrometry, both C_{60} and C_{70} are clearly observed (Fig. 5).

Samples collected from the liquid nitrogen (LN) cold trap are similar to those collected at the fiber wool trap. The only difference is that there are fewer soot particles and more soluble material than at the fiber wool trap. When dissolved, the solvent turned to a light yellowish color. When analyzed with time-of-flight mass spectrometry, C_{60} and C_{70} are observed in the solution. However, when the sample from the LN trap was directly analyzed by FT-ICR mass spectrometry, we notice that the mass spectrum showed a whole array of peaks that we interpret as higher fullerenes up to C_{118} (Fig. 6). This is similar to the mass spectrum of negative ion clusters in a benzene flame (3).

Discussion

Hot-filament CVD. The presence of fullerenes in the HF-CVD chamber may explain the origin of the five-fold twinning defect occasionally observed in CVD diamond chambers (*21*). TEM studies of diamond crystals exhibiting five-fold symmetry reveal polyhedral core particles (*22*). Lattice matching is important for CVD diamond growth. Fullerene provides a good nucleation site for the CVD diamond (*23*) along with an inherent five-fold symmetry. We speculate that fullerene may be the source of these core particles. If this is indeed the case, other elements like gold (*24*), which also exhibit this same defect structure, may be candidates for fullerene-like allotropes.

It is also possible that the co-production of fullerene under diamond growth conditions is a significant factor in diamond nucleation. Fullerenes have been used as precursors of diamond growth in the absence of hydrogen. Gruen *et al.* (*25*) used a fullerene vapor in a pure Ar carrier gas to successfully grow diamond films. They found that *in situ* optical measurements reveal a strong C_2 emission. These C_2 fragments could be a growth species for the CVD diamond.

Hydrogen abstraction from single CH_x species has been the preferred mechanism for diamond growth. In the hot-filament method that we employed, substrate temperature and processing pressure are similar to the MW-CVD method. It is conceivable that fullerene may form readily, but quickly fragment due to the diamond growth environment. If this is the case, then diamond film growth due to C_2 addition may be more important in the CVD process than previously suspected.

Microwave enhanced CVD. UV-vis, IR, and mass spectroscopies of the solute and raw soot material formed via the MW-CVD method indicate a complex assortment of various carbons and hydrocarbons. For example, in the UV-vis spectra of soot material dissolved in ether or hexane, the absorption band at 250 nm is routinely observed. Since corannulene has a pronounced absorption peak at 250 nm (*26*), it is possible that corannulene is a precursor of fullerenes in the MW-CVD method. However positive identification of the individual species will require extraction, purification, and further analysis.

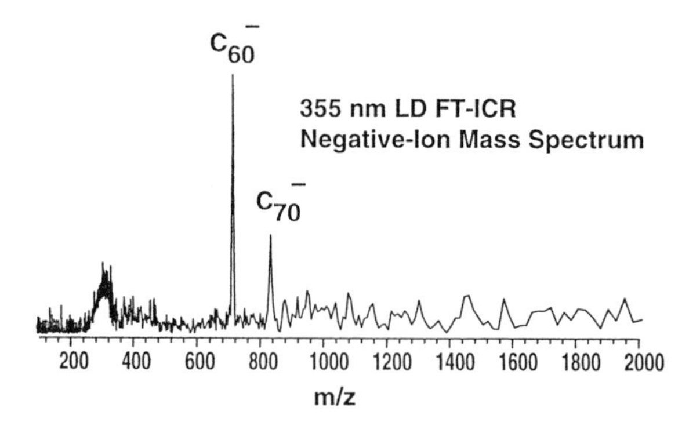

Figure 5. FT-ICR negative-ion mass spectrum of the sample (dissolved in toluene) collected at the fiber wool trap in the microwave enhanced CVD chamber.

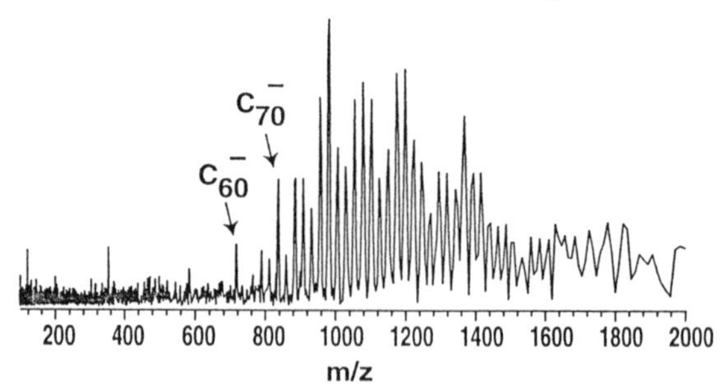

Figure 6. FT-ICR negative-ion mass spectrum of raw sample collected at the LN trap in the microwave enhanced CVD chamber.

The polymer-like films produced in the MW-CVD chamber are also intriguing. It is significant that simple molecules like acetylene can so readily be converted into complex polymer-like films in the microwave plasma. In contrast to its compositional complexity is its simple UV-vis absorption spectrum. The ability to control the absorption edge of these films through processing parameters may be useful for new applications.

Summary

We have demonstrated that fullerenes can be synthesized in both HF-CVD and MW-CVD methods. The observation of fullerene under HF-CVD diamond-forming conditions suggests a close relationship between diamond nucleation and fullerene formation. We believe that the formation mechanism of fullerene in the CVD process could be similar to that in the combustion method (*27*), where acetylene addition at the early stage and hydrogen abstraction at the late stage are important. We have shown that CVD is a versatile technique capable of producing not only fullerene but also polymer-like hydrocarbon films.

Acknowledgment

The authors at UCF acknowledge partial support from the College of Arts and Sciences, University of Central Florida. The ASU portion of this study was suppoted by NSF Grant no. EAR-9219376 (to PRB). The FSU portion was supported by NSF Grant no. CHE-93-22824, NIH Grant no. GM-31683, and the National High Magnetic Field Laboratory at Florida State University.

References

1. H. W. Kroto, J. R. Heath, S. C. O'Brien, R. E. Curl and R. E. Smalley, *Nature* **318**, 162 (1985)
2. W. Kratschmer, L. Lamb, K. Fostropoulos, and D. R. Hoffman, *Nature* **347**, 354 (1990)
3. Ph. Gerhardt, S. Loffler, and K. H. Homann, *Chem. Phys. Lett.*, **137**, 306 (1987).
4. J. B. Howard, J. T. McKinnon, Y. Markovsky, A. L. Lafleur, and M. E. Johnson, *Nature* **352**, 139 (1991).
5. R. E. Haufler, et al, *J. Phys. Chem.* **94**, 8634 (1990)
6. R. Taylor and G. J. Lanley, in "Recent Advances in the Chemistry and Physics of Fullerenes and Related Materials" Ed. by K. M. Kadish and R. S. Ruoff, p. 68, The electrochemical Society, Inc. (1994).
7. K. Yoshie, S. Kasuya, K. Eguchi, and T. Yoshida, *Appl. Phys. Lett.*, **61**, 2782 (1992).
8. K. Inomata, N. Aoki, and H. Koinuma, *Jpn. J. Appl. Phys.*, **33**, L197 (1994).
9. G. Peters and M. Jansen, *Angew. Chem. Int. Ed. Engl.*, **31**, 223 (1992).
10. D. H. Parker, P. Wurz, K. Chatterjee, K. R. Lykke, J. E. Hunt, M. J. Pellin, J. C. Hemminger, D. M. Gruen, and L. M. Stock, *J. Am. Chem. Soc.*, **113**, 7499 (1991).

11. Q. L. Zhang, S. C. O'Brien, J. R. Heath, Y. Liu, R. F. Curl, H. W. Kroto, and R. E. Smalley, *J. Phys. Chem.* **90**, 525 (1986).
12. A. L. Lafleur, J. B. Howard, J. A. Marr, and T. Yadav, *J. Phys. Chem.* **97**, 13539 (1993).
13. J. R. Heath, in *Fullerenes: Synthesis, Properties, and Chemistry of Large Carbon Clusters* (eds. G. S. Hammond and V. J. Kuck) 1-23, (American Chemical Society, Washington, DC, 1991).
14. R. E. Smalley, *Acc. Chem. Res.* **25**, 98 (1992).
15. L. Chow, A. Horner, and H. Sakouri, B. Roughani, and S. Sundaram, *J. Mater. Res.* **7**, 1606 (1992).
16. L. Chow, H. Wang, S. Kleckley, T. K. Daly, and P. R. Buseck, *Appl. Phys. Lett.*, **66**, 430 (1995).
17. C. P. Chang, D. L. Flamm, D. E. Ibbotson, and J. A. Mucha, J. *Appl. Phys.*, **63**, 1744 (1988).
18. L. Chow, H. Wang, S. Kleckley, A. Schulte, and K. Casey, *Solid State Comm.* **93**, 999 (1995).
19. Y. Huang, L. Pasa-Tolic, S. Guan, A. G. Marshall, *Anal. Chem.*, **66**, 4385 (1994).
20. R. L. Hettich and P. R. Buseck, *Carbon*, **34**, 685 (1996).
21. J. Narayan, *J. Mater. Res.* **5**, 2414 (1990).
22. Qi Lichang, Xuan Zhenwu, Yang Peichun, Pu Xin, and Hou Li, *J. Crystal Growth*, **112** 580 (1991).
23. R. Meilunas, R. Chang, S. Liu, M. Kappes, *Appl. Phys. Lett.,* **59**, 3461 (1991).
24. S. Ino and S. Ogawa, *J. Phys. Soc. of Japan*, **22** 1365 (1967).
25. D. Gruen, S. Liu, A. Krauss, J. Luo, and X. Pan, *Appl. Phys. Lett.*, **64**, 1502 (1994).
26. A. L. Lafleur, J. B. Howard, J. A. Marr, and T. Yadav, *J. Phys. Chem.* **97**, 13539 (1993).
27. C. J. Pope, J. A. Marr, and J. B. Howard, *J. Phys. Chem.*, **97**, 11001 (1993).

Chapter 7

Chemical Vapor Deposition of Organic Compounds over Active Carbon Fiber To Control Its Porosity and Surface Function

Yuji Kawabuchi, Chiaki Sotowa, Keiichi Kuroda, Shizuo Kawano, D. Duayne Whitehurst, and Isao Mochida

Institute of Advanced Material Study, Department of Molecular Science and Technology, Graduate School of Engineering Sciences, Kyushu University, Kasuga, Fukuoka 816, Japan

Chemical vapor deposition (CVD) of some organic compounds was examined to control the porosity and surface function of active carbon fiber (ACF). In this system, the deposition takes place only on the pore wall of the ACF, when the precursor organic compound and deposition temperature around 700°C were selected carefully. The surface of the ACF was modified by carbon derived from heterocyclic compounds (pyridine, pyrrole, furan and thiophene) through CVD. The moderately activated ACF modified by pyridine, pyrrole and thiophene showed molecular sieving activity, that modified by furan did not. Only furan was decomposed at this temperature. Thermal stability is a key factor to get molecular sieving performance after CVD. Pyridine produced amorphous carbon within the pore, which appears to maintain the pyridine ring structure, creates basic sites over the surface of the ACF. Thus catalytic oxidation of SOx over ACF of high surface area was accelerated.

Pore size and surface functionality of active carbon fiber (ACF) were modified by design by selective chemical vapor deposition (CVD) where the deposition temperature was selected carefully (1). In this system, particular organic compounds, like benzene, are useful due to their thermal stability (2). The pore structure of ACF was another key factor which influences strongly its performance after CVD. The ACF of low activation, which carries micropores in majority, showed molecular sieving activity, when the carbon deposition automatically ceased at a certain level. Mean while the ACF of high activation, which carries mesopores, never did show molecular sieving activity, since most carbon was deposited in the mesopores, which plugged the micropores. In this paper, CVD onto ACFs of some thermally stable aromatic compounds containing heteroatoms was tried in order to control their porosity and surface functionality. The treated ACFs were evaluated with respect to both gas separation of CO_2/CH_4 and catalytic oxidation of SO_x for its removal from flue gas.

EXPERIMENTAL

ACF and its characterization. Some of physical and chemical properties of two commercial pitch based active carbon fibers (ACF-1 and 2) used in the present study are listed in Table 1. After outgassing at 150°C for 4h, BET surface areas were measured by N_2 adsorption at -196°C, using a Simazu ASAP 2000 apparatus.

CVD by organic compounds pyrolysis over active carbon fiber. Pyrolysis of heterocyclic compounds (pyridine, pyrrole, furan, and thiophene) were performed by flowing a helium stream containing controlled amounts of the organic compounds at 150ml/min over the ACF (200mg) suspended in a quartz basket in a micro balance (CAHN 1000). A thermocouple, in a glass tube, was inserted directly below the quartz basket. The sample was heated in the He flow to the fixed temperature at a programmed rate of 10°C/min and maintained at the temperature for 1h. Then, a prescribed amount of organic compound was supplied by a micro feeder into the He gas flow to be sent to the heated ACF for a fixed period. The weight uptake of the sample was recorded continuously by the balance.

Identification of the organic species produced in the pyrolysis of organic compounds. The organic species produced from organic compounds as substrates were analyzed at the outlet of the reactor, using both GC-FID (Yanaco, G-2800) and GC-MS (Simazu, GC-17A-QP-5000). The latter detector was mainly used to detect heavy products (>50 a.m.u.). Products deposited on the reactor wall were recovered and analyzed by GC.

Adsorption of CO_2 and CH_4. Adsorptions of CO_2 and CH_4 on the ACFs were carried out separately using a volumetric adsorption apparatus. Each ACF was evacuated to 10^{-2} torr at 150°C for 1h prior to the adsorption study. The gas volumes adsorbed by ACF within a given time interval at 30°C were determined from the change of pressure in the closed system. The initial pressure of each gas to be adsorbed was fixed at 760torr. The amount adsorbed vs. time was plotted to determine the rate of the adsorption. CO_2 adsorption capacity is defined in the present paper as the amount of CO_2 adsorbed in 2min. CO_2 selectivity is defined as the ratio of CO_2 adsorbed in 2min to CH_4 adsorbed in 2min. These parameters are more reflective of kinetic properties than equilibrium properties, as are appropriate parameters for PSA applications.

SO_2 removal. SO_2 removal was carried out at 30°C, using a fixed bed flow reactor (3). The weight of ACF was 0.25g. The total flow rate was 100ml/min. The model feed gas contained 1000ppm SO_2, 5vol% O_2 and 10vol% H_2O in nitrogen. Aq. H_2SO_4 was recovered at the outlet of the reactor. SO_2 concentrations in the inlet and the outlet gases were observed continuously by a flame photometric detector (FPD).

Temperature-Programmed-Decomposition of SO_2 adsorbed on ACFs. Temperature-Programmed-Decomposition (TPDE) spectra of the ACFs were measured by a using an apparatus made of quartz equipped with a mass spectrometer (ANELVA AQA-200). A mixed gas containing SO_2 (1000ppm) and O_2 was adsorbed on ACF of 0.1g at room temperature. The ACF of 0.1g was heated to 500°C by 10°C/min while the evolved gases such as SO_2 were analyzed by the mass spectrometer.

RESULTS

CVD on ACF (ACF-1) of low activation

Pyrolysis of organic compounds over ACF. The weight increases of ACF during the pyrolysis of organic compounds at 725°C are illustrated in Figure 1, where the concentration of organic compounds were 2vol% in He flow. Pyridine increased the weight of ACF slowly upto 110mg/g within 40min when the weight gain ceased. Pyrrole, furan, and thiophene also showed similar profiles of weight increase and saturation level. However, the time to achieve the saturation with pyridine was longer than with pyrrole but shorter than with thiophene.

Table 1 Some properties of active carbon fiber

Sample	carbon deposition amount (%)	Elemental analyses (%) C	H	N	O	Ash	S.A.[2] (m/g)	P.V.[3] (ml/g)	SO_2[4] (mg/g)
ACF[1]-1 as-received	0	88.7	0.9	0.7	9.7	—	700	0.35	—
ACF[1]-2 as-received	0	93.9	0.7	0.3	4.6	0.5	2150	1.08	9
Treated ACF-2 (Point1)	8	95.3	0.7	1.5	2.4	0.1	1880	0.94	—
Treated ACF-2 (Point2)	16	94.5	0.6	2.3	2.4	0.2	1590	0.78	19
Treated ACF-2 (Point3)	25	93.4	0.6	3.0	2.9	0.1	1270	0.62	—
Treated ACF-2 (Point4)	46			———			1.3	0	—

1) activated carbon fiber from coal tar pitch, fiber diameter 10-15 μm

2) surface area

3) pore volume

4) SO_2 (1000ppm) adsorption capacity at 30℃, 760torr using gravimetric system

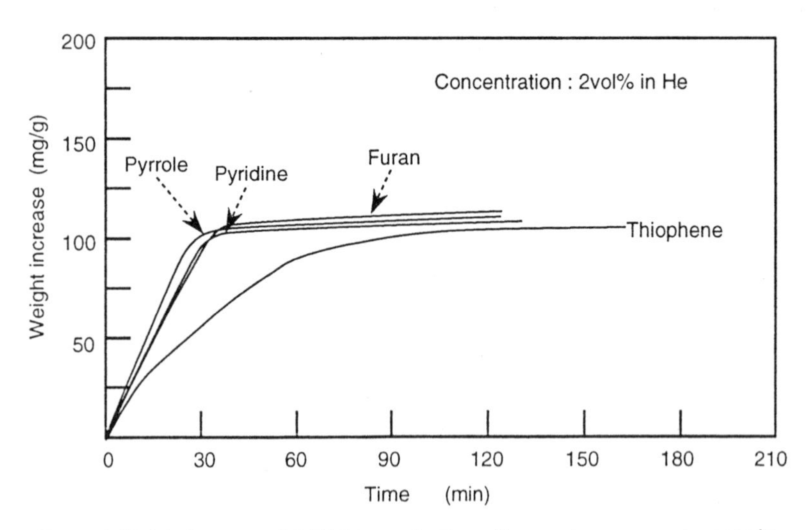

Figure 1 Weight increase of ACF-1 by contacting with organic compounds at 725℃.

Products of the pyrolysis reaction. Table 2 shows GC-FID and GC-Mass (heavy products; >50 a.m.u. detected by GC-Mass) analysis of products at the outlet of the reactor. Only pyridine was detected in the gas phase after the pyrolysis of pyridine at 725°C while a small amount of dimer was found on the reactor wall. Thiophene stayed also unchanged in this temperature, although very small amounts of its dimer were also detected as in the case of pyridine. Furan was reactive, producing mainly butadiene, cyclopentadiene, and benzene (>50 a.m.u.) at concentrations of over 2%

Adsorption profiles of CO_2 and CH_4 on as-received and treated ACFs. Figure 2 illustrates adsorption profiles of CO_2 and CH_4 on as-received and treated ACFs. As-received ACF adsorbed both CO_2 and CH_4 very rapidly within 1 min, and adsorptions of both CO_2 and CH_4 were apparently saturated by 2min. After 2min, the as-received ACF adsorbed 37ml/g of CO_2 and 16ml/g of CH_4, giving a CO_2 selectivity (CO_2/CH_4) of 2.3. The ACF, which gained weight by CVD of pyridine to 110mg/g of the saturation level, showed an excellent CO_2/CH_4 adsorption selectivity of about 50, while CO_2 adsorption capacity was decreased only slightly to 86% of that on the original ACF. The adsorption of CH_4 was markedly reduced by CVD to less than 1ml/g, thus dramatically improving the selectivity. Pyrrole and thiophene provided excellent selectivity of CO_2/CH_4 adsorption as pyridine. The ACFs treated by furan were found to have poor CO_2/CH_4 kinetic adsorption selectivities. The CO_2 adsorption was very slow and no saturation was observed by 5min.

CVD on ACF (ACF-2) of high activation

Pyrolysis of pyridine over ACF. The weight increases of the ACF during the pyrolysis of pyridine at 725°C are illustrated in Figure 3, where the concentration of pyridine was 2vol% in He flow. Up to 600°C, pyridine did not react, no weight gain was found. At 700°C , the weight of ACF increased slowly. At 725°C, the weight of the ACF increased up to 460mg/g within 90min when the weight gain ceased, as shown in Figure 3. ACFs defined as points 1-4 in Figure 3 where the weight gains of the ACF were 8, 16, 25 and 46%, were analyzed for nitrogen content which increased after CVD treatment by pyridine. Carbon deposition on this particular ACF decreased its surface area and pore volume (see Table 1). This ACF never did show molecular sieving activity, even though the carbon deposition ceased at a defined level.

Table 2 Products of the pyrolysis reaction

organic compaund	reaction temperature (℃)	products at the outlet of the reactor
N	725	N > N N trace (mol ratio)
O	725	O 71%; CH_2=CH–CH=CH_2 17%; 8%; CH_2 CH CH CH–CH 3%; CH_3 1% (T.I.C. ratio ; > 50 a.m.u.)
S	780	S > S S trace (mol ratio)

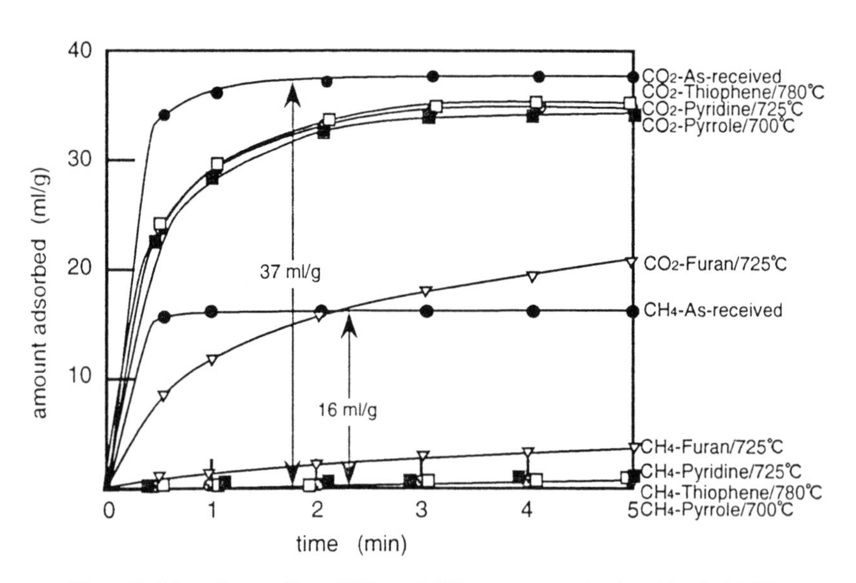

Figure 2 Adsorption profiles of CO_2 and CH_4 over as-received and treated ACFs.

Initial pressure : 760 torr Final pressure : 500~600 torr

Catalytic activity on SO_2 removal. Figure 4 illustrates effects of carbon deposition from benzene and pyridine on SO_2 breakthrough profiles over pitch-based ACF at 30°C, in the presence of 10% H_2O. The treatment by pyridine was very effective in decreasing the level of steady state SO_2 concentration. The modified ACF showing 8% of carbon deposition from pyridine (Point 1) showed complete removal of SO_2 for at least 15h at a W/F of $2.5x10^{-3}$ $g{\cdot}min{\cdot}ml^{-1}$. However, the ACFs modified by benzene showed rather poor activities compared to the ACF modified by pyridine, even though the same increase of weight was obtained.

TPD of adsorbed SO_2. Figure 5 illustrates the TPDE profiles of as-received ACF and pyridine treated ACF, after they were exposed to 1000ppm SO_2 in He carrier gas containing O_2. As-received ACF liberated SO_2 in the temperature ranges of 30-80°C and 120-400°C with the peaks at 50 and 240°C, respectively. The peaks are attributed to adsorbed SO_2 and SO_3 (H_2SO_4) (4), respectively. ACF modified by pyridine also showed two similar peaks, and liberated much more SO_2 than the as-received ACF. The adsorption of SO_2 was enhanced very much, in the pyridine-modified ACF.

DISCUSSION

In previous studies (2, 5, 6), we suggested the following scheme of carbonization which narrows the pore width. Benzene which was the one of the best organic compound for achieving a molecular sieving capability, is adsorbed on the pore wall, and by catalytic action of the ACF it is condensed into non-volatile substances within a short time. The adsorbed species are carbonized into isotropic carbon as a results of minor rearrangement of adsorbed species into graphitic layer stacking. The adsorption leading to such a carbonization is restricted to within the pores and does not occur on the outer surface of ACF. Under the flow conditions, adsorption on the outer surface is low and the residence time is not long enough for the carbonization to occur. Thus, carbonization within the pores continued to thicken the carbon coating on the walls as long as benzene could diffuse into the pores. When the pore width was reduced by the carbon coating and become less than the molecular thickness of benzene (0.37nm), the carbonization stopped automatically. The pore width of 0.37nm is in-between the molecular sizes of CH_4 (0.38nm) and CO_2 (0.33nm), and can distinguish them with true molecular sieving ability (7).

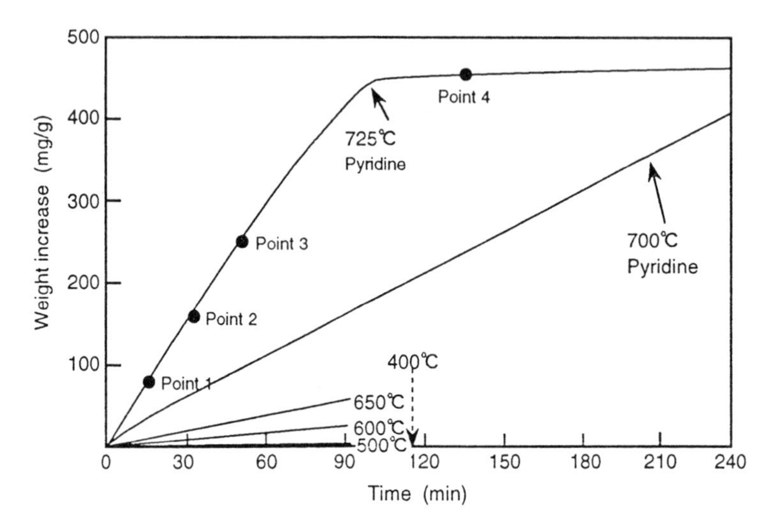

Figure 3 Decomposition profiles of pyridine over ACF-2 at some temperature.

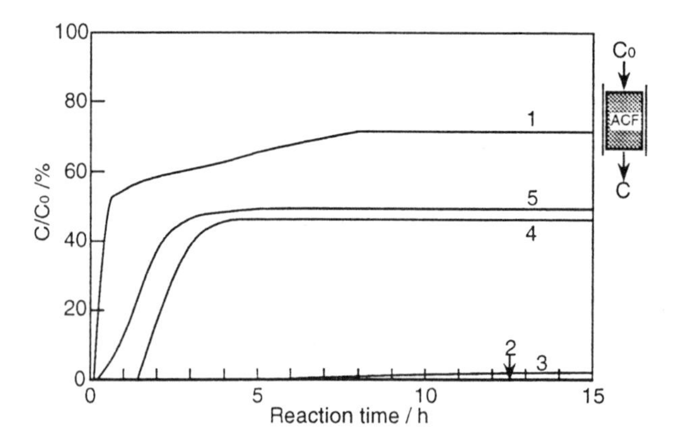

Figure 4 Breakthrough profiles of SO_2 over Pitch-ACFs.

SO_2 1000ppm, O_2 5 vol%, H_2O 10 vol%
Reaction Temp. 30℃, W/F = 2.5×10^{-3}g min ml^{-1}

1: As-received ACF-2
2: Treated ACF-2 by pyridine-8wt%
3: Treated ACF-2 by pyridine-16wt%
4: Treated ACF-2 by benzene-8wt%
5: Treated ACF-2 by benzene -16wt%

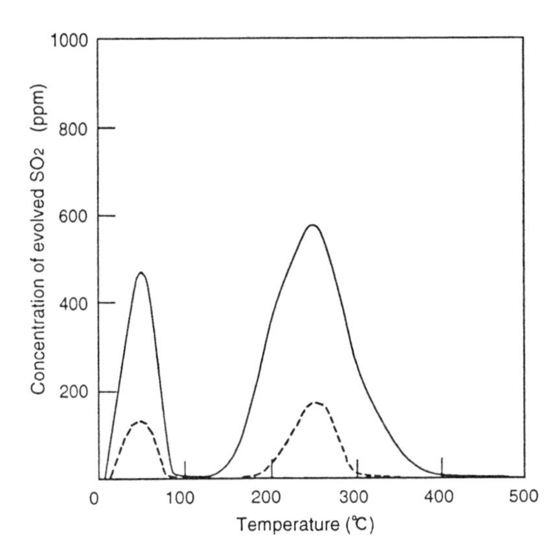

Figure 5 TPDE spectras of SO_2 evolution from as-received ACF and treated ACF.
Weight: 100mg, Carrier gas: He, Flow rate: 100ml/min

.... As-received ACF-2

———— ACF carbon deposited from pyridine (16%)

In the present study, pyridine, pyrrole, furan, and thiophene were used as the carbon precursor in CVD, because they possess similar molecular shape and thickness to benzene, although heteroatoms are included in their rings. Pore size in an activated pitch based ACF could be successfully controlled at the nano-scale by carbon deposition from pyridine, thiophene and pyrrole at a specific range of temperature (around 700°C) to obtain molecular sieving activity sufficient to separate CO_2 from CH_4 . Although furan also showed a similar profile of weight increase of ACF by carbon deposition, the molecular sieving selectivity was not improved at all. The kinetic adsorption capacity was also significantly diminished. Only furan among the compounds examined at this temperature range was decomposed to a variety of products. Thermal stability of aromatic compounds is a key factor in getting molecular sieving activity by CVD. Similar treatment failed to introduce molecular sieving activity to another fiber of high activation. ACF microporous structure is also another key obtaining fine control of pore size. The wall of mesopores can be coated by the deposited carbon to plug the micropores.

The present study also reports introduction of the significant catalytic activity to the pitch-based ACF (S.A.2000m^2/g) after the CVD treatment by pyridine. The activity observed in the present study allowed the complete removal of 1000ppm SO_2 at room temperature by W/F of 2.5×10^{-3} $g \cdot min \cdot ml^{-1}$. The ACF modified by pyridine showed higher catalytic activity than the ACF modified by benzene, even though the same amount of carbon was deposited. Pyridine provided at 725°C basic functionality on the surface. Thus, catalytic SOx removal is very much enhanced.

TPDE profile of SO_2 from ACF after the adsorption of SO_2 and O_2 exhibited two adsorbed species. More SO_2 was found over modified ACF than those of as-received or benzene treated ACFs. Pyridine derived grain accelerates the adsorption and oxidation activity against SO_2. The larger amount of deposited carbon provides an increased number of nitrogen atoms, however the surface area tends to decrease. Thus, the balance between amount of nitrogen and surface area must be optimized for the highest activity.

Literature Cited.

1. Kawabuchi, Y. ; Kawano, S. ; Mochida, I. *Carbon*, **1996**, 34, 711.
2. Kawabuchi, Y.; Kishino, M.; Kawano, S; Whitehurst, D. D.; Mochida, I. *Langmuir*, **1996**, 12, 4281.
3. Kisamori, S.; Kuroda, K.; Kawano, S.; Mochida, I.; Matsumura, Y.; Yoshikawa, M. *Energy & Fuels*, **1994**, 8, 1337.
4. Yamamoto, K.; Seki, M.; Kawazoe, K. *Nippon Kagaku Kaishi*, **1972**, 1046.
5. Kawabuchi, Y.; Kishino, M.; Kawano, S; Whitehurst, D. D.; Mochida, I. *Proceeding of the European Carbon Conf.*, Newcastle, UK, pp. 570 (1996).
6. Kawabuchi, Y.; Sotowa, C.; Kishino, M.; Kawano, S.; Whitehurst, D.D.; Mochida, I. *Chem. Lett.*, **1996**, 941.
7. Kawabuchi, Y. ; Kawano, S. ; Mochida, I. *Carbon*, **1995**, 33, 1611.

Chapter 8

Recombination of Oxygen and Nitrogen Atoms on Silica and High-Temperature Coating Materials

Young C. Kim, Sung-Chul Yi, and Sei-Ki Moon

Department of Chemical Engineering, Hanyang University, Seoul 133–791, Korea

The recombination of oxygen and nitrogen atoms on the surfaces of two coating materials of the Space Shuttle Orbiter (SSO), a reaction cured glass (RCG) and a spinel (C742), was investigated. The recombination probability, γ, i.e., the probability that atoms impinging on the surface will recombine, was measured in a diffusion reactor. Value of γ for oxygen atom on C742 (3×10^{-2}) was much higher than on RCG (4×10^{-4}) at the temperature of SSO re-entry (*ca.* 1000K). These results agree with γ estimated from the surface temperatures of the coating materials during the actual SSO re-entry; 1×10^{-2} for C742 and 5×10^{-4} for RCG. Atom recombination occurs at surface active sites which cover a small fraction of the surface and chemisorbed atoms. Above 700K, the chemisorbed atoms react mainly with gas-phase atoms impinging on the active sites. The higher value of γ on C742 indicates a higher number density of active site and a larger surface area than RCG. Interestingly, γ above 1000K on both surface decreases due to the desorption of the chemisorbed atoms from the active sites. It suggests the probability of designing less active surface by inducing the desorption at lower temperature.

A new concept for space vehicles, such as a Space Shuttle Orbiter (SSO), is the repeated use of vehicles (*1,2*). In the past, the vehicles for one time flight have entered and free-fallen in the earth's atmosphere (*3*). This abrupt passage of vehicle through air generated such a high surface temperature that an ablative heat shield was evaporated from the vehicles. One of the most important consideration in the development of an SSO is the design of a heat shield that can survive repeated re-entries into the earth's atmosphere. The SSO has a capability to maneuver and to control its descending trajectory. The actual trajectory is determined by flying close to the thermally-allowed surface temperature. For vehicle descending with a slower speed, a corresponding long re-entry time requires thicker insulation tiles. Since the payload of the vehicle is central to its economical viability, a faster descending trajectory is preferred.

The highest surface temperature in the trajectory occurs between 60km and 100km in altitude due reactions on the surface of the tiles. This flight through the upper atmosphere causes dissociation of dioxygen and dinitrogen in the shock wave generated in front of the vehicle. The atoms recombine on the exterior surface of the SSO, releasing the large heat of reaction (498 kJ mol^{-1} for oxygen) which contributes up to half of the overall heat transfer to the vehicle. The maximum surface temperature rise from this source would occur if all atoms arriving on the surface recombine and release the heat of recombination to the surface. At the opposite extreme, the temperature rise would be minimum for an inactive surface.

The effect of atom recombination on the surface temperature of the SSO has been directly demonstrated during the actual re-entry. The coating material on insulation tiles used in the SSO is called a reaction cured glass (RCG) (*4-7*), which consists of *ca.* 90% silica. In flights of STS-2, -3, and -5, some of the RCG tiles were over-coated with a spinel, mixture of Fe, Co, and Cr oxides (Ferro Co., C742). These tiles were placed side by side during actual SSO re-entry and the surface temperature of these tiles was monitored. Under re-entry condition, the surface temperature of C742 tiles (1400K) was much higher than that of RCG (1150K), indicating that the RCG surface is considerably less active toward recombination of atoms.

In this work, the apparent recombination coefficient, γ_A, i.e., the probability that atoms colliding on the surface will recombine, was measured in a well-defined system built. Our results were compared with the recombination probability γ_A estimated from the surface temperatures measured during the re-entry of the SSO (*5*).

Experimental

The apparent recombination probability of oxygen and nitrogen atoms on surfaces of tiles was measured in a diffusion tube first described by Smith (*8*). A detailed description of the apparatus and technique to measure γ_A has been described elsewhere (*9*). Atoms generated in a microwave discharge diffused down a closed-end diffusion tube, the wall of which was covered with the surface under study. The depletion of atoms down the tube due to atom recombination on the wall was measured with a Pt/Pt-10% Rh thermocouple probe traveling along the axis of the tube. The temperature rise of the probe after turning on the atom generator, ΔT, was assumed to be proportional to the atom concentration in the gas phase (*8*). When the rate of atom recombination was first order with respect to the atom concentration in the gas phase, ΔT decreased exponentially with the distance away from the atom generator. From the slope of the plot of $\ln \Delta T$ *vs.* the distance from the atom generator, the apparent recombination probability was obtained (*9*). The steeper slope indicates a higher γ_A of the wall. Experiments were carried out from 194 to 1250K by using either a cooling bath or heater around the diffusion tube.

Two SSO coating materials (RCG and C742) have been tested. The surface of RCG was prepared in the same way used in the preparation of the insulation tiles of the SSO. A mixture of 94% Glass (Corning 7930, 96% silica), 4% boron oxide and 2% SiB_4 was ground to less 325 mesh in ethyl alcohol (*4-7*). The mixture was stirred for one day before it was applied inside a fused quartz tube (18mm O.D., 21mm I.D., General Electric, type 204) that fitted closely inside the diffusion tube(25mm O.D., 22mm I.D., 90cm in length). The sample tube was dried at room temperature (RT) and the surface was oxidized and fused at 1500K for 1.5 hours in air. The C742 surface was prepared by applying a thin layer of the spinel over a RCG surface. The same batch of spinel powder in polyvinyl acetate used in the SSO (*10*) was applied and dried at RT. The C742 sample tube was also heated at 1500K for 1.5 hours.

Dioxygen (Liquid Carbonic, 99.995%) and dinitrogen (Liquid Carbonic, 99.999%) were further purified by passage through molecular sieves refrigerated at 194K. Dry air was synthesized by mixing the purified dioxygen and dinitrogen.

Results

Plots of $\ln\Delta T$ *vs.* axial distance of the thermocouple tip from the atom generator divided by the radius of the sample tube gave straight lines in all experiments. This confirmed that the rate of atom recombination was first order with respect to the atom concentration in the gas phase. The apparent recombination probability γ_A was calculated from the slopes of these straight lines. The true recombination probability γ was calculated by dividing γ_A by a surface roughness factor F. The factor F was the ratio of the real surface area to the geometric surface area. In this study, the sample surface were assumed to be as rough as a fused quartz (F = 2.4) (*11*).

In Fig. 1, the recombination probability of oxygen was plotted from 194K to 1250K. At low temperature, LT, (194K), γ increased with increasing temperature. At medium temperature, MT, (300~700K), γ decreased slightly as temperature increased. At high temperature, HT, (700~1000K), γ increased again with increasing temperature. Finally, above *ca* 1000K, γ decreased. At all the temperatures, the recombination probability on C742 was more than an order of magnitude larger than on RCG. Temperature dependence of γ for nitrogen atom was similar to that of oxygen, as shown Fig. 2. The recombination probability measured in air fell between the values of γ for oxygen and nitrogen.

Discussion

Comparison with actual flight experiments. Attention is now turned toward a comparison of recombination probability obtained in the diffusion tube and that calculated from surface temperatures during actual re-entry. The surface temperatures of the RCG and C742 tiles were continuously monitored during the re-entry flight. The actual flight conditions were obtained from a combination of radar, onboard

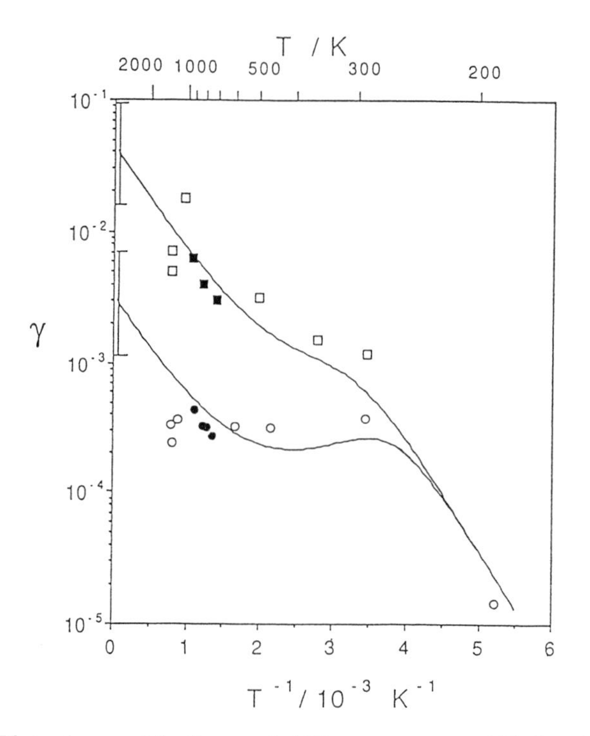

Figure 1. Plot of recombination probability, γ, versus 1/T for the oxygen atom recombination.

□ C742
○ RCG

Filled data points at high temperature are used for data fitting.
Solid lines are the calculated values with the parameter shown in Table II.

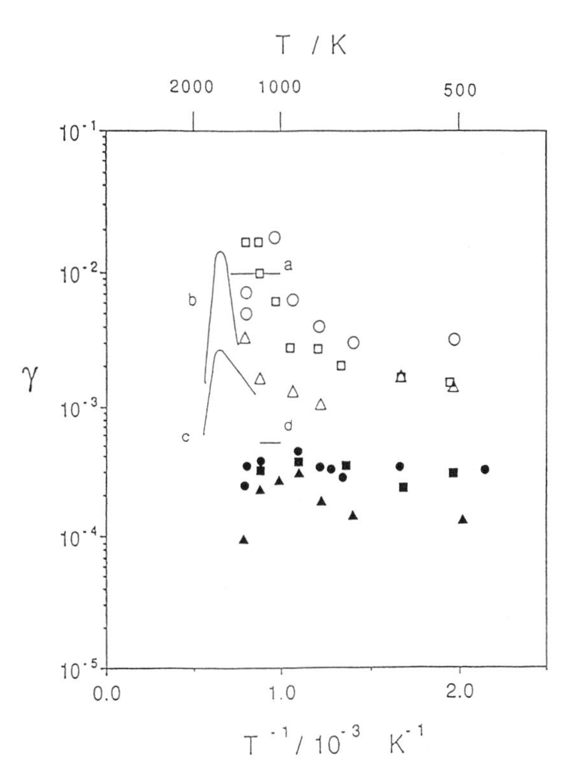

Figure 2. Recombination probability, γ, on the surfaces of materials used on the Space Shuttle Orbiters.

a	Flight experiment on C742 (Reference 5)
b	Arc jet experiment on C742 (Reference 11)
○	Oxygen recombination on C742 (This work)
△	Nitrogen recombination on C742 (This work)
□	Recombination in air on C742 (This work)
c	Flight experiment on RCG (Reference 11)
d	Arc jet experiment on RCG (Reference 5)
●	Oxygen recombination on RCG (This work)
▲	Nitrogen recombination on RCG (This work)
■	Recombination in air on RCG (This work)

instrumentation, and meteorological instruments. A recombination probability was calculated from the surface temperatures through extensive computational analysis of the flow-field and boundary layer around the SSO (*8*), which included the transport of mass, momentum, energy, and concentrations of chemical species. Rakich *et al.* (*5*) obtained good fits between the calculated and the measured surface temperatures on the RCG (850~1150K) and the C742 (1000~1400K) when values of γ were 5×10^{-4} and 1×10^{-2}, respectively. As shown in Fig. 2, values of γ of oxygen in the diffusion tube on both RCG and C742 are in good agreement with the results from the actual flight data in a factor of two. Under re-entry conditions, the concentration of nitrogen atoms is much smaller than that of oxygen since the thermal dissociation energy of dinitrogen (950 kJ mol^{-1}).

In a ground simulation of re-entry by using an arc-jet wind tunnel (*12*), the activity of RCG and C742 coating materials for oxygen and nitrogen recombination has been also estimated. The value of γ was estimated from the rise of surface temperature due to recombination of atoms which are generated by the arc. Stewart *et al.* (*12*) reported the recombination probability of oxygen and nitrogen atoms on RCG and C742 by using an arc-jet facility at NASA Ames Research Center. Arc-jet experiments are restricted only to temperature above 1000K. Their results for each coating material appear to be generally consistent (Fig. 2), showing the fall-off and a higher activity of C742. To understand the difference in activity between RCG and C742, the mechanism of atom recombination should be understood in the hopes of development of a less active material.

Mechanism of Atom recombination. In order to explain the complex temperature dependence of γ of atom on the surface of silica, a mechanism of atom recombination on surface was proposed (*11*). The surface is covered with a small fraction of active sites, denoted by *, which hold chemisorbed atoms irreversibly. The chemisorbed atoms A* recombine with atoms A arriving at the active sites. In addition to the direct arrival of atoms from the gas phase, atoms reversibly adsorbed on the rest of the surface diffuse along the surface to the active sites. Here is a proposed sequence of elementary steps for adsorption, desorption, surface diffusion, and recombination at the active sites:

$A + * \longrightarrow A*$	irreversible adsorption on active site
$A \longrightarrow A_r$	reversible adsorption anywhere else
$A_r \longrightarrow A_r$	surface diffusion
$A_r \longrightarrow A$	desorption
$A_r + * \longrightarrow A_2 + *$	Langmuir-Hinshelwood recombination
$A + * \longrightarrow A_2 + *$	Eley-Rideal recombination

The proposed mechanism that explains is covered by active sites, which strongly hold chemisorbed atoms. Atoms impinging on the surface reach to the active site either by direct impingement on the active sites or by surface diffusion.

At sufficiently low temperature (LT), all atoms impinging on the surface are assumed to be adsorbed and to active sites before desorption. At high temperature

(HT), the effect of surface diffusion of adsorbed atoms is negligible since the adsorbed atoms desorb before reaching the active sites, so atoms can reaching the gas phase on the active sites. The recombination probability at HT is thus expressed as:

$$\gamma = \varphi \exp(-E/RT) \tag{1}$$

where φ is the fraction of the surface area covered by active sites, exp(−E/RT) is the reaction probability of an arriving atom with the chemisorbed atom and E is the activation energy required for recombination.

In Fig. 1, the filled data points in the HT region before the fall-off were analyzed with the two adjustable parameters, φ and E. The values obtained from a least squares fit are shown in Table Ⅰ. Value of φ for RCG (3×10^{-3}) indicates that less than one percent of the surface is covered by active sites, while the higher φ for C742 (5×10^{-2}) indicates that C742 contains more surface active sites per unit area. The activation energy (16 kJ mol^{-1}) is almost same as that on silica (*11*).

Let us analyze the data from LT at HT quantitatively. Henry (*13*) solved the diffusion-reaction equation in the model, where the active sites of radius a are distributed uniformly by a distance 2b. The recombination probability is

$$\gamma = \varphi + \left[\frac{2aX_D}{b^2}\frac{I_1(b/X_D)K_1(a/X_D) - I_1(a/X_D)K_1(b/X_D)}{I_0(a/X_D)K_1(b/X_D) - I_1(b/X_D)K_0(a/X_D)}\right]\exp(-E/RT) \tag{2}$$

where the first term, $\varphi = (a/b)^2$, is the contribution of direct impingement on the active sites. The second term is due to the surface diffusion, where I_i and K_i are the modified Bessel functions of *i*th order. The term surface diffusion distance of a adsorbed atom before desorption, X_D, is

$$X_D = (a/2)\,\beta^{1/2} \exp((1-\rho)E_d/2RT) \tag{3}$$

where $\beta (= \nu_D / \nu_d)$ and $\rho (= E_D / E_d)$ are the ratios of the frequency factors and activation energies, respectively, ν_D and ν_d are the frequency factors for surface diffusion and desorption from the surface, respectively.

With assumed values (*11*) of $\beta = 10^{-2}$, $\rho = 0.5$, and the values obtained from HT data, E = 16 kJ mol^{-1} and $\varphi = 3 \times 10^{-3}$ (for RCG) and 4×10^{-2} (for C742), results from LT to HT were analyzed by a least-squares fit of Eq. 2. As shown in Table Ⅱ, that yields the value of the only adjustable parameter, E_d, which is close that on silica (51 kJ mol^{-1}) (*11*). The calculated curve shown in Fig. 1 fits the data from LT to HT reasonably well.

It is interesting that γ on RCG and C742 decreases above 1000K and 1100K, respectively. All our results are reversible with respect to the surface temperature. The fall-off of γ at very high temperature is attributed to the thermal desorption of the chemisorbed atoms. As a result, the possibility that two atoms meet each other on the surface diminishes. Greaves and Linnett (*14*) also observed the slight decrease of γ in oxygen recombination on a silica glass above 900K. The fall-off of γ on a silica

Table I. Activation energy and pre-exponential factor of atom recombination

Surface	E / kJ mol^{-1}	φ
C742	17±5	5.3×10^{-2}
RCG	15±2	2.9×10^{-3}

Error bars of φ are shown in Fig. 1.

Table II. Estimation of E_d by curve fitting

Coating	RCG	C742
ρ	0.5	
β	10^{-2}	
E / kJ mol^{-1}	16	
φ	3×10^{-3}	5×10^{-2}
E_d / kJ mol^{-1}	53±1	44±1

above 1250K has been reported for nitrogen recombination (*15*). The fall-off on both coating materials was also detected in the arc-jet experiments (Fig. 2). By promoting the desorption of chemisorbed atoms from the active sites at lower temperature we could make less active coating material.

Since γ at HT is directly proportion to the number density of active sites (Eq.1), identification and elimination of active sites is important in developing a active surface for a better SSO flight performance. The C742 spinel ($Cr(FeCo)O_4$) was found to be more active than RCG. Dickens and Sutcliffe (*16*) investigate the oxygen recombination on various surface of oxides including Co_3O_4, Fe_2O_3, Cr_2O_3, and silica. The rate of recombination is first order with respect to gaseous oxygen atoms at the temperature investigated (RT~625K). It is investigating that the Co_3O_4 and Fe_2O_3 show the same temperature dependence of γ found in this study. This suggests that atom recombination on both oxide surfaces occurs though the surface diffusion mechanism involving the active sites. Higher γ on Co_3O_4 and Fe_2O_3 (5×10^{-2}) than silica (2×10^{-3}) at 625K indicates a higher number density of surface active sites for Co_3O_4 and Fe_2O_3 than silica. Silica, which is the main constituent of the RCG, was found to be the least active among the oxide investigated. The nature of active sites and chemisorption should be investigated further.

Surface defects and impurities during the preparation and annealing could provide the active sites. In addition, surfaces of the coating material may be not as smooth as that of glass. Since the γ_A is proportion to the surface area, we have to examine physical conditions, such as the roughness and the defects by a scanning electron microscope (SEM) and an atomic force microscope (AFM).

Conclusion

Although the previous arc-jet and flight data are useful for improved design of atmospheric re-entry vehicles and for flight trajectory calculations, they will be of little use in the design of improved (less active) coating materials. The diffusion tube method permits a more precise measurement of the apparent recombination probability over the wide range of temperature, thus providing a better insight into the kinetics of the reaction. Knowledge of the mechanism and the surface morphology could help in the design of less active sites of the surface. For instance, our model suggests that elimination of the active sites can drastically reduce catalytic recombination of atoms at re-entry temperature.

The most interesting feature of these reaction is the sudden decrease in the recombination probability at high temperature. Identification of the cause of this phenomenon could suggest modifications in the composition of the RCG coating aimed at enhancing the drop in the rate of atom recombination on the coating surface at re-entry temperature.

Acknowledgment

This work was supported by a grant No. KOSEF 941-1100-037-2 from the Korean Science and Engineering Foundation.

References

(1) Howe, J. T., *J. Spacecraft and Rockets*, **1985**, *22*, 19.
(2) Scott, C. D., *J. Spacecraft and Rockets*, **1985**, *22*, 489.
(3) Ried, R. C., Jr., In *Manned Spacecraft*: *Engineering and Design and Operation*; Purser, P. E.; Faget, M. A.; Smith, N. F. Eds.; Fairchild Pub.: New York, 1964.
(4) Stewart, D. A.; Rakich, J. V.; Lanfranco, M. J., *AIAA Paper* 82-1143, **1981**.
(5) Rakich, J. V.; Stewart, D. A.; Lanfranco, M. J., *AIAA Paper* 82-0944, **1982**.
(6) Goldstein, H. E.; Leiser, D. B.; Katvala, V., U.S. Patent No. 40 093 771, 1978.
(7) Goldstein, H. E. *et al*, In *Borate Glasses*: *Structure, Properties, Applications*, Pye, L. D. Ed.; Plenum Press : New York, 1978 ; 623
(8) Smith, W. V., *J. Chem. Phys.*, **1943**, *11*, 110.
(9) Kim, Y. C., *Ph. D Thesis*, Stanford University, **1991**.
(10) Stewart, D. A.; Rakich, J. V.; Lanfranco, M. J., *NASA CP-2283*, **1987**.
(11) Kim, Y. C. ; Boudart, M., *Langmuir*, **1991**, *7*, 2999.
(12) Stewart, D. A.; Rakich, J. V.; Lanfranco, M. J., *AIAA Paper*, **1982**, 248.
(13) Henry, C. R., *Surf. Sci.*, **1989**, *223*, 519.
(14) Greaves, J. C.; Linnett, J. W., *Trans. Faraday Soc.*, **1959**, *55*, 1355.
(15) Rosner, D. E., *NASA CR-134124*, Washington: National Aeronautics and Space Administration, 1973.
(16) Dickens, P. G. ; Sutcliffe, M. B., *Trans. Faraday Soc.*, **1964**, *60*, 1272.

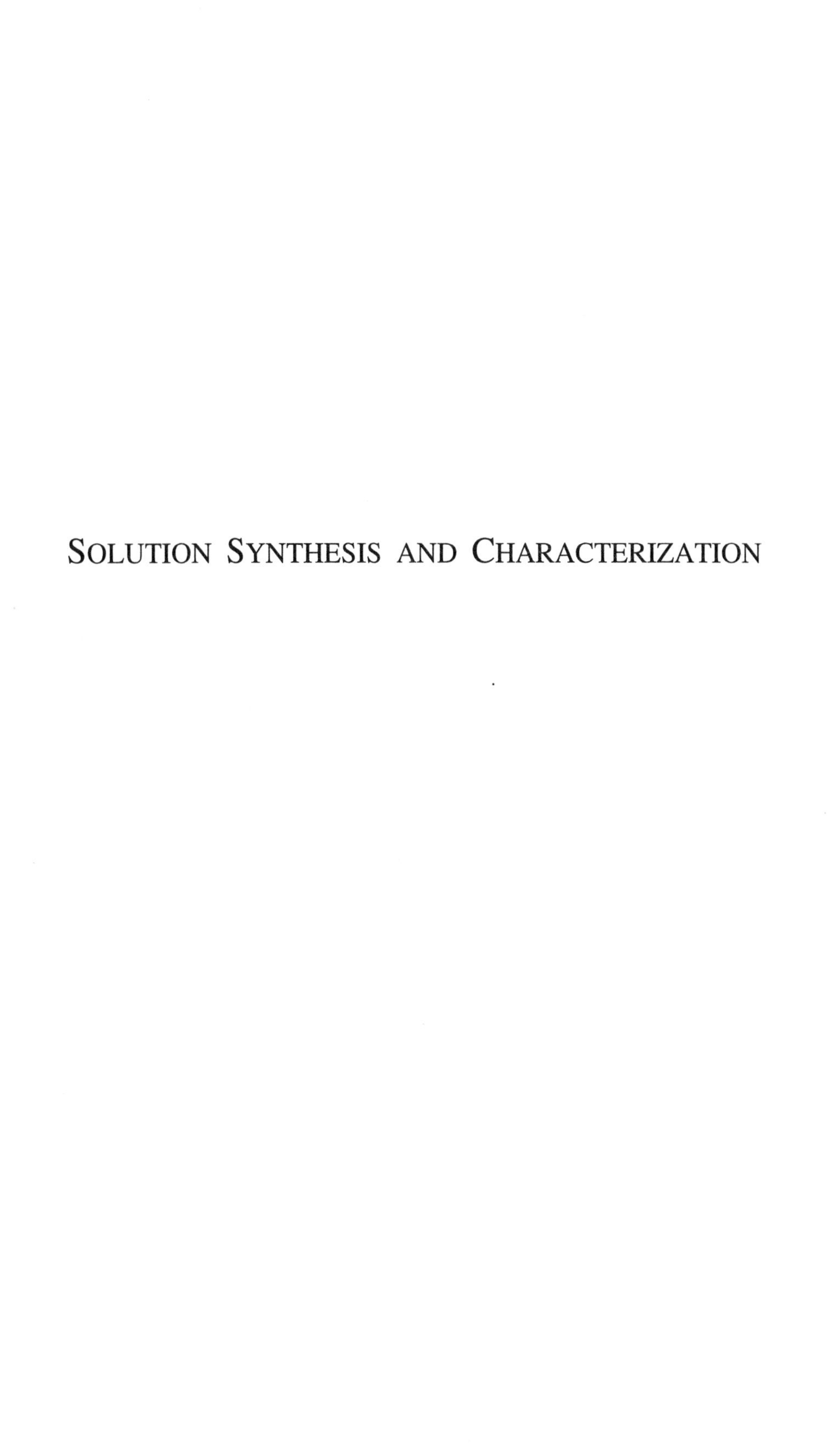

Solution Synthesis and Characterization

Chapter 9

Application of Chemical Principles in the Solution Synthesis and Processing of Ceramic and Metal Particles

James H. Adair, Jeffrey A. Kerchner, Nelson S. Bell, and Melanie L. Carasso

Department of Materials Science and Engineering, University of Florida, P.O. Box 116400, Gainesville, FL 32611–6400

There is an increasing appreciation for the role that solution synthesis plays in the preparation of metal and ceramic powders and films. The objective of this manuscript is to review a general approach based upon rigorous chemical principles that may be employed to produce a powder or film of a desired phase, size distribution, and particle morphology. The approach is based on the development of a fundamental understanding of the phase stability and solution chemistry of the material system, via the calculation of phase stability and speciation diagrams. The colloidal principles involved in the production of a well-dispersed powder within a particular synthesis scheme are also reviewed. Finally, the principles which control the morphology of the synthesized particles are presented. Each of these steps which comprise the general approach for tailoring the desired chemical and physical characteristics will be discussed with respect to exemplary material systems including calcium oxalate monohydrate, zirconia, and α-alumina.

Chemical principles will be used to provide an integrated approach to the solution synthesis of inorganic particles with controlled size and shape. The trends and approaches for chemical synthesis of particles will be reviewed, followed by presentation of the integrated approach. In recent years there has been a growing emphasis on the synthesis of particles with controlled size and morphology for a variety of applications, particularly those which benefit from nanometer-size particles (1-13). These applications include electronic devices, particularly multilayer capacitors in which the trend is toward submicron layers (11). There is also a growing appreciation for the importance of anisotropically-shaped, nanometer-size particles in flat panel optical displays (14,15). Control over particle features such as size and morphology depends on a number of interrelated processes involving the solid phase(s) present during and after development of a mature particle, the complex ionic equilibria in solution, and the nature of the solid-solution interface, including surface adsorbates and surface charge.

There have been a number of investigators who have made significant inroads into understanding and controlling the properties of fine particles synthesized from solution (1-13). The technological and economic significance of precipitation processes is considerable. The Bayer process, for example, in which Gibbsite is extracted from bauxitic ore, annually produces over 50,000 million tons of refined aluminum (hydrous) oxide (16). The emphasis of much of the recent work has been directed toward the synthesis of nanometer-size materials, in which the yield of particles from reverse micellar techniques is quite low. However, similar criticisms can be directed toward vapor phase synthesis of nanometer particles. In addition, the degree of aggregation is almost impossible to control in vapor phase synthesized materials, in contrast to solution synthesized particles for which dispersion techniques are well established during both synthesis and particle recovery (17).

There are six integrated steps that may be employed in the development of a particle system with tailored size and morphology. These are: (i) know the material; (ii) determine synthesis conditions to prepare the desired phase; (iii) develop methods to control particle size within the context of the material system; (iv) develop techniques to control particle morphology within the context of the material system; (v) control the state of agglomeration using colloid chemical principles; and (vi) control yield of the material by selection of the starting materials and their concentrations.

The use of this systematic approach provides a scientifically rigorous way to produce particles with the desired properties, tailored for a specific application. Each of the steps will be discussed in detail, with selected examples provided to illustrate the benefits as well as the limitations inherent in such an approach.

(i) Know the material

This first step is obvious but often overlooked; simply know the material. Table I provides a systematic overview of the important properties and specific information which should be known before one embarks upon the synthesis of a particular material. The requirement for detailed knowledge of the chemical, physical, colloidal, electronic and optical properties of the materials involved is often ignored at the outset of a synthesis. In the eventuality that there is no need to produce particles of controlled size and morphology, there is little need for an appreciation of all the critical properties of a material. However, for those applications which require particles with specific characteristics, it is essential to understand as much about the material as possible to optimize the opportunity for discrete control of particle features, as well as to ensure that problems such as poor yield from a particular precursor or irreversible agglomeration are avoided or minimized.

Precipitation from solution is a straightforward process for many material systems. The need to understand critical chemical properties such as phase equilibria and solubility product is obvious. Less obvious is the need to understand the physical, colloidal, electronic, and optical properties. The physical properties such as crystal structure are important because the atomic structure on various crystallographic habit

Table I. Material properties which affect particle synthesis

Chemical Properties	Phase Equilibria, Solubility Product, Solubility, Complex Species in Solution, Competing Phases
Physical Properties	Crystal Structure, Solid State Solutions, Space Group
Colloidal Properties	Surface Groups, Hydrophilicity, Isoelectric Point, Potentiometric Titration Behavior, Hamaker Constant
Electronic Properties	Dielectric Constant, Dielectric Dispersion, Ferroelectricity, Piezoelectricity, Ferromagnetism
Optical Properties	Color, Band Gap, Absorption Spectra

planes can ultimately control the morphology of particles (18). It is useful to know the colloidal properties of a particular material as a function of solution properties to anticipate and avoid synthesis conditions which yield agglomerated particles. The electronic and optical properties are related to the attractive van der Waals interactions which promote the formation of agglomerated particles (19). Understanding of these properties can be used to predict with considerable rigor the dispersion schemes required to prevent agglomeration of the synthesized particles.

(ii) Determine synthesis conditions to prepare the desired phase.

The conditions that may be used to produce a desired phase form the boundary conditions for any synthesis scheme in which one wishes to produce particles of controlled size and morphology. Assuming that thermodynamic equilibrium is achieved during a synthesis, there are two ways to determine whether a particular set of conditions (e.g. solution pH, ionic strength, temperature, initial concentrations, etc.) will produce the desired phase of the material. The most common approach, after a careful assessment of the known literature, is to acquire the necessary, usually soluble, precursor chemicals and proceed with a synthesis, followed by careful characterization of the resulting material to assess whether the desired phase has been produced. This iterative procedure is accompanied by a relatively qualitative development of a knowledge database until particles of the desired phase and properties are ultimately produced.

Thermodynamic equilibrium may be approached either from the direction of precipitation (i.e., from supersaturated solution conditions), or from undersaturated solution conditions (20). In the latter case, particles of the desired phase are used to determine the solution conditions where the desired phase is stable or decomposes. The particles used in the second approach to deduce phase stability conditions can be prepared by a variety of synthesis methods which provide the correct phase but not necessarily the desired physical properties for the particles (1). In this way a processing map which provides the solution and processing conditions where the desired phase is the stable, homogeneous phase can be efficiently developed.

(iii) Develop methods to control particle size within the context of the material system.

Control over the size of particles depends upon the mechanism of particle formation. From a broad viewpoint, there are two fundamental mechanisms of particle formation. These are (a) *in-situ* transformation of a virtually insoluble precursor to the ultimately stable phase, and (b) dissolution of any solid precursor materials followed by recrystallization from solution to the ultimate phase (21). For particles synthesized by *in-situ* transformation, the driving force is the difference in free energy between the solid precursor and the final product, similar to any solid-state transformation. In contrast, it has been proposed that the difference in solubility between a sparingly soluble precursor and the final product is the effective supersaturation, or driving force for the formation of particles in a dissolution-recrystallization process (2l). Classical notions of precipitation fall into the latter class of dissolution-recrystallization, because supersaturation in the solution is the fundamental feature which dictates particle size. For particles synthesized directly from solution or via dissolution-recrystallization from solution, Overbeek has provided a detailed overview of the impact of the growth mechanism on the breadth of the particle size distribution (22).

The particles shown in Figure 1 serve to highlight the role that supersaturation in the solution phase plays in the control of particle size and particle morphology for precipitation processes. The degree of saturation in a solution, S, is defined as ratio of the product of the activities of the precipitating species to the solubility product of the final product:

$$S = \frac{\text{product of activities of reactants}}{\text{solubility product}}$$

If S is greater than 1, the solution phase is *super*saturated with respect to the precipitating material. However, as noted above, thermodynamic equilibrium may be approached from either a supersaturated state or in *under*saturated solution conditions, where the ratio is less than 1. For such supersaturated and undersaturated solutions, precipitation or dissolution, respectively, of the equilibrium phase takes place until thermodynamic equilibrium is achieved, with a degree of saturation equal to 1.

Typically, particle morphology becomes more defined as saturation decreases below about 100, as shown for the calcium oxalate monohydrate particles in Figure 1. The particles precipitated at a supersaturation greater than 3000 are agglomerates composed of very fine primary particles. The high initial supersaturation leads to a large number of these fine particles, and this results in agglomerated masses of the primary particles unless steps are taken to provide adequate dispersion (see step (v), below). In contrast, the particles produced at supersaturations of 89 and 65 are well-defined crystallites for which the habit planes are obvious. At relatively low supersaturations, the growth rates for the various habit planes are such that the slowest growing habit planes can manifest themselves in the final form of the particle (18, 20). Thus, the degree of saturation in a solution dictates not only the size of particles, but also the morphology. The supersaturation values indicated in Figure 1 were calculated using a freeware computer program. Numerous user-friendly, computer programs are available to calculate the activities of species and consequently the degree of saturation for a variety of material systems with varying degrees of rigor, depending upon the model used for the activity coefficient (13, 23).

Material systems with inherently low solubility typically undergo *in-situ* transformation from sparingly soluble precursor materials to (usually) polycrystalline agglomerates composed of nanometer-size primary particles. This is demonstrated in Figure 2 for zirconia (ZrO_2) in aqueous solutions. Matijevic and co-workers (24) have shown that the complex chemistry of a solution can be exploited to control the supersaturation of particles. For example, the coarser ZrO_2 particles in Figure 2 were obtained by complexing the Zr in the precursor solution with ethylenediaminetetraacetate (EDTA). The particles were hydrothermally prepared at 190°C and pH 9, where hydrolysis of EDTA takes place. This results in controlled decomposition of the Zr-EDTA complex ion to release soluble Zr, probably as the $Zr(OH)_5^0$ species, in the solution phase without producing the large supersaturations

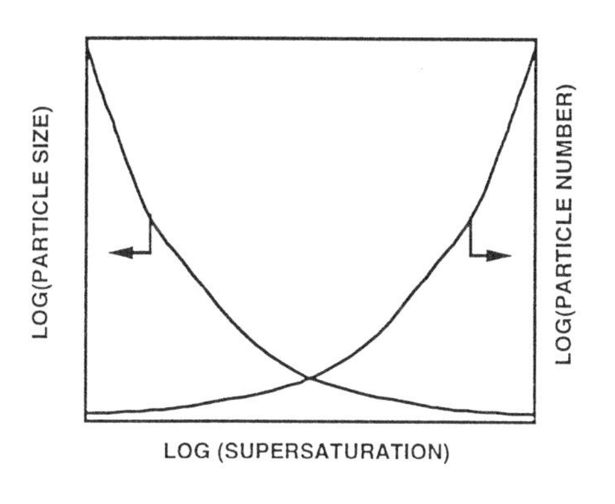

Figure 1a.

Figure 1b.

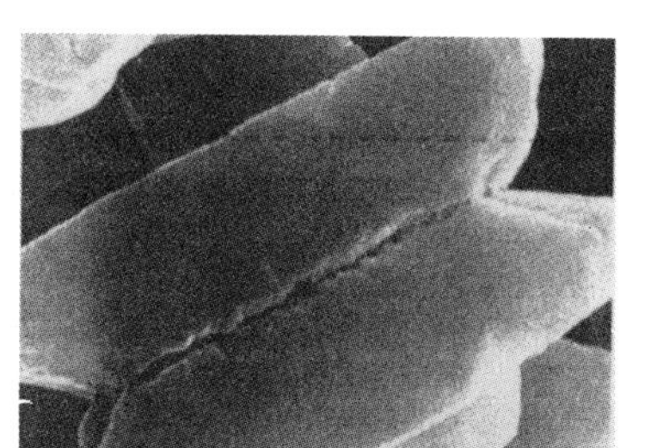

Figure 1c.

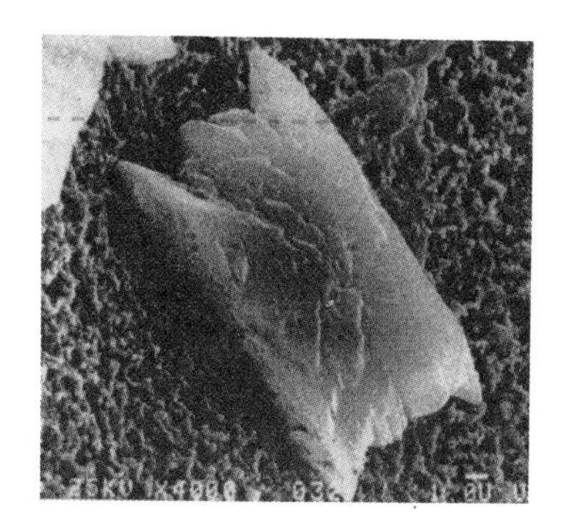

Figure 1d.

Figure 1. (a) Supersaturation as a function of particle size and number, (b) COM particles at $S > 3000$, (c) $S = 89$, (d) $S = 65$.

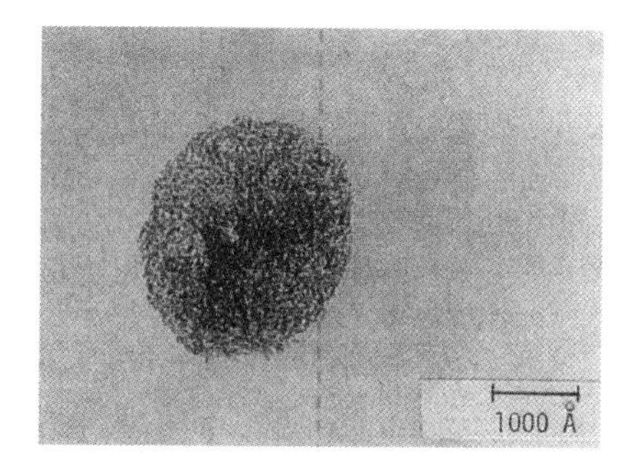

Figure 2a.

Figure 2b.

Figure 2. Zirconia particles synthesized (a) in aqueous solution and (b) in the presence of EDTA.

normally expected for the Zr-H_2O system (25). Thus, by careful manipulation of the complex chemistry of the system, particles of the desired particle size can be obtained.

Another common way to control particle size, as long as initial supersaturation is regulated, is by the addition of either homoepitaxial or heterogeneous particles to act as "seeds" for the nucleation of the desired phase. As the seed concentration is increased above a critical concentration, known as the intrinsic nucleation concentration, the addition of seeds can be effective in controlling particle size. The interested reader is referred to an article describing the seeding procedure used for this system (26).

(iv) Develop techniques to control particle morphology within the context of the material system.

Control over the shape of particles has already been mentioned in the discussion on particle size control. Here some of the basic paradigms for particle morphology control will be mentioned. Particles must be synthesized at relatively low initial supersaturations to produce particles with controlled morphology. The relatively low yields inherent at modest initial supersaturations may be overcome by the use of complexing agents. The formation of a soluble complex ion with one of the constituents of the precipitating solid ensures that there is an isolated reservoir of material which does not directly contribute to the supersaturation. Thus, as ions are removed from solution via the precipitation reaction, metal ions are released from the metal-complexing agent moiety to the solution. This process depends on the metal-complex moiety providing metal species to the solution to maintain a relatively constant and modest supersaturation level (24,26).

Another criterion for particle morphology control is that the slowest growing crystallographic habit planes for a particular solution condition are present on the final particle. There are numerous morphological theories that have been developed in attempts to explain particle morphologies (18). However, all of these theories have fallen short in *a-priori* predictions of particle shape. Another approach to take in developing a fundamental understanding of particle morphology control in a system consisting of particulate phase and solvent is to produce particles of specific morphologies and then identify the resulting habit planes. Various computer programs can be used in such an approach to predict equilibrium particle morphology and atomic planes, as a function of process conditions (23). This can result in the formation of a database for morphological forms, which can be used ultimately to predict particle morphologies for a particular material system under arbitrary process conditions. Complexing agents may be used to control the shape of particles by complexation on specific habits, with a consequent reduction in growth rate. Buckley (18a) has listed numerous examples of complexing agents with resulting effects on particle morphology.

A final processing parameter that affects particle morphology is the hydrodynamic behavior within the reaction vessel. The α-Al_2O_3 polyhedron particles in Figures 3a and 3b were produced at a shear rate estimated to be less than 1 s^{-1} (36 rpm), while the platelet particles in Figures 3c and 3d were produced in an autoclave at a shear rate in excess of 12 s^{-1} (240 rpm). Stirring or shear rate is typically varied in the development of precipitation processes to aid in deducing the particle growth mechanism(s). Walton (20a) has discussed the effect of shear rate on diffusion controlled growth manifest at low stirring rates, where a concentration gradient may be present near the growing interface. In contrast, interface-controlled growth is typically found at high shear rates, where surface reactions such as adsorption of species control growth rate. In the present work on the Al-1,4-butanediol system, the effect of stirring has a profound effect on the particle morphology because of the transition from diffusion to interface control as a function of stirring rate (16, 26).

(v) Control the state of agglomeration using colloid chemical principles.

Control over the state of agglomeration is the final step in producing a powder of the desired physical features. Certain consolidation processes for ceramic and metal powders, particularly pressing operations, intentionally agglomerate particles to ensure good flow properties. However, in general it is desirable to avoid agglomerates, particularly agglomerates in which the primary particles are bound together with solid bridges, as is usually the case during precipitation processes. Such agglomerates, also often designated "hard aggregates," can not be broken down to their primary particles by common chemical dispersion schemes. The greater the supersaturation, the larger the yield of powder from a precipitation process. Agglomeration between primary particles is more extensive at high number particle concentrations and at the high ionic strengths inherent to a large supersaturation. Thus, the high yields desirable for most precipitation processes result in a large number concentration of small particles in which the collision frequency between particles and, hence, the probability of agglomeration, is high.

The tendency for precipitated particles to agglomerate may be eliminated or at least minimized by incorporating the basic principles of colloid and interfacial chemistry into a precipitation scheme. It is well established that the electric potential near the solid-solution interface, known as the zeta potential, for a particle in solution can be attenuated by manipulating solution features such as solution pH and by adding electrically charged adsorbing ionic species (19). The pH value at which the zeta potential changes polarity (i.e., reaches zero) is known as the isoelectric point (IEP). For example, as shown in Figure 4, EDTA at relatively modest concentrations causes the IEP for a 3 mole percent Y_2O_3 doped ZrO_2 (known as 3 m/o Y-TZP from the designation tetragonal zirconia polycrytals) to shift from ~ pH 9 to ~ pH 4, consistent with the acid dissociation constants of EDTA. In concert with the shift in IEP, the zeta potential at pH 9 becomes much greater in magnitude, with a value of -50mV attained at an EDTA concentration of $5x10^{-3}$M. Polymeric dispersants are typically less effective than such electrostatic dispersants in precipitating solutions because polymeric additives often have a gross effect on the nucleation and growth kinetics of the solid phase.

Since the tendency to agglomerate is reduced at higher surface potentials, it is expected that TZP produced in the presence of ~ 10^{-3}M EDTA will be much less agglomerated than a TZP powder produced at pH 9 without EDTA, close to the IEP. Hydrothermally derived 3 m/o Y-TZP powders produced at 190°C in the presence and

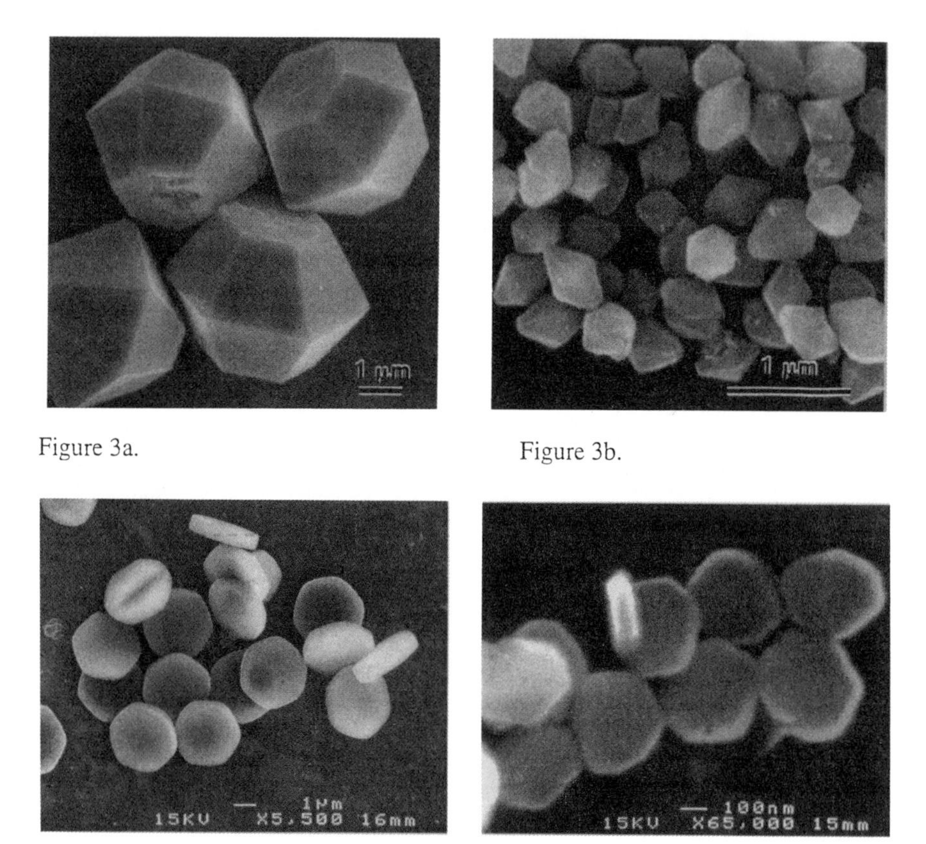

Figure 3a. Figure 3b.

Figure 3c. Figure 3d.

Figure 3. Effect of stirring rate on glycothermally synthesized alumina particles. Polyhedra (a), (b) synthesized at 36 rpm; platelets (c), (d) synthesized at 240 rpm.

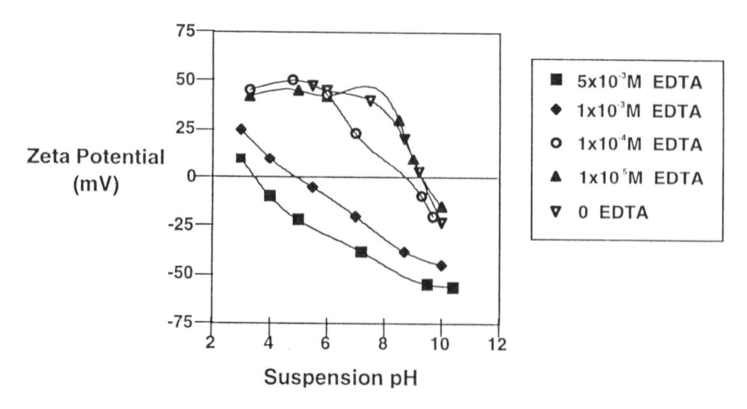

Figure 4. The effect of EDTA on the zeta potential of 3 m/o Y-TZP.

absence of EDTA are shown in Figure 5 (Adair, unpublished data). The presence of EDTA resulted in a TZP powder with well-dispersed particles, in contrast to the agglomerated masses in the TZP produced in the absence of EDTA dispersant. It is important to note that more is not better in the case of an electrostatic dispersant. Modest levels of EDTA were added to disperse the precipitating TZP particles because high levels of this complexing agent could result in increased solubility of the TZP particles, to a point where a homogenous solution results. As noted in the particle size control section, such complex formation may be used to produce larger particles if desired. Modest levels of EDTA were also used in the TZP synthesis because complexing agents can also affect particle morphology.

(vi) Control yield of the material by selection of the starting materials and their concentrations.

The production of particles with reasonable yields is a critical part of any acceptable particle synthesis process. It is possible to use dissolution-recrystallization reactions to produce powders at high yields without compromising particle properties for a specific application. Some of the features required to provide reasonable yields of product are given in Table II (7). Inexpensive precursor compounds, easy to handle chemicals, and compounds which react cleanly to form the product are desirable because of high yield avoiding the need for rigorous post-synthesis washing. There is a trade-off between expense and washing in the case of chloride precursor compounds. Chloride is typically difficult to remove from powders (27), but it is also the least expensive of all precursors. The use of aqueous-based materials is advantageous from environmental and safety considerations.

Conclusions

At the outset of the development of the synthesis of particles from solution there is an integrated approach which may be taken to optimize the production of particles with the desired features for a particular application. An understanding of how thermodynamic equilibrium is approached leads to an understanding of the phase stability of the desired solid phase. Fundamental principles of solution and colloid chemistry provide guidelines for control of particle size, particle shape and agglomeration. To complement the application of these fundamental principles, there are a number of computer programs that ease the trauma associated with predictions of phase stability, ionic equilibria, and predictions of agglomeration tendency.

Acknowledgments

Partial support for this work was provided by NIH Grant No. POG 5P01 DK20586-17 and Cabot Performance Materials, Boyertown, PA.

Figure 5a.

Figure 5b.

Figure 5. 3 m/o Y-TZP powders hydrothermally synthesized without (a) and with (b) EDTA dispersant.

Table II. Recommended features for synthesis processes[a]

Recommendation	Avoid	Use
Use inexpensive inorganic precursors	Alkoxides	Chlorides
	Metal Organics	Nitrates Oxides Hydroxides
Select most reactive precursor	Carbonates Oxides	Hydroxides (Hydrous) Oxides
Use easy-to-handle chemicals	Fuming Chlorides	Chloride Aqueous Solutions
	Fuming Nitrates	
Use precursors which react cleanly to form desired product	Sulfates Acetates	Chlorides (wash required) Nitrates

[a] Courtesy of W.D. Dawson, CMI Inc., Columbus, OH.

Literature Cited

1. Johnson, D.W. in *Advances in Ceramics*; Messing, G.L., Mazdiyasni,K.S., McCauley, J.W., and Haber, R.A., Eds.; Ceramics Powder Science, The American Ceramic Society, Inc., Columbus, OH, **1987**, Vol. 21; pp. 3-20.
2. Kato, A. in *Materials Science Monographs*; Vincenzini, P., Ed.;,High Tech Ceramics, Part A; Elsevier Science Publishers B.V., Eds.: Amsterdam, **1987**, Vol. 38; pp. 459-469.
3. Osseo-Asare, K.; Arriagada, F.J. in *Cer.Trans.*, **1990**, *12,* The American Ceramic Society, Inc.; Westerville, OH, pp. 3-16.
4. Chow, G.M.; Markowitz, M.A.; Singh, A. *J. Materials.*, **1993**, *11,* pp. 62-65.
5. Benedetti, A.; Fagherazzi, G.; Pinna, F. *J. Am. Ceram. Soc.*, **1989**, *72*, pp. 467-469.
6. Hirano, S. *Am. Ceram. Soc. Bull.*,**1987,** *66*, pp. 1342-1344.
7. Dawson, W.J. *Am. Ceram. Soc. Bull.*, **1989,** *67,* pp.1673-1678.
8. Nancollas, G. H. *Cer. Trans.,* **1987**, *1*; The American Ceramic Society, Inc., Westerville, Ohio, pp. 8-22.
9. Randolph, A.D.; Larson, M. A. *Theory of Particulate Processes: Analysis and Techniques of Continuous Crystallization*; Academic Press, Inc.: San Diego, CA, **1988**; pp. 109-134.
10. (a) Fendler, J. H. *Chem. Rev.* **1987**, *87*, 877. (b) Baral, S.; Fendler, J. H. *J. Am. Chem. Soc.* **1989**, *111*, 1604. (c) Zhao, X. K.; Herve, P. J.; Fendler, J. H. *J. Phys. Chem.* **1989**, *93*, 908. (d) Zhao, X. K.; Xu, S.; Fendler, J H. *ibid.* **1990**, *94*, 2573. (e) Zhao, X. K.; Fendler, J. H. *ibid.* **1988**, *12*, 3350.
11. Höppener, R.; Daemon, A. *Proceedings of the 14th Capacitor and Resistor Technology Symposium,* **1994,** Components Technology Institute Inc., Huntsville, AL, pp. 122-129.
12. Komarneni,S.; Menon, V.C.; Li, Q.H. Cer. Trans., **1996**, *62*, The American Ceramic Society, Westerville, OH, pp.37-46.
13. Lencka, M.M.; Riman R.E. *Chem. Mater.*, **1993**, **5**, pp. 61-70.
14. Matsumoto, S. *Electronic Display Devices*, **1990**, John Wiley, New York, NY.
15. Li, T.; Kido, T.; Adair, J.H. *Proceedings of the Ninth IEEE International Symposium on Applications of Ferroelectrics,* **1995**, IEEE Catalog No. 94CH3416-5, IEEE Service Center, Piscataway, NJ, pp. 791-793.
16. (a) Cho, S.B., Venigalla, S.; Adair, J.H. *J. Am. Ceram. Soc.*, **1996**, *79* , pp. 88-96. (b) Cho, S.B., Venigalla, S.; Adair, J.H. *Cer. Trans.,* **1995**, *54*, The American Ceramic Society, Inc., Westerville, OH, pp. 139-150. (c) Inoue, M., Tanino, H., Kondo, Y.; Inui, T. *J. Am. Ceram. Soc.*, **1989**, *72*, pp. 352-353.
17. Flint, J.H.; Haggerty, J.S. *Cer.Trans.*, **1988**, 1, The American Ceramic Soc., Westerville, OH, pp. 244-252.
18. (a) Buckley,H.E. "Crystal Growth," John Wiley and Sons, Inc., NY, pp.105-120, **1951**. (b) Hartman, P. *J. Cryst. Growth*, **1980**, *49*, 166. (c) Webb, W.W. and Forgeng,W.D. *J. Appl. Phys.*, **1957**, *28 [12]* 1449. (d) Kuznetsov,V.A. *Soviet Phys. - Crystallo.,* **1966**, *10 [5]* 561. (e) Davey R.J. and Dobbs, B. *Chem. Engr. Sci.*, **1987**, *42 [4]* 631. (f) Koga, T., and Nakatani, S. U.S. Patent 4,202,871,

1980. (g) Hartman, P. *J. Cryst. Growth*, **1980**, *49*, 166. (h) Kuznetsov, V.A. *in "Hydrothermal Synthesis of Crystals,"* A.N. Lobachev (ed.),Consultants Bureau, NY, pp.52-61, **1971**. (i) James,J.A. and Kell, R.C. in "Crystal Growth," B.R. Pamplin (ed.), Pergamon Press, NY, pp.557-75, **1975**. (j) Laudise, R.A. in *"Crystal Growth: An Introduction,"* P. Hartman (ed.), American Elsevier Publishing Co., Inc., NY, pp.172-80, **1973**. (k) Weissbuch, Addadi, L. Lahav, M., and Leiserowitz, L. *Science*, **1991**, *253*, 637.

19. (a) Hunter, R.J. *Zeta Potential in Colloid Science*, **1981**, Academic Press, New York, NY. (b) Hunter, R.J. *Foundations of Colloid Science,* Vol. I, **1986**, Oxford Press, London, U.K. (c) *ibid.*, Vol. II, **1989**.
20. (a) Walton, A.G. *The Formation and Properties of Precipitates*, **1979**, Robert E. Krieger Publ. Co., Huntington, NY. (b) Nielsen, A.E. *Kinetics of Precipitation*, **1964**, The MacMillan Co., New York, NY. (c) Matijevic, E. in *Chemical Processing of Advanced Materials*, Hench, L.L. and West, J.K., Eds.; John Wiley and Sons, Inc., New York, NY, 1992; pp. 513-527.
21. (a) Adair, J.H.; Denkewicz, R.P. Arriagada, F.J.; Osseo-Asare, K. *Cer. Trans.*, **1988**, *1*, The American Ceramic Society, Inc., Westerville, OH, pp. 135-145. (b) Adair, J.H.; Denkewicz, R.P. *ibid.*, **1990**, *12*, pp. 25-32. (c) Denkewicz, R.P.; TenHuisen, K.; Adair, J.H. *J. Mat. Res.*, **1990**, *5*, pp. 2698-2705. (d) Rossetti, Jr., G.A.; Watson, D.J.; Newnham, R.E.; Adair, J.H. *J. Crystal Growth*, **1992**, *116*, pp. 251-259.
22. Overbeek, J.Th.G. *Adv. Colloid Interface Sci.*, **1982**, *15*, pp. 251-277.
23. (a) Finlayson, B. *Calcium Metabolism in Renal Failure and Nephrolithiasis*, **1977**, John Wiley and Sons, New York, pp. 337-382. (b) B. Finlayson, J.H. Adair, and R.V. Linhart, *Finlayson's EQUIL v.1.3*, copyright pending, **1996**. (c) H. Krarup, R.V. Linhart, and J.H. Adair, *OPAL*, copyright pending, **1996**. (d) Dowty, D.; Richards, R.P. ATOMS© and SHAPE© v4.0, **1991-1993**, Kingsport, TN.
24. (a) Matijevic, E. *J. Colloid Int. Sci.*, **1973**, *43*[2], p. 217. (b) Sapieszko, R. S. Matijevic, E. *NACE*, **1980**, *36*(10), pp. 522-530. (c) Matijevic, E.; Wilhelmy D.M. *J. Colloid Interface Sci.*, **1982**, *86*(2), pp. 476-484.
25. Stambaugh E.P.; Adair, J.H.; Sekercioglu, I.; Wills, R.R. U.S. Patent 4,619,817, **1986**.
26. (a) Bell, N.S.; Cho, S.B.; Adair, J.H. *This Proceedings*. (b) Kerchner, J.A.; Moon, J.; Chodelka, R.E.; Morrone, A.; Adair, J.H. *ibid.*
27. Adair, J.H. *Engineered Materials Handbook*, **1991**, *4*, CASM International, Materials Park, OH, pp. 90-94.

Chapter 10

Precursors for Aqueous and Liquid-Based Processing of Ferroelectric Thin Films

Allen W. Apblett, Galina D. Georgieva, Larry E. Reinhardt, and Edwin H. Walker

Department of Chemistry, Tulane University, 6823 St. Charles Avenue, New Orleans, LA 70118–4698

Several highly water-soluble metal carboxylate precursors for barium titanate, strontium titanate, barium strontium titanate, and neodymium molybdate have been synthesized. A new, facile preparative route for $SrTiO(Ox)_2$ (Ox=oxalate) has been developed that involves passing a solution of $K_2TiO(Ox)_2$ through an ion-exchange resin to convert it to the proteo-derivative, $H_2TiO(Ox)_2$. Treatment of the latter with $SrCl_2$ results in precipitation of $SrTiO(Ox)_2$ with a ratio of strontium:titanium of 1:1.001. Finally, reaction of $SrTiO(Ox)_2$ with refluxing methoxyacetic acid produces a water-soluble ceramic precursor that may be used to prepare thin films of strontium titanate by metallo-organic deposition. Mixing this precursor with the analogous barium compound affords preceramic compounds for barium strontium titanates. A liquid precursor for barium titanate may also be prepared by dissolving barium acetate in liquid titanium 2-[2-(2-methoxy)ethoxy]ethoxyacetate. Alternatively, a barium titanium gluconate which is extremely water-soluble may also be readily prepared and used as a precursor for barium titanate. As well, it has been found that a water-soluble precursor for neodymium molybdate may be readily prepared via reaction of neodymium gluconate with molybdenum trioxide.

The synthesis of ceramics by liquid phase processing provides a convenient method for the preparation of ceramic films and bodies (*1*). These methods potentially contribute the advantages of low processing temperatures, the ability to prepare composite materials or complex shapes, and strict control over the stoichiometry of the elements. One such technique for the preparation of ceramic thin films is metallo-organic deposition (MOD) (*2,3*), a non-vacuum, solution-based method of depositing thin films. A suitable metallo-organic precursor dissolved in a suitable solvent is coated on a substrate by spin-coating, screen printing, or spray- or dip-coating. The soft metallo-organic film is then pyrolyzed in air, oxygen, nitrogen or other suitable atmosphere to convert the precursors to their constituent elements, oxides, or other compounds. Shrinkage generally occurs only in the vertical dimension so conformal coverage of a substrate may be realized. Metal carboxylates are often used as precursors for ceramic oxides since they are usually air-stable, soluble in organic solvents, and decompose readily to the metal oxides. MOD processes for the generation of many oxide-based materials have already been developed: e.g. $BaTiO_3$ (*4*), indium tin oxide (*5*), SnOx (*6*), $YBa_2Cu_3O_7$ (*7*) and ZrO_2 (*8*). The usual carboxylate residue used is a long slightly-branched alkyl chain (e.g. 2-ethylhexanoate or neodecanoate) that confers the necessary solubility in organic solvents. Barium titanate has been synthesized previously using a variety of metallo-organic precursors (*9*). These include routes based solely on metal alkoxides (*9,10*) or metal carboxylates (e.g. the Pechini (or citrate) process (*11*)) and mixed carboxylate/alkoxide precursors (*12*).

The increasing demand for environment-friendly processes places stringent requirements on precursors for ceramic materials. In particular, the avoidance of organic solvents necessitates the development of preceramic compounds that are either water-soluble or which are amenable to solventless processing. We report herein several metal carboxylate precursors for a variety of ferroelectric ceramics that are either liquids or which are water-soluble and are therefore suitable for such processing techniques.

Experimental

All reagents were commercial products and were used without further purification with the exception of 2-[2-(2-methoxyethoxy)ethoxy] acetic acid, MEEAH which was dried over activated 5A molecular sieves for ca. 24 hours. Water was distilled and deionized in a Modulab UF/UV Polishing apparatus before use. Infrared spectra were obtained as KBr pellets or neat films on KBr plates using a Mattson Cygnus 100 FT-IR spectrometer. ^1H and $^{13}C\{^1H\}$ NMR were obtained on a Bruker AC 200 MHz spectrometer and chemical shifts are reported relative to tetramethylsilane. Thermogravimetric studies were performed using 20-30 mg samples under a 100 ml/minute flow of dry air in a Seiko TG/DTA 220 instrument or a TA Instruments Hi-Res TGA 2950 Thermogravimetric Analyzer. The temperatures were ramped from 25 °C to 1025 °C at a rate of 2 °C per minute or from 25 °C to 650 °C at a rate of 5 °C/min. The metal content of single metal carboxylates was determined gravimetrically upon ignition in air. Metal contents of bimetallic compounds were determined by ICP spectroscopy on a Perkin Elmer Optima instrument or by X-ray fluorescence spectroscopy on a Spectro X-Lab energy dispersive XRF spectrometer. Bulk pyrolyses at various temperatures were performed in ambient air in a temperature-programmable muffle furnace using 1-2 g samples, a temperature ramp of 5 °C/minute and a hold time of 6-12 hours. X-ray powder diffraction patterns were obtained using copper K_α radiation on a Scintag XDS 2000 diffractometer equipped with an automated sample changer and a high resolution solid state detector. Jade, a search/match software package, was used in the identification of XRD spectra. Microanalysis for C, H, and N content was performed by Oneida Research Services or Tulane's Coordinated Instrument Facility. Chloride content was determined by dissolution in water and measurement using a chloride ion-selective electrode.

Preparation of $SrTiO(Ox)_2(H_2O)_4$ A solution of $K_2TiO(Ox)_2(H_2O)_2$ (3.54g, 10.0 mmol) in 150 ml of distilled water was passed through a proton-charged cation exchange column (20 g of resin). The effluent from the column was found to be very acidic (pH=2) and it was drained directly into a solution of $SrCl_2{\cdot}6H_2O$ (2.67g, 10.0 mmol) in 100 ml of water. Immediate reaction occurred to yield a voluminous white precipitate. This was isolated by filtration, washed with water (50 ml) and dried in the air. The yield was 2.63g (66 %). XRF analysis indicated a ratio of Sr:Ti of 1:1.001.

Preparation of $SrTiO(O_2CCH_2OMe)_4(HO_2CH_2OMe)_2(H_2O)_3$ A mixture of $SrTiO(Ox)_2(H_2O)_4$ (1.09g, 2.73 mmol) and 25 ml of methoxyacetic acid were heated at reflux (ca. 202 °C). The $SrTiO(Ox)_2(H_2O)_4$ dissolved gradually to give a yellowish solution over a period of twenty hours with a small quantity of solid remaining undissolved. The reaction mixture was filtered and removal of the excess acid from the filtrate in vacuo yielded a slightly-sticky pale yellow solid (1.61g, 80 %). ^1H NMR ($CDCl_3$): δ(ppm) 3.38(s, 12H, CH_3), 3.40(s, 6H, CH_3), 4.03(s, 4H, CH_2), 4.10(s, 4H, CH_2), 4.61(s, 4H, CH_2), 6.64(s, 4H, OH). ^{13}C NMR($CDCl_3$): δ(ppm) 58.7, 59.0, 59.2, 69.3, 70.0, 170.2, 171.8, 174.7. IR (thin film between KBr plates, cm^{-1}): 3447(s, br), 3119(s, sh), 2998(s, sh), 2950(s, sh), 2836(s), 1754(s), 1653(s, sh), 1576(s, br), 1506(w), 1457(s, sh), 1429(s), 1381(m), 1340(s), 1324(s, sh), 1206(s, sh), 1166(w), 1115(s, sh), 1109(s), 1040(w), 1024(w), 986(m), 935(m), 914(m), 833(w), 813(w), 799(w), 764(w), 742(w), 717(w), 710(w), 686(m), 678(m), 673(m), 665(w), 657(m), 645(w), 637(w), 521(w), 607(m), 594(w), 580(w), 562(w), 549(w), 537(w), 517(w), 510(w), 469(w), 425(m).

Preparation of $Ti(MEEA)_4$ $Ti(^iOPr)_4$ (13.70 g, 48.19 mmol) was added to a stirred solution of 2-[2-(2-methoxyethoxy)ethoxy]acetic acid (34.48 g, 193.5 mmol) in dry ethanol (50 mL) under an atmosphere of dry nitrogen. Over a period of 30 minutes, the reaction mixture changed color from light yellow to dark yellow. At this point, the ethanol was removed by rotary evaporation to yield a viscous liquid which was then vacuum-dried at 100 °C. The yield of the final viscous yellow liquid was 36.95 g (100.4%). ^1H NMR (CD_3CN) δ 1.22 (t, 0.57 H, CH_3)* , 3.28 (s, 3H, CH_3O), 3.4-3.7 (multiplet, 9.4 H, OCH_2CH_2O) 4.06 (s, 2H, CH_2CO_2) 4.1 (q, 0.38H, CH_2CH_3*. $^{13}C\{^1H\}$ NMR (neat with a D_2O insert for locking) δ 14.9 (CH_3CH_2O)*, 59.2 (CH_3O), 72.8 (O_2CCH_2) 171.7 (O_2C), 172.5 (O_2C)* Starred peaks are due to the ethyl ester of 2-[2-(2-methoxyethoxy)ethoxy]acetic acid. The ceramic yield was 9.32 % indicating a titanium content of 5.59%. IR (neat film on KBr plates, cm^{-1}) 3461(s,br), 2885(s), 1907 (m,br 2526(w), 1749(s), 1558(s), 1456(s), 1330(m), 1203(s), 1107(s), 1026(m), 854(m), 669(m)

Preparation of $BaTi(MEEA)_4(CH_3CO_2)_2$ A solution of $Ba(O_2CCH_3)_2$ (0.6369 g, 2.494 mmol) in 5 ml of distilled water was added to $Ti(MEEA)_4$ (2.1345 g, 2.492 mmol). Removal of the water in vacuo yielded 2.56 g (92.4 %) of a viscous yellowish liquid. $^{13}C\{^1H\}$ NMR (neat with a D_2O insert for locking) δ14.9 (CH_3CH_2O)*, 21.9 (CH_3CO_2), 59.2 (CH_3O), 61.2 (CH_3O)*, 69.1 (OCH_2CH_2O)* 71.1, 71.4, and 71.5 ppm (OCH_2CH_2O), 72.7 (O_2CCH_2), 171.2 (O_2CCH_2O), 172.7 (O_2C)*, 175.7 (O_2CCH_3). Starred peaks are due to the ethyl ester of 2-[2-(2-methoxyethoxy)ethoxy] acetic acid. IR (neat film on KBR plates, cm^{-1}) 3448(br), 2920(s), 2198(w), 1961(w), 1751(s), 1601(s), 1454 (sh),1429(s), 1325 (w), 1294 (w), 1249 (w), 1201 (w) 1111(s), 1028 (m), 976 (w), 931(w), 902 (w), 850(m), 601 (w), 582 (w), 516(w), 462(w)

Preparation of $Ba(gluconate)_2$ $BaCO_3$ (4.94 g, 25 mmol) was reacted with gluconic acid (20.65 g of a 47.5 wt % solution, 50 mmol) in 50 ml of water at reflux for approximately 1 to 2 hours. Upon completion, the solution was allowed to cool and then filtered to remove a small trace of brown insoluble material. The solution was then slowly evaporated on a rotary evaporator to yield 14.81 g (97.3%) of

$Ba(gluconate)_2(H_2O)$ as a brownish-yellow solid. IR (KBr pellet, cm^{-1}) 3844 (w), 3828 (w), 3812 (w), 3788 (m), 3721 (w), 3700 (w), 3661 (w), 3638 (w), 3366 (s, br), 3270 (s, br), 2942 (w), 2903 (w), 2820 (w), 2390 (m, br), 2348 (w), 2288 (m, br), 1958 (w), 1902 (w), 1858 (w), 1834 (w), 1782 (w), 1761 (w), 1744 (w), 1726 (w), 1593 (vs), 1402 (s), 1377 (s), 1325 (s), 1267 (m), 1223 (m), 1134 (m), 1088 (vs), 1007 (s), 970 (w), 920 (w), 883 (m), 796 (m), 686 (m), 658 (w), 635 (m), 561 (m), 534 (w). A sample of material fired for 6 hours at 630°C had a ceramic yield of 32.44%. XRD analysis on the resulting powder revealed a match for $BaCO_3$. The barium content of the precursor was therefore found to be 22.57% while the barium content calculated for $[HOCH_2(HCOH)_4COO]_2Ba(H_2O)_{4.5}$ is 22.56%. The presence of 4.5 equivalents of water per barium was confirmed by the weight-loss of 13.3 % observed for dehydration in TGA experiments.

Preparation of Titanium Gluconate Chloride The titanium gluconate precursor was prepared as follows: $TiCl_3$ (19.29 g of 20 wt % solution, 25 mmol,) was allowed to stir for 36 hours with gluconic acid (41.42 g of a 47.5 wt % solution, 100 mmol) in 100 ml of water. The reaction was monitored for a color change (purple to yellow) which accompanied the oxidation of Ti (III) to Ti (IV) and the formation of the soluble precursor. The solution was filtered to remove a small trace of TiO_2 and was slowly evaporated on a rotary evaporator (temperature ca. 45°C) to yield a dark, hygroscopic solid (22.14 g). IR (KBr pellet, cm^{-1}) 3912 (w), 3895 (w), 3877 (w), 3862 (w), 3844 (w), 3829 (w), 3810 (w), 3788 (w), 3719 (w), 3698 (w), 3682 (w), 3659 (w), 3395 (s, br), 2932 (w), 2386 (m, br), 2348 (w), 2299 (m, br), 1742 (m), 1628 (m), 1410 (w), 1379 (w), 1335 (w), 1231 (w), 1132 (w), 1080 (w), 1038 (w), 882 (w), 802 (w), 635 (w), 584 (w). Themogravimetric analysis indicated total organic removal at 582°C with a ceramic yield of 8.21%. XRD analysis confirmed a match for TiO_2 (Anatase). The titanium content of the precursor was then calculated to be 4.92%. The chloride content of the precursor was determined to be 6.88% which corresponds to 1.89 chloride ions per titanium while the carbon and hydrogen content was found to be 32.65% and 5.63%, respectively. Thus, the approximate formula of this material is $TiC_{26.4}H_{54.8}Cl_{1.9}O_{30.4}$, but ^1H and ^{13}C NMR indicates it is not a pure compound. Attempts to purify the material results in precipitation of an insoluble white solid.

Synthesis of Gluconate Precursor for $BaTiO_3$ Barium gluconate (1.52 g, 2.5 mmol) was dissolved in 20 ml of water and separately, the titanium gluconate precursor (2.43g, 2.5 mmol) was dissolved in 20 ml of water. Upon combining both solutions, no precipitation was observed. The resulting brown colored solution was evaporated on a rotary evaporator (temperature approximately 45°C) to form a dark, hygroscopic solid in essentially quantitative yield (2.93 g). IR (KBr pellet, cm^{-1}) 3746 (w), 3418 (s, br), 3391 (s, br), 2938 (w), 2363 (w), 2336 (w), 1734 (m), 1636 (m), 1539 (w), 1522 (w), 1416 (w), 1395 (w), 1321 (w), 1290 (w), 1231 (w), 1130 (w), 1080 (w), 1049 (w), 873 (m), 812 (w), 793 (w), 664 (w), 627 (w), 586 (w), 496 (w). A sample of material fired for 6 hours at 630°C had a ceramic yield of 14.33%

Preparation of Neodymium Molybdenum Gluconate $Nd_2(CO_3)_3$ (1.01 g, 1.93 mmol) was mildly refluxed with gluconic acid (5.06 g of a 49 % aqueous solution, 12.6 mmol) in 50 ml of water for approximately 2 hours. The solution was allowed to cool and then filtered. MoO_3 (0.866 g, 6 mmol) was then added (under the same mild refluxing conditions) to the neodymium gluconate solution. The reflux was continued for 3 hours at which point the MoO_3 had completely dissolved. Upon cooling, the solution was filtered and evaporated on a rotary evaporator to yield an off-white solid (4.11 g). IR (KBr pellet, cm^{-1}) 3954 (w), 3935 (w), 3914 (w), 3856 (w),

3821 (w), 3754 (w), 3717 (w), 3673 (w), 3443 (s, br), 3422 (s, br), 3082 (w), 3034 (w), 2998 (w), 2922 (w), 2891 (w), 2853 (w), 2820 (w), 2789 (w), 2762 (w), 2670 (w), 2641 (w), 2610 (w), 2550 (w), 2527 (w), 2369 (w), 2342 (w), 2286 (w), 2249 (w), 2211 (w), 2178 (w), 2135 (w), 2020 (w), 1946 (w), 1919 (w), 1890 (w), 1616 (s), 1578 (s), 1418 (s), 1395 (s), 1325 (w), 1263 (w), 1119 (w), 1086 (w), 1034 (m), 926 (m), 891 (m), 799 (w), 660 (w), 637 (s), 588 (w). A sample of material fired for 6 hours at 630°C had a ceramic yield of 36.1%

Results and Discussion

Recently, a facile preparative route for a water-soluble precursor for highly-stoichiometric barium titanate was reported (*13*). This involved the reaction of the very successful barium titanate precursor , $BaTiO(Ox)_2$ (Ox=oxalate, $C_2O_4^{2-}$), with refluxing methoxyacetic acid. This yields $BaTiO(O_2CR)_4(HO_2CR)_2$, {R= CH_2OMe} a viscous liquid that is soluble in a variety of organic solvents and water. In this investigation, the related strontium titanate precursor, $SrTiO(O_2CR)_4(HO_2CR)_2$, {R= CH_2OMe} was prepared in an analogous manner. First, $SrTiO(Ox)_2 \cdot X(H_2O)$ was prepared in a two-step process from $K_2TiO(Ox)_2$.This compound was converted to the corresponding proteo-derivative $H_2TiO(Ox)_2 \cdot X(H_2O)$ by passing an aqueous solution of it through a proton-charged cation exchange resin column. The eluent from the column (pH=2) was drained directly into an aqueous solution of $SrCl_2$ to afford $SrTiO(Ox)_2 \cdot 4(H_2O)$ as a voluminous precipitate in a 66% yield. XRF analyses of this material indicated a strontium to titanium ratio of 1:1.001. This compound could be used as a precursor to strontium titanate powders but, due to its insolubility, it cannot be processed into films or other useful morphologies. In order to circumvent this problem, the oxalate complex was converted to the corresponding methoxyacetic acid derivative by reflux in an excess of methoxyacetic acid (ca. 202 °C). This substitution is facilitated by the thermal instability of oxalate which decomposes above 180 °C. Removal of the excess acid by vacuum distillation yields a sticky solid with the formula, $SrTiO(O_2CR)_4(HO_2CR)_2(H_2O)_3$. The extra equivalents of coordinated acid play an important role in stabilizing the compound and attempts to remove them result in decomposition to an insoluble material. Clear evidence for the coordinated acid is provided by both infrared and NMR spectroscopy. The former shows a very broad absorption at 3119 cm^{-1} due to the hydroxyl stretch of the carboxylic acid. The carboxylate moieties of the coordinated neutral and deprotonated acids give rise to two characteristic IR absorptions at 1754 and 1653 cm^{-1} for $\nu_{as}(COOH)$ and $\nu_{as}(COO^-)$, respectively. The ^1H NMR spectrum of $SrTiO(O_2CR)_4(HO_2CR)_2(H_2O)$ closely resembles that of the previously reported barium analog (13). It contains two resonances, corresponding to the methoxy protons at 3.38 and 3.40 ppm in a 2:1 ratio; three signals for the CH_2 protons at 4.03, 4.10 and 4.61 ppm in a 1:1:1 ratio, as well as a broad signal for the carboxylate protons of the neutral acid ligands and the associated water molecule at 6.64 ppm. The ^{13}C NMR spectra of the $SrTiO_3$ precursor contains three resonances for the methoxy carbon atoms at 58.7, 59.0 and 59.2 ppm. However the number of resonances corresponding to methylene carbon atoms is two (69.3 and 70.0 ppm) rather than three. Nevertheless, there is some indication that the CH_2 carbon resonance at 69.3 ppm is constituted of two overlapping signals. The spectra also contain three additional low field resonances due to the carboxylate carbon atoms. These resonances appear at 170.2, 171.8 and 174.7 ppm. The NMR spectral data indicate the presence of three types of methoxyacetate ligands in the ratio of 1:1:1 as evidenced from the ^1H NMR integrated intensities. One type of ligand is clearly the coordinated neutral acid. The other two types of ligands may be assigned as the carboxylate anions associated with strontium for one set and titanium for the other but there is no definitive evidence that this is the case. $SrTiO(O_2CR)_4(HO_2CR)_2(H_2O)_3$ is very soluble in water and a variety of organic solvents such as ethanol, acetonitrile, and chlorocarbons. Therefore, like the barium analog (*13*) it may be used to prepare strontium titanate thin films by spin or dip-coating and firing at 700 °C. Furthermore, since the two titanate precursors are miscible with each other it is also possible to prepare films of $Ba_xSr_{1-x}TiO_3$ in the same manner simply by mixing the precursors. This precursor could also be readily doped with other trace metal ions by mixing in their methoxyacetate salts into the strontium titanate precursor.

One drawback of the methoxyacetate precursors is that their excellent processing characteristics are dependent on the presence of excess methoxyacetic acid since its removal yields fairly insoluble glassy solids. Obviously, the chelating ability of the methoxyacetate ions alone is not sufficient to prevent polymerization of the titanyl complexes. Potentially, then, this problem could be circumvented by including additional ether linkages in the organic chain of the carboxylate. For example, we have demonstrated that metal salts of 2-[2-(2-methoxy)ethoxy]ethoxyacetate ($MeOCH_2CH_2OCH_2CH_2OCH_2CO_2^-$, MEEA) are liquids at room temperature (*14*). Such a liquid metal carboxylate salt of titanium is readily prepared by reaction of four equivalents of the carboxylic acid with titanium isopropoxide in dry ethanol. The product yielded by this reaction is a yellow liquid that is soluble in polar organic solvents such as ethanol, acetonitrile, and chloroform. As well, quite surprisingly, it dissolves in water without hydrolysis. The ^1H and ^{13}C NMR spectra of the neat liquid indicates that it contains a small amount of the ethyl ester of 2-[2-(2-methoxy)ethoxy] ethoxyacetate which is formed as a minor side product of the reaction. Aside from these peaks, there is a single set of resonances attributable to the carboxylate anion indicating that there is only one carboxylate environment (at least on the NMR time-scale since rapid exchange would give the same result).

Barium acetate can be dissolved in $Ti(MEEA)_4$ to afford a precursor for barium titanate. The barium salt will dissolve slowly in the neat titanium carboxylate but the most rapid way to prepare the precursor is to dissolve both $Ba(CH_3COO)_2$ and $Ti(MEEA)_4$ in a small amount of water and then dry in vacuo to yield a pale yellow liquid. Dissolution of barium acetate has little effect on the ^{13}C NMR resonances of the titanium compound aside from the appearance of a new set of peaks due to the acetate ions (δ= 21.9 and 175.7 ppm). However, the line widths do broaden somewhat due to the increased viscosity of the liquid. Thermal gravimetric analysis (Figure 1) of this precursor in air indicates that the coordinated water and organic ligands are lost gradually over the temperature range of 75-450 °C. The mass of the residue corresponds to formation of barium carbonate and titania. However, the X-ray diffraction pattern (Figure 2) of the compound obtained by heating the precursor to 500 °C only shows weak, broad reflections that are attributable to barium carbonate and no crystalline titanium-containing phases are observed. Upon heating to 600 °C for four hours, the product converts almost completely to tetragonal barium titanate with traces of barium carbonate present. The conversion to barium titanate is complete at 700 °C. Presumably, other barium salts besides the acetate could be used to introduce the barium into the $Ti(MEEA)_4$ including barium nitrate or barium 2-[2-(2-methoxyethoxy)ethoxy]acetate (which is also a liquid). As well, other barium titanium oxide phases could be readily prepared by varying the barium to titanium ratio.

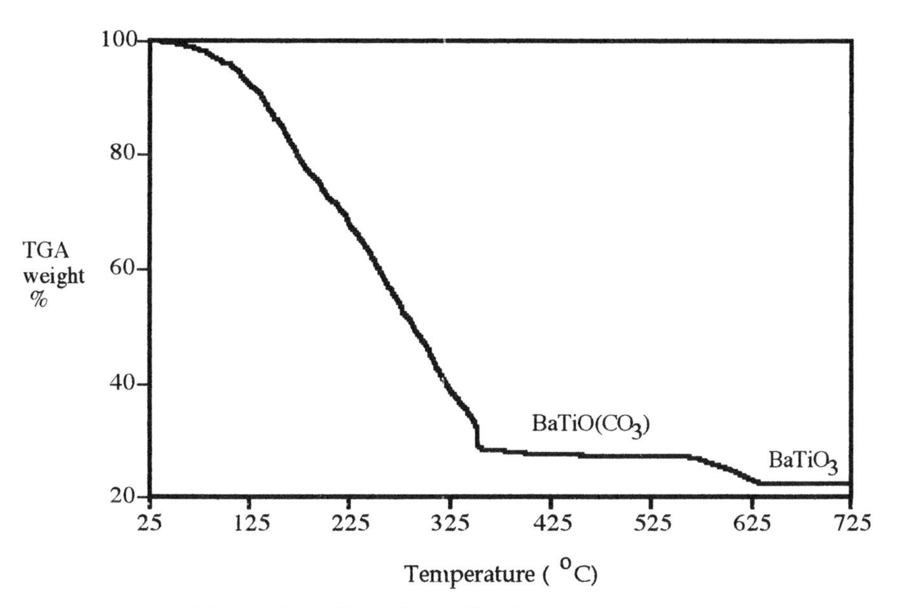

Figure 1 TGA Trace for $BaTi(MEEA)_4(CH_3CO_2)_2$

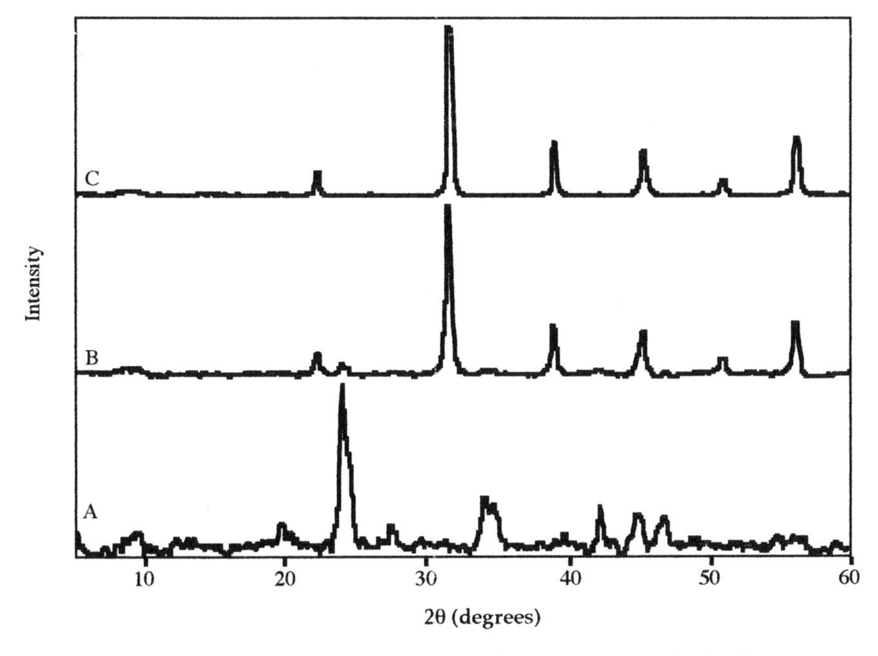

Figure 2 XRD Patterns of Powders Obtained from Pyrolysis of $BaTi(MEEA)_4(CH_3CO_2)_2$ at (A) 500 °C, (B) 600 °C, and (C) 700 °C

Gluconic acid, is another carboxylic acid that is capable of conferring significant water solubility on metal carboxylate salts. Gluconic acid also has the significant advantage that it is extremely inexpensive. However, due to a facile lactonization reaction, it is commercially available only as an aqueous solution. Unfortunately, when such solutions are reacted with $Ti(OEt)_4$, hydrolysis reactions effectively compete with the deprotonation reaction and insoluble gels are obtained. However, a highly water-soluble titanium gluconate chloride can be prepared by using an aqueous system which contains hydrolytically stable titanium ions - $TiCl_3$ in hydrochloric acid. Stirring a mixture of four mole equivalents of gluconic acid and $TiCl_3$ in the air results in gradual oxidation of the purple Ti^{3+} ions to Ti^{4+} and eventual formation of a light yellow solution. Removal of the volatiles in vacuo without heating yields a grayish-yellow, extremely hygroscopic solid. This material is very heat sensitive and, above 60°C, converts to a dark brown insoluble material. The success of this preparative route for titanium gluconate is likely attributable to chelation of the Ti^{3+} ions by gluconate before oxidation.

^{13}C NMR spectroscopy of the solid in D_2O indicates that it is a complex mixture of free gluconic acid, lactonized gluconic acid, and several gluconate anion environments (there are at least five resonances in the carboxylate region). The material also contains 1.89 chloride ions per titanium ion, as indicated by analysis of an aqueous solution of the compound using a chloride ion selective electrode (this number may be judged the lower limit of chloride content since the ISE will not detect any strongly bonded Cl^-). Since titanium (IV) chloride bonds undergo ready hydrolysis, the chloride ions are likely present as counterions to cationic titanium gluconate complexes or, possibly strongly hydrogen-bonded hydrochloric acid. Attempts to purify the titanium gluconate by precipitation from aqueous solution using alcohols or acetone results in decomposition and formation of a white solid that is completely insoluble in water but which still contains gluconate anions according to infrared spectroscopy and elemental analysis. However, the unpurified product is stable indefinitely and it readily redissolves in water. It appears that the impurities stabilize the titanium complex, possibly by prevention of polymerization via complexation of the metal ions. This contention is supported by the fact that the precipitated solid will redissolve in the mother liquor if the solvent used to induce the precipitation is allowed to evaporate.

A gluconate precursor for barium titanate is readily prepared by mixing the correct amounts of the individual barium and titanium gluconates in aqueous solution. Removal of the water in vacuo from this solution yields a grey solid. This compound shows an amorphous X-ray diffraction pattern which is quite different from the patterns of the individual barium and titanium gluconate salts. This, suggests that the two compounds have reacted to produce a material that is structurally quite different. The thermal gravimetric analysis trace for this precursor in air shows two main weight loss steps. The first occurs in the range 25 to 170 °C and corresponds to dehydration. The second weight loss begins at 170 °C and is the result of pyrolysis of the gluconate anion. This occurs gradually as the temperature rises and then becomes extremely rapid at 449 °C. The ceramic yield at the end of this step (500 °C) is 14.3% . Since no weight loss is observed from this point up to 1000 °C, it may be concluded that complete conversion to the oxides has occurred at this temperature. The XRD pattern of the bulk precursor heated to 600 °C indicates that the product is a mixture of barium titanate and barium chloride. The formation of the latter phase is an unfortunate consequence of the chloride content of the titanium precursor. Surprisingly, however, this material converts to barium titanate upon heating to 800 °C. Since no detectable weight loss is observed, it may be presumed that the amount of chloride present was quite small. Thus, this precursor shows some promise for water-based processing of barium titanate thin films. In particular, if the chloride could be removed from the titanium precursor (possibly by ion-exchange), a much improved $BaTiO_3$ precursor could be prepared.

It is possible to combine the water solubility of gluconate salts, the known propensity of glucose to complex the molybdate anion (*15*), and the high reactivity of molybdenum trioxide with diols (*16*) to design unique, water-soluble, bimetallic precursors for lanthanide molybdate ferroelectric phases such as $Nd_2(MoO_4)_3$. These have typically been prepared in the past by elevated temperature (700 °C) reactions of the metal oxides (*17*), a process that is not amenable to the synthesis of thin films. In this investigation, the direct preparation of $Nd_2(MoO_4)_3$ by reaction of Nd^{3+} with MoO_4^{2-} was discovered to be unworkable due to the lack of overlap of the pH stability ranges of the cation and the anion. Thus, hydrolysis reactions compete with the desired precipitation reaction and precipitates are obtained which contain both sodium and high amounts of neodymium relative to molybdenum. Furthermore, even if the precipitation reaction worked, it would provide little opportunity for processing of ferroelectric thin films.

A bimetallic single-source precursor for neodymium molybdate was readily prepared by a single pot reaction in which neodymium gluconate was prepared by reaction of neodymium carbonate with gluconic acid in refluxing water. After the neodymium carbonate had completely reacted, molybdenum trioxide was added to the refluxing reaction mixture. Under these conditions, MoO_3 quickly dissolved as a soluble complex which we propose is the molybdate ester of the gluconate. To date, we do not have any conclusive structural information but one possible structure is given in Figure 3.

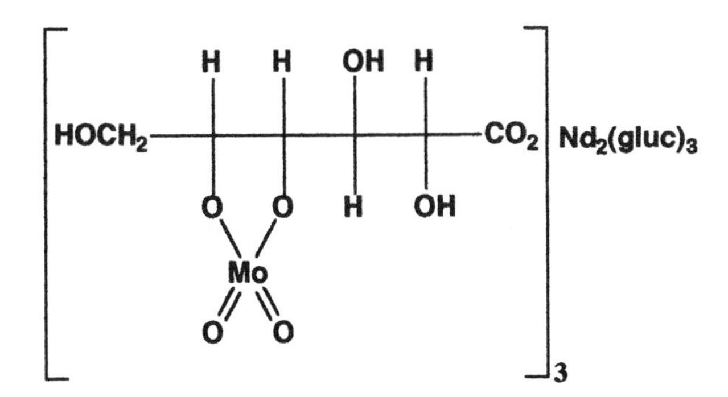

Figure 3 Possible Structure of $Nd_2(MoO_4)_3$ Precursor

The thermal gravimetric analysis trace for this precursor in air (Figure 4) shows two main weight loss steps. The first occurs in the range 25 to 175 °C and corresponds to dehydration. The second weight loss begins at 175 °C and is the result of pyrolysis of the gluconate anion. A gradual loss of weight is observed up to 590 °C where it becomes quite rapid. The ceramic yield at the end of this step (630 °C) is 34.6 %. The XRD pattern of the bulk precursor heated to 630 °C shows that the material is isostructural to $Gd_2(MoO_4)_3$ (*18*) which is also a ferroelectric material (*17*). The high water-solubility and low-temperature conversion of the neodymium molybdenum gluconate to neodymium molybdate make it an excellent precursor for water-based preparation of $Nd_2(MoO_4)_3$ thin films.

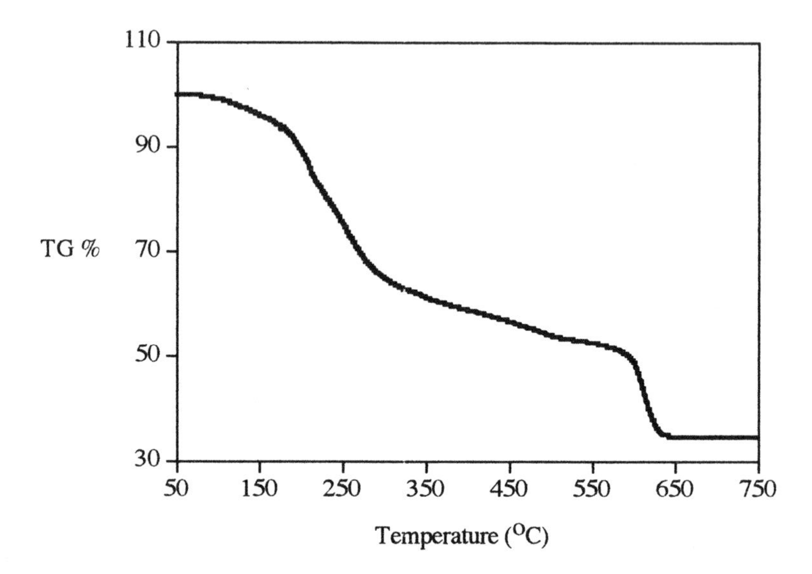

Figure 4 TGA Trace for Neodymium Molybdenum Gluconate in Air

Conclusions

Metal methoxyacetate, 2-[2-(2-methoxy)ethoxy]ethoxyacetate, and gluconate salts are excellent precursors for either water-based or solventless processing of ferroelectric thin films. As well, neodymium gluconate reacts with molybdenum trioxide to afford a water-soluble bimetallic precursor for neodymium molybdate. All of these compounds are readily prepared and convert to the desired oxide ceramics below 700 °C. Initial results for thin film preparation are promising and this investigation is ongoing.

Acknowledgments

We thank the Board of Regents Support Fund for support of this research through Contract # LEQSF(1993-96)-RD-A-26. We also express gratitude to the Department of Defense for support of the molybdate portion of this research through Grant # 93DNA-2.

References

1. Livage, J. in *Transformation of Organometallics into Common and Exotic Materials: Design and Activation*; NATO ASI Series E, No. 141, R. M. Laine ed. (Martinus, Nijhoff, Dordrecht, 1988).

2. Mantese, J. V.; Micheli, A. L; Hamdi, A. H.; Vest, R. W. *M.R.S. Bull.*, (XIV), 1173 (1989).

3. Vest, R. W. In *Ceramics Films and Coatings* Wachtman, J. B.; Haber, R. A. Eds; Noyes Publications: Park Ridge, N.J., 1993; pp 303-347.

4. Xu, J. J.; Shaikh, A. S.; Vest, R. W. *IEEE Trans UFFC* 36, 307 (1989).

5. Xu, J. J.; Shaikh, A. S.; Vest, R. W. *Thin Solid Films* 161, 273 (1988).

6. Maruyama T.; Kitamura, K. *Jpn. J. Appl. Phys.* 28, L312 (1989).

7. Hamdi, A. H. ; Mantese, J. V. ; Micheli, A. L.; Laugal, R. C. O.; Dungan, D. F.; Zhang, Z. H.; Padmanabhan, K. R. *Appl. Phys. Lett.* 51, 2152 (1987).

8. Hebert, V.; His, C.; Guille, J.; Vilminot, S. *J. Mat. Sci.* 26, 5184 (1991).

9. Phule, P. P.; Risbud, S. H. *J. Mater. Sci.* 25, 1169 (1990) and references therin.

10. Xu, Z.; Chae, H. K.; Frey, M. H.; Payne, D. A. *Mat. Res. Soc. Symp. Proc.* 271, 339 (1992).

11. Pechini, M. P. *US Patent 3,330,697* (1967)

12. Chandler, C. D. ; Hampden-Smith, M. J.; Brinker, C. J. *Mat. Res. Soc. Symp. Proc.* 271, 89 (1992).

13. Apblett, A. W.; Georgieva, G. D.; Raygoza-Maceda, M. I. *Ceram. Trans.,* 43 , 73-74 (1994).

14. Apblett, A. W.; Long, J. C.; Walker, E. H.; Johnston, M. D.; Schmidt, K. J.; Yarwood, L. N. *Phosphorus, Sulfur, Silicon, and Related Elements,* 93-94 , 481-482 (1994).

15. Brixner, L. H.; Bierstedt, P. E.; Sleight, A. W.; Licis, M. S. *Mat. Res. Bull.,* 6 , 545-554 (1971).

16. Wilson, A. J.; Penfold, B. R.; Wilkins, C. J. *Acta Cryst.* **C39**, 329-330 (1983).

17. Geraldes, C. F.; Castro, M. M.; Saraiva, M. E. *J. Coord. Chem.* 17 , 205-219 (1988).

18. Powder Diffraction File #24-0428. JCPDS (1995).

Chapter 11

Nucleation and Formation Mechanisms of Hydrothermally Derived Barium Titanate

Jeffrey A. Kerchner[1], J. Moon[2], R. E. Chodelka[1], A. A. Morrone[1], and James H. Adair[1]

[1]Department of Materials Science and Engineering, University of Florida, Gainesville, FL 32611–6400
[2]Department of Materials Science, Massachusetts Institute of Technology, Cambridge, MA 02139

Barium titanate powders were produced using either an amorphous hydrous Ti gel or anatase precursor in a barium hydroxide ($Ba(OH)_2$) solution via a hydrothermal technique in order to discern the nucleation and formation mechanisms of $BaTiO_3$ as a function of Ti precursor characteristics. Isothermal reaction of the amorphous Ti hydrous gel and $Ba(OH)_2$ suspension is believed to be limited by a phase boundary chemical interaction. In contrast, the proposed $BaTiO_3$ formation mechanism from the anatase and $Ba(OH)_2$ mixture entails a dissolution and recrystallization process. $BaTiO_3$ crystallite nucleation, studied using high resolution transmission electron microscopy, was observed at relatively low temperatures (38°C) in the amorphous hydrous Ti gel and $Ba(OH)_2$ mixture. Additional solution conditions required to form phase pure crystallites include a CO_2-free environment, temperature >70°C and solution pH ≥13.4. Analysis of reaction kinetics at 75°C was performed using Hancock and Sharp's modification of the Johnson-Mehl-Avrami approach to compare observed microstructural evolution by transmission electron microscopy (*1*).

Barium titanate ($BaTiO_3$) has become one of the most commonly used ferroelectric materials since its discovery in 1945 (*2,3*). Recently, hydrothermal synthesis has been exploited in the commercial synthesis of high purity, aggregate-free, crystalline $BaTiO_3$ (*4*). Knowledge of the nucleation and formation mechanisms will allow precise control of final particle size and morphology for use in the multilayer capacitor, thermistor, and transducer industries (*5,6*). In contrast to hydrothermal synthesis, conventional synthesis by solid state reaction of $BaCO_3$(s) and TiO_2(s) requires calcination temperatures above 1200°C and produces aggregated particulate which require subsequent milling (*7*).

Hydrothermal synthesis is defined as the treatment of aqueous solutions or suspensions at elevated temperature and pressure. The dynamic nature of the interaction between TiO_2(s), Ba^{2+}(aq), and OH^-(aq) in the solution phase determines the nucleation and growth mechanisms of $BaTiO_3$ particles. However, the closed nature of the hydrothermal system hinders *in-situ* observation of nucleation. In addition, possible phase transformation or crystallization of the Ti precursor prior to reaction with the $Ba^{2+}(aq)$ solute species makes determination of particle formation mechanisms difficult. The crystal structure, as well as the diffusion distance for $Ba^{2+}(aq)$ species (i.e., the surface area), and solubility of the Ti precursor may significantly influence reaction kinetics of hydrothermally derived $BaTiO_3$. Generally, the reactivity of Ti precursors with respect to the formation of $BaTiO_3$ is recognized to follow the hierarchy: amorphous hydrous Ti gel > anatase (TiO_2) > rutile (TiO_2) (*8*). Ensuing variations in the available nucleation sites, particle size, and particle size distribution of the $BaTiO_3$ powders occur due to the differing reaction pathways. Regardless of formation mechanism, four processes must chronologically transpire, the slowest of which is rate limiting (*9*). These four steps include transport of the solute through suspension, surface adsorption and dehydration of solute species, surface diffusion, and finally, crystallite formation exhibiting long-range order and growth.

A fundamental understanding of the $BaTiO_3$ rate limiting mechanisms under the specified hydrothermal conditions, in addition to knowledge of the dominant processing variables, will allow future control of both particle size and morphology. Though processing variables, such as the solution pH, solids loading (i.e., total molarity), and Ba:Ti precursor molar ratio, may affect the observed formation mechanisms, these variables are held constant in order to determine the influence of Ti precursor characteristics on $BaTiO_3$ formation kinetics. Due to the closed nature of the hydrothermal reaction vessel, the solid state reaction based Hancock and Sharp modification was employed in order to discern the formation mechanisms and corroborate the results obtained by direct observation of microstructural evolution by transmission electron microscopy (*1*).

Materials and Methods

Titanium Gel Synthesis and Characterization. The titanium gel (lot number 6814-320) was prepared at Cabot Performance Materials, Boyertown, PA, by neutralization of $TiCl_4$ with NH_4OH at pH 3.5 to pH 4 and exhibited a $300 m^2/g$ surface area, as reported by Cabot. The as-received Ti gel was stored in the frozen state to minimize aging upon receipt because preliminary experiments with ambient stored gels produced inconsistent results as a function of aging time. The phase and morphology of the as-received Ti gel were examined by X-ray diffractometry (XRD), transmission electron microscopy (TEM), and high resolution transmission electron microscopy (HRTEM). In order to vary the Ti precursor characteristics, three heat-treatments, which included heat treatment at 130°C for 24 hours in a conventional laboratory furnace, calcination at 600°C for 1 hour, and hydrothermal crystallization at 150°C for 30 min, were conducted with the as-received gel. The phase composition and morphology of each heat-treated Ti precursor were examined using XRD and TEM. Thermogravimetric analysis/differential thermal analysis (TGA/DTA) was performed to corroborate the thermal treatment results and determine the as-received gel solids loading.

$BaTiO_3$ Synthesis Conditions. Prior to investigation of crystallization kinetics, a study of the conditions required to synthesize phase-pure $BaTiO_3$ (i.e., without barium carbonate contamination) was performed using the as-received hydrous Ti gel and barium octa-hydroxide ($Ba(OH)_2 \cdot 8H_2O$) from Solvay Performance Chemicals, Milano, Italy. The Ti molarity of the as-received slurry was calculated based on thermogravimetric analysis (TGA) assuming $TiO_2(s)$ was the sole calcination product at 800°C. The as-received gel was added to the CO_2-free aqueous suspension containing $Ba(OH)_2$ at 25°C to produce a suspension with a Ba:Ti molar ratio equal to 1.1:1. The dual precursor suspension was homogenized with an ultrasonic horn for 5 minutes under argon atmosphere in a sealed Teflon reaction vessel. The necessary solution pH to produce phase pure $BaTiO_3$ was evaluated by variation of the $Ba(OH)_2$ concentration (either 0.1M or 0.2M) at an isothermal reaction temperature of 90°C for 6 hours under an argon atmosphere, followed by XRD analysis. Based on these results, a feedstock concentration of 0.2M was heated under various isothermal conditions (including 60°C, 70°C, and 80°C) for 12 hours using 23ml acid digestion bombs. After hydrothermal reaction, the powders were washed with pH adjusted (pH~10), CO_2-free deionized water to remove the excess Ba present in solution, yet prevent incongruent dissolution of the barium ions from the $BaTiO_3$ particle surface (*10,11*).

$BaTiO_3$ Kinetic Analysis. The $BaTiO_3$ kinetic analysis was performed using the as-produced anatase precursor (i.e., without drying or washing), whereas the as-received hydrous Ti gel was suspended in CO_2-free deionized water before introduction to the $Ba(OH)_2$ solution. Reactions were performed with a 1.1:1 Ba:Ti molar ratio and 0.2M $Ba(OH)_2$ concentration at 75°C by addition of the Ti and Ba precursor suspensions into the hot sealed Teflon reactor under argon atmosphere. The time zero designation indicates the extraction time of the first sample upon stabilization at the desired solution temperature. Quantitative evaluation of the fractional crystallinity and observation of microstructural evolution were performed by XRD and HRTEM, respectively.

Results and Discussion

Ti Precursor Characteristics. The phase of the as-received hydrous Ti gel was both XRD and TEM amorphous. Heat treatment of the as-received gel at 130°C produced a mixture of an amorphous phase with anatase. Increase in the reaction temperature to the 150°C hydrothermal treatment produced anatase crystallites, whereas calcination at 600°C produced a rutile phase. Further investigation proceeded with only the as-received gel and hydrothermally treated precursor, hereafter referred to as the anatase precursor. Observation of the precursor morphology by TEM indicated a fine gel structure and a 10nm granular nature for the as-received hydrous gel and anatase precursor, respectively.

Dissolution-recrystallization has been proposed for the formation mechanism of TiO_2 under hydrothermal conditions (*12*). In such a case, the nature of the aqueous species at a given solution pH determines the phase of the precipitate. Crystalline titanium dioxide (TiO_2) exhibits three different crystal structures, rutile, anatase, and brookite, depending upon the connection modes of the $[TiO_6]$ octahedra building block. The stability of the structure decreases as the number of the shared edges or faces increases. Thus rutile is the most stable (i.e., least reactive) and brookite the least stable of the three polymorphs (*13*). Recently, Cheng et al. demonstrated that the phase and even morphology of TiO_2 can be manipulated via hydrothermal treatment (*14*). As reported by Cheng and colleagues, granular anatase is formed at a reaction pH ranging from pH 1 to pH 8. At a solution value greater than pH 2, $[Ti(OH)_4^0](aq)$ is the dominant aqueous Ti species (*15*). Upon crystallization to TiO_2, a Ti-O-Ti linkage forms via a dehydration reaction from $[Ti(OH)_4^0](aq)$. Every hydroxyl (OH^-) group in the $[Ti(OH)_4^0](aq)$ tetrahedra is a chemically active site, thus the probability of edge sharing between $[TiO_6]$ octahedra building blocks increases on formation of $TiO_2(s)$. Therefore, the formation of anatase is preferred over rutile. Subsequently, as the reaction medium pH decreases below pH 2, the dominant aqueous species of Ti changes to $[Ti(OH)_2^{2+}](aq)$. Less available hydroxyl groups limits the possibility of edge sharing, and therefore rutile formation is thermodynamically favored at a low pH (*14*).

As predicted, the hydrothermal treatment of the amorphous Ti gel at pH 7 resulted in the formation of nanosize granular anatase. However, the heat treatment and calcination of the as-received hydrous gel produced different phases as a function of treatment temperature. Results of TGA/DTA confirm the transformation from the amorphous to the metastable anatase and eventually to the stable rutile phase.

$BaTiO_3$ Synthesis Conditions. The solution pH is one of the most critical reaction parameters in the control of the phase composition and formation mechanism of solution synthesized particles since it mainly determines the chemical nature of the aqueous species (*16-18*). Therefore, the solution pH necessary to synthesize phase-pure $BaTiO_3$ was investigated. The Ba concentration is significant since the pH of the reaction medium was controlled solely by variation of the $Ba(OH)_2$ concentration. The 0.1M and 0.2M $Ba(OH)_2$ concentrations resulted in a solution pH of pH 12.4 and pH 13.1, respectively. XRD analysis demonstrated the need for highly alkaline solution conditions (pH>13) to form phase-pure cubic $BaTiO_3$. Below pH 13, the reaction products were mostly X-ray amorphous except for a barium carbonate phase. The XRD confirmation that a solution pH greater than pH 13 (i.e., 0.2M $Ba(OH)_2$) is required to form phase pure $BaTiO_3$ is consistent with theoretical predictions (*11*).

In addition to solution pH, the lowest isothermal synthesis temperature for $BaTiO_3$ formation was investigated to allow measurement of the nucleation and growth processes. Of the three investigated hydrothermal temperatures, only the 80°C treatment resulted in the production of a well-crystallized phase-pure $BaTiO_3$. Barium titanate formation was observed at 70°C, though complete crystallization was not achieved within 12 hours. Lastly, the sample treated at 60°C contained unreacted amorphous Ti gel and a $BaCO_3$ contaminant phase. Therefore, the following kinetic analysis was performed at reaction temperatures above 70°C as well as a 0.2M $Ba(OH)_2$ concentration.

$BaTiO_3$ Kinetic Analysis. Utilization of the Hancock and Sharp modification of the Johnson-Mehl-Avrami analysis permits formation mechanism evaluation of an otherwise closed hydrothermal system(*1,19*). Through application of the following equation,

$$\ln[-\ln(1-f)] = \ln r + m \ln t \qquad (1)$$

the slope of the linear regression denotes the constant m, which approximates the system geometry in solid state systems. Additionally, f indicates the fraction crystallized, t represents the time at the isothermal temperature in seconds, and r represents a constant which is dependent on the nucleation frequency and linear grain growth rate in solid state reactions (*1*). Samples withdrawn as a function of time at 75°C were analyzed by XRD to evaluate the fractional crystallinity of the reaction products as illustrated in Figure 1. Observation of $BaTiO_3$ isothermal crystallization from the anatase precursor shown in Figure 2b indicates decreased reaction kinetics and an induction period in comparison to the as-received amorphous Ti gel, Figure 2a. The maximum fractional crystallinity for $BaTiO_3$ formation from the as-received gel and anatase precursor is 0.652 and 0.475, respectively. The transformation was not complete, however it is sufficient to permit use of the Hancock and Sharp technique, which is valid for a fractional crystallinity range of 0.15 to 0.5 (*1*).

Application of the kinetic analysis to the obtained fractional crystallinity is presented in Figure 2. The $BaTiO_3$ reaction kinetics from the amorphous gel exhibit a single-stage rate law where a linear regression yields m = 1.09. An m value equal to 1.09 implies a phase-boundary controlled mechanism where the initial crystallization is controlled by an interfacial chemical reaction (*19*). Conversely, the crystallization kinetics for $BaTiO_3$

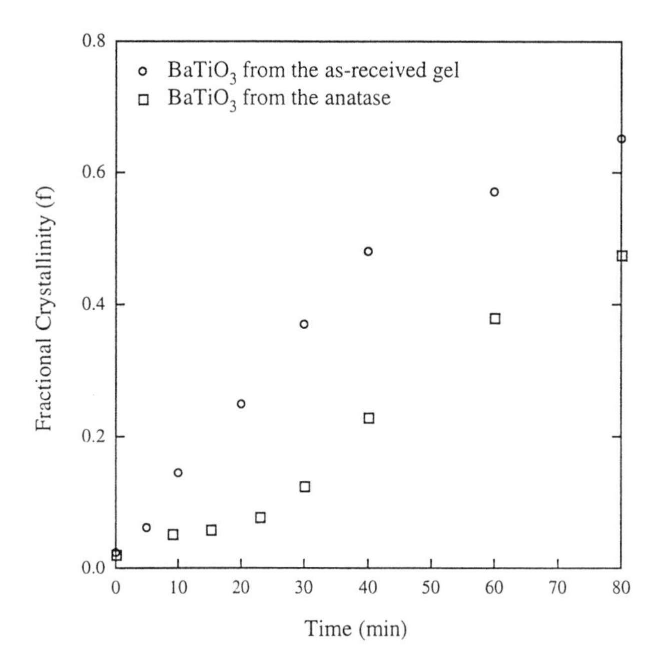

Figure 1. Fraction of $BaTiO_3$ crystallized as a function of time at 75°C for both the as-received TiO_2 gel and anatase reactions in barium hydroxide aqueous solution.

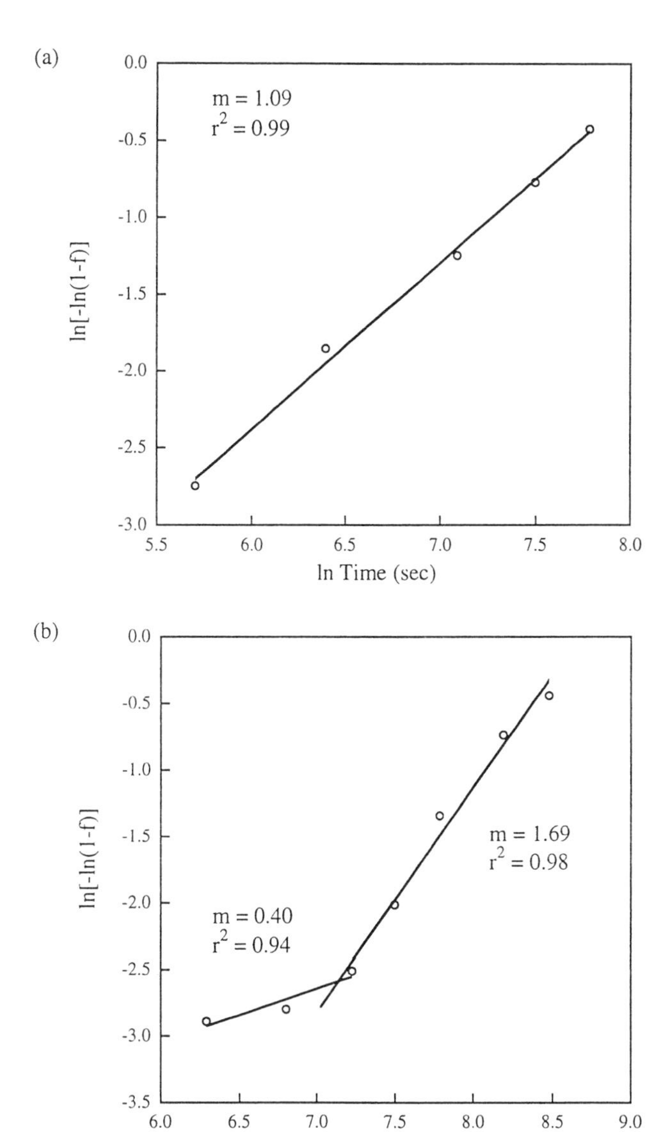

Figure 2. Hancock and Sharp kinetic analysis of $BaTiO_3$ formation mechanisms for the (a) amorphous Ti gel and (b) anatase precursor. The m is a constant determined by the system geometry and r^2 is the correlation coefficient.

formation from the anatase precursor are described by a two-stage rate control where the first stage is characterized by $m = 0.4$. Transition to the second stage occurs at approximately $f = 0.06$ and is identified by $m = 1.69$. The formation kinetics of both stages do not directly correlate to the mechanisms of standard reaction rate equations for various solid state geometries. Dissimilarity between the solution technique with the standard state equations is not unexpected, as some standard state reactions (e.g., $BaCO_3$ formation) do allow direct comparison (*1*). However, the abrupt transition from slow to rapid reaction rates may indicate that a precipitation-like mechanism prevails during the initial stage of the reaction whereby time is required to produce a solution condition favorable to crystallite formation.

As-received Hydrous Gel Reaction. In order to verify or refute the calculation-based kinetic analysis, morphological evolution was investigated using HRTEM and TEM. In the case of $BaTiO_3$ formation from the amorphous Ti gel, approximately spherical 5nm-10nm size $BaTiO_3$ nuclei embedded within the Ti gel matrix were observed as shown in the HRTEM in Figure 3a. The weak diffraction ring patterns produced by the selected area diffraction in Figure 3b confirm the presence of $BaTiO_3$. The observation of crystal lattice fringes which span the nuclei diameter possibly implies a dissolution-recrystallization formation mechanism, as coring of the nuclei would be visible if an in-situ transformation reaction occurred. The TEM micrographs in Figure 4 illustrate particle formation and growth at the following reaction times and corresponding fractional crystallinity: Figure 4a, 3 minutes and f= 0.07, Figure 4b, 9 minutes and f= 0.17, Figure 4c, 35 minutes and f= 0.40, and the selected area diffraction pattern (SADP) at 35 minutes reaction time (Figure 4d). Spherical 25nm particles (f= 0.07) evolve from the amorphous Ti gel and exhibit growth to a particle size of 50nm at 35 minutes, after which time the particle size remains nearly constant.

Based on the results of the kinetic study and observed microstructural evolution, a phase boundary controlled mechanism is proposed in which the rate controlling step entails a chemical reaction at the gel surface. The amorphous gel exhibits a fine, porous network structure which is believed to be thoroughly infiltrated by the aqueous solute and solvent. The interconnected network structure is experimentally observed by both the large surface area obtained via a nitrogen adsorption technique (~$300m^2/g$) and TEM photomicrographs of the precursor at early stages of reaction (*20*). Transport of Ba^{2+} solute species through the porous network structure is extremely rapid in comparison to bulk diffusion through a freshly formed $BaTiO_3$ product layer. Therefore, it is reasonable to assume that, due to the relatively high solubility of Ba^{2+} ions (with respect to the titanium species), the Ba^{2+} species cover the entire surface of the Ti gel throughout the reaction as illustrated in Figure 5 (*10,21*). To permit chemical interaction between the TiO_2(s) and Ba^{2+}(aq) the Ti-O-Ti bonds in the highly cross-linked hydrous gel must break and undergo rearrangement in addition to the dehydroxylation or dehydration reactions. Local nucleation and initial growth are marked by Ba^{2+} incorporation to the nuclei lattice. Latter growth occurs via aggregation with unstable smaller $BaTiO_3$ primary crystallite clusters, resulting in a raspberry-like appearance of $BaTiO_3$ particles, as observed in this investigation by HRTEM and reported by Chien and colleagues (22).

Anatase Precursor Reaction. The earliest $BaTiO_3$ nucleation from the hydrothermally treated Ti precursor could not be visually discerned from the aggregated anatase. However, electron diffraction patterns suggest the presence of barium titanate. The HRTEM photomicrograph of a sample extracted on ramping prior to attainment of the isothermal temperature illustrates 10nm nuclei in Figure 6. Due to the overlap and (approximately 10°) misorientation of the nuclei (observed by the split diffraction spots

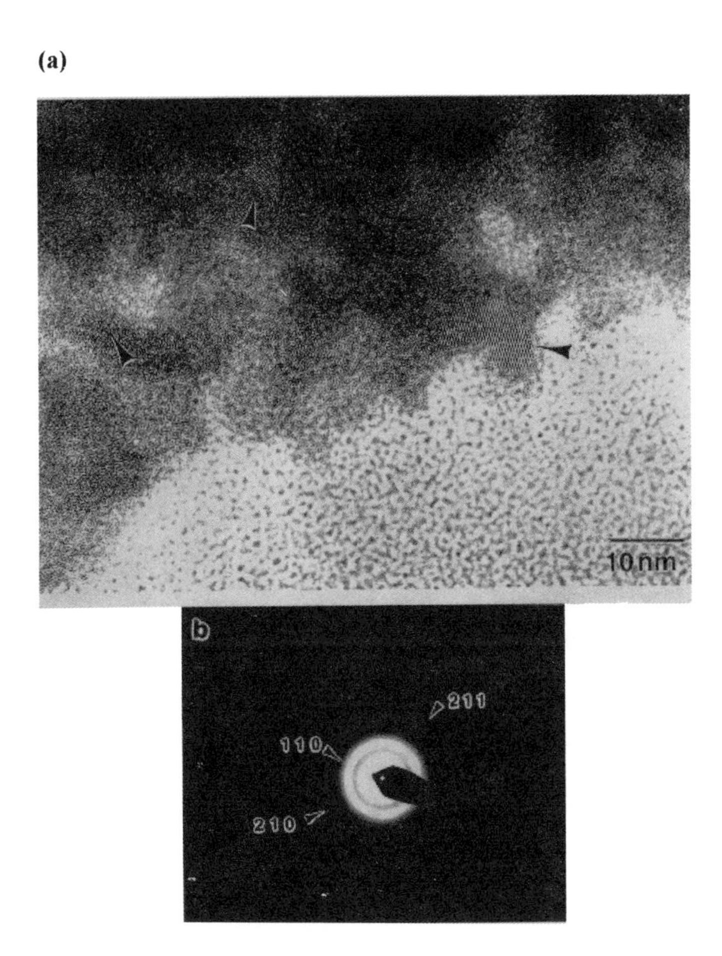

Figure 3. HRTEM photomicrograph illustrating $BaTiO_3$ nucleation within the as-received gel on ramping to the isothermal temperature, and corresponding selected area diffraction pattern .

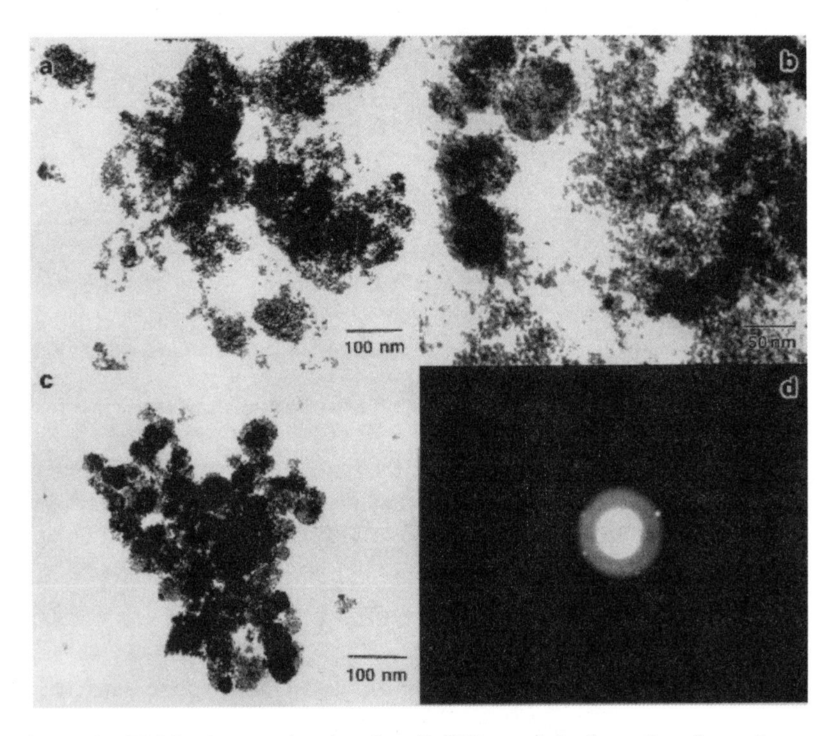

Figure 4. TEM micrographs showing $BaTiO_3$ particle formation from the as-received gel and $Ba(OH)_2$ reaction as a function of reaction time: (a) 3 min; (b) 9 min; (c) 35 min; and (d) selected area diffraction pattern, respectively.

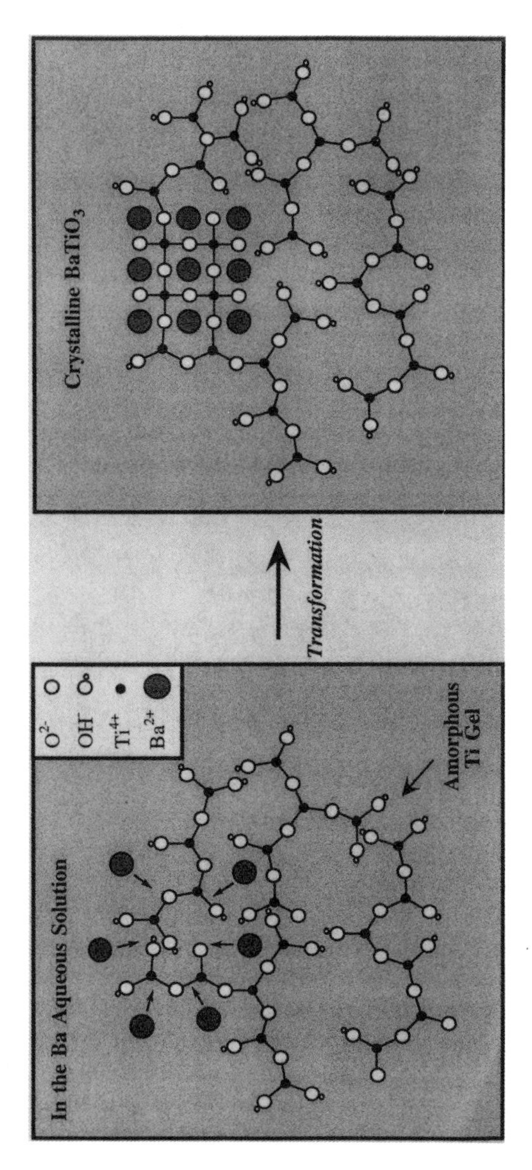

Figure 5. Proposed $BaTiO_3$ formation from the as-received gel precursor and 0.2M $Ba(OH)_2$.

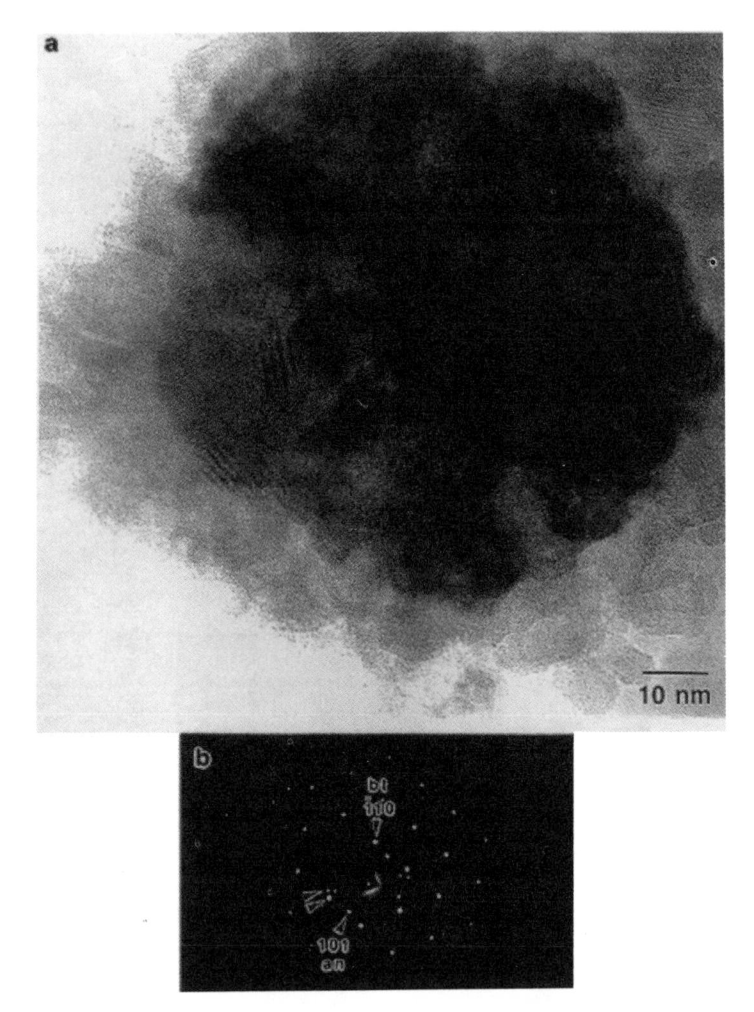

Figure 6. HRTEM photomicrograph of the anatase derived $BaTiO_3$ nuclei including selected area diffraction pattern. The double arrows indicate a split reflection spot of $BaTiO_3$ due to lattice misfit of the anatase and $BaTiO_3$ denoted by 'an' and 'bt', respectively.

for the $0\bar{1}1$ in the SADP), Moire fringes are produced, as noted by the thick alternating dark and light bands. Similar to the $BaTiO_3$ nuclei produced via the amorphous gel reaction, particle coring is not observable. The TEM micrograph shown in Figure 7 illustrates the aggregated nature of the 10-20nm anatase precursor with f= 0 (Figure 7a), the rough surface of 100nm size crystallites with f= 0.10 (Figure 7b), and the eventual growth into 200nm size particles with f= 0.40 (Figure 7c).

In comparison of the kinetic analysis and microstructural evidence, the following formation mechanism for $BaTiO_3$ from the hydrothermally treated Ti precursor is proposed. Nucleation of $BaTiO_3$ initiates heterogeneously on the smallest particles of the aggregated titania via dissolution-recrystallization. As proposed in the Ti gel reaction, Ba species are expected to surround the TiO_2 due to the high relative solubility. Although the solubility of the anatase at 25°C is approximately 10^{-7}M, which is too low to provide a sufficient level of supersaturation for nucleation, this value does not consider local solubility due to the surface energy effect of small particles (*23*). The slow initial kinetics, as observed in Figure 1, probably represent an incubation-like period for dissolution. Upon achievement of the required supersaturation level, $BaTiO_3$ nucleates on the anatase via a reaction between $[Ti(OH)_4^0](aq)$ and $[BaOH^+](aq)$ and grows by a solute addition reaction, continuously consuming the anatase. The rate limiting step is the dissolution of the titania. Therefore, overall reaction kinetics are slower than the system involving the more reactive amorphous Ti gel.

Several $BaTiO_3$ crystallites can simultaneously, heterogeneously nucleate on the same anatase particle, which results in lattice misorientation and the presence of Moire fringes. The hexagonal lattice pattern of (111) $BaTiO_3$ clearly reveals a phase boundary between the barium titanate and anatase. In addition, the TiO_2 lattice fringes were observed to cross the phase boundary without interruption, which implies that $BaTiO_3$ grows heterogeneously at the expense of the anatase. Unlike the abundant nucleation sites present in the amorphous hydrous gel reaction, nucleation occurs only at the solution interface of the smallest anatase particles. As a result, the average $BaTiO_3$ particle size (~0.2μm) derived from the anatase precursor is larger than the $BaTiO_3$ particles (~0.05μm) obtained from the amorphous Ti gel at identical fraction crystallinity (f= 0.40).

Conclusions

Both hydrothermal treatment and solid state heating caused modification of the as-received hydrous gel microstructure and phase. Investigation of the Ti precursor variations by assessment of the reaction rates and TEM photomicrographs determined the Ti precursor to significantly influence the formation mechanisms of $BaTiO_3$ at 75°C, as predicted. Nucleation of $BaTiO_3$ occurs at relatively low temperatures (i.e., <100°C) in both Ti precursor systems and was observed at 38°C in the amorphous gel and $Ba(OH)_2$ suspension. Increased reaction kinetics without an induction period were observed during the amorphous hydrous Ti gel and $Ba(OH)_2$ reaction, as compared to the anatase precursor. At similar stages in the crystallization process (i.e., 0.4 fraction crystallinity) the particulate derived from both the gel and anatase exhibited an equiaxed morphology, but differed in particle size (0.05μm and 0.2μm, respectively).

The amorphous gel reaction produced relatively spherical 5nm to 10nm $BaTiO_3$ nuclei which grew within the porous amorphous gel matrix. Based on the Hancock and Sharp kinetic analysis, the proposed $BaTiO_3$ formation mechanism entails a phase boundary limited reaction where the chemical interaction at the gel surface is the rate controlling step. In the case of $BaTiO_3$ formation from the anatase precursor, 10nm size crystallites which nucleated atop the aggregated anatase were observed. It is believed

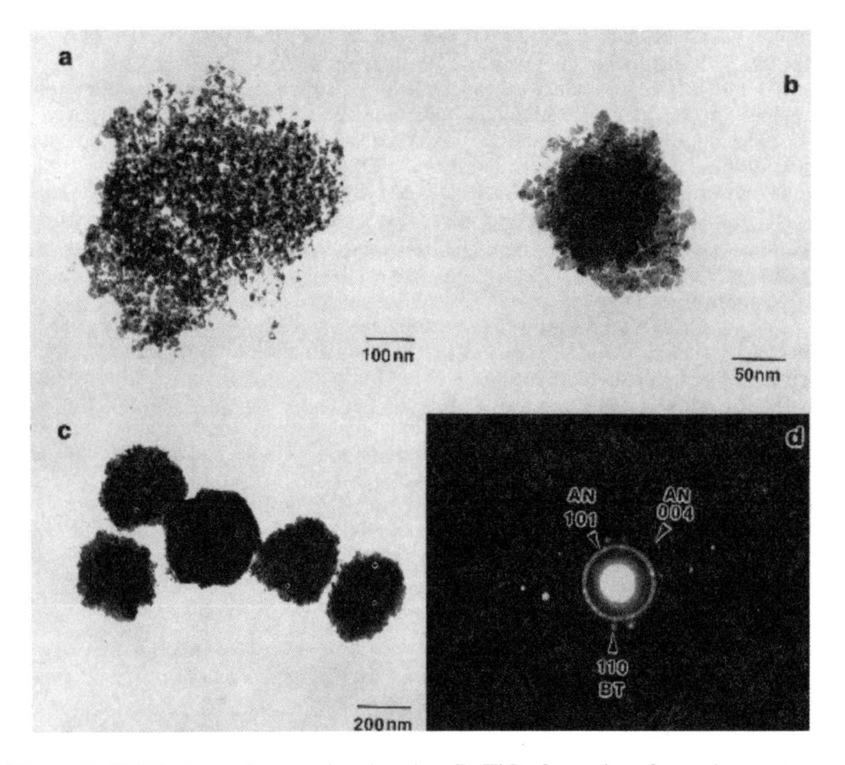

Figure 7. TEM photomicrographs showing $BaTiO_3$ formation from the anatase precursor as a function of reaction time: (a) 5 min; (b) 25 min; (c) 60 min; and (d) SADP from 60 minutes. The 'AN' and 'BT' represent diffraction spots for the anatase and $BaTiO_3$, respectively.

that $BaTiO_3$ nucleates heterogeneously on the anatase particles via a dissolution and recrystallization process and grows at the expense of the neighboring anatase. The current study, performed at mild hydrothermal conditions (i.e., <100°C), may facilitate a better understanding of $BaTiO_3$ formation mechanisms at elevated temperatures. However, care must be exercised in the application of the low temperature analysis since the predominant mechanisms may change as a function of temperature and pressure.

Acknowledgments

The authors would like to thank the research sponsors, Cabot Performance Materials, for the technical advisement and funding. In addition, the authors greatly appreciate the HRTEM expertise of Dr. Augusto Morrone at the Major Analytical Instrumentation Center.

Literature Cited

1. Hancock, J. D.; Sharp, J. H. *J. Am. Ceram. Soc.* **1972**, *55*, 74.
2. Wul, B.; Goldman, I. M. C. R. Acad. Sci. URSS. **1945**, *46*, 139; *49*, 177.
3. Wul, B.; Goldman, I. M. C. R. Acad. Sci. URSS. **1946**, *51*, 21.
4. Riman, R. E. , In *Surface and Colloid Chemistry in Advanced Ceramic Processing*; Pugh, R. J. and Bergstrom, L., Ed.; Surfactant Science Series; Marcel Dekker Inc.: New York, NY, 1994, Vol. 51; pp 29-69.
5. Whatmore, R. W.; In *Fundamentals of Ceramic Engineering*, Vincenzini, P. Ed.; Elsevier Science Publishers: London, UK, 1991; pp 223-254.
6. Jaffe, B.; Cook, W.; Jaffe, H. Piezoelectric Ceramics, Academic Press, Inc.: New York, NY, 1971; pp 74-107.
7. Galagher, P. K.; Thomson J. *J. Am. Ceram. Soc.*, **1965**, *48*, 64.
8. Pfaff, G. *J. Euro. Ceram. Soc.*, **1991**, *8*, 35.
9. Mersmann, A. In *Crystallization Technology Handbook*; Mersmann, A. Ed.; Marcel Dekker Inc.: New York, NY,1995; pp 1-78.
10. Utech, B.; M.S. Thesis, The Pennsylvania State University, 1990.
11. Lencka, M. M.; Riman, R. E. *Chem. Mater.*, **1993**, *5*, 61.
12. Kondo, M.; Shinozaki, K.; Ooki, R.; Mizutani, N. *J. Ceram. Soc. Japan*, **1994**, *102*, 742.
13. Evans, R. C. An Introduction to Crystal Chemistry, Cambridge University Press: New York, NY, 1964; 2nd Ed.
14. Cheng, H.; Ma, J.; Zhao, Z.; Qi, L. *Chem. Mater.*, **1995**, *7*, 663.
15. Moon, J.; Ph.D. Dissertation, University of Florida, 1996.
16. Henry, M.; Jolivet, J. P.; Livage, J. In *Chemistry, Spectroscopy and Applications of Sol-Gel Glasses, Structure and Bonding*; Reisfeld, R., Jorgensen, C. K., Ed.; Springer-Verlag: Berlin, Germany, 1992, Vol. 77; pp 153-206.
17. Livage, J.; Henry, M.; Jolivet, J. P. In *Chemical Processing of Advanced Materials*; Hench, L. L., West, J. K., Ed.; John Wiley and Sons: New York, NY, 1992, pp 223-237.
18. Livage, J.; Henry, M.; Jolivet, J.; Sanchez, C. *MRS Bull.* **1990**, *15*, 18.
19. Sharp, J. H.; Brindley, G. W.; Narahari Achar, B. N. *J. Am. Ceram. Soc.*, **1966**, *49*, 379.
20. Kutty, T.; Padmini, P. *Mater. Res. Bull.*, **1992**, *27*, 945.
21. Sharma, R. K.; Chan, N. H; Smyth, D. M. *J. Am. Ceram. Soc.*, **1981**, *64*, 448.
22. Chien, A. T., et al. *J. Mater. Res.*, **1995**, *7*, 1784.
23. Eckert, J. O., Jr.; Hung-Houston, C. C.; Gersten, B. L.; Lencka, M. M.; Riman, R. E. *J. Am. Ceram. Soc.*, in press.

Chapter 12

High-Temperature Synthesis of Materials: Glycothermal Synthesis of Alpha Aluminum Oxide

Nelson S. Bell, Seung-Beom Cho[1], and James H. Adair

Department of Materials Science and Engineering, University of Florida, North–South Drive, Gainesville, FL 32611–6400

A technology for the glycothermal synthesis of α–Al_2O_3 particles of controlled size and shape has been developed. This technique allows for the production of anisometrically shaped particles with controlled morphology and size distribution. Morphological control of particle shape is described as a function of reaction conditions in which platelet morphologies are produced at high shear rates and polyhedra are formed at low shear rates. Particle size control is also investigated via the use of seed materials which promote epitaxial growth, thereby reducing the surface energy contribution to the nucleation barrier. α–Al_2O_3 seed materials can be used to form particle sizes as small as 100-200 nm without agglomeration. α–Fe_2O_3 seeds have similar effects in comparison with the α–Al_2O_3 seeds, but require higher values of seed number concentration to achieve equivalent particle size effects. Morphological control is unaffected by seeding additions.

In the continuing effort to improve the quality of ceramic powders, preferred characteristics include small and uniform particle size (<1.0 μm) and a non-agglomerated state of dispersion. Several attempts have been reported involving chemical techniques that produce ceramic powders with nanosize particles (1-7). These chemical techniques have been shown to be advantageous by eliminating the need for high temperature calcination reactions and milling procedures to achieve fine particle sizes. Processes that produce the final powder at low temperature and fast reaction time result in a more cost effective material.

Solution synthesis is a technique for forming crystals of a desired phase by precipitation from a supersaturated solution containing the solute components. Unlike other methods for forming submicron and nanoscale particles, solution synthesis can generate chemically homogeneous materials by control of the molecular components during the reaction. Nancollas (8) identifies reaction variables in solution processing that affect the crystallite morphology, size and reaction kinetics. These include: the

[1]Current address: Department of Ceramics, Rutgers University, P.O. Box 909, Piscataway, NJ 08855–0909

solids loading, stirring dynamics, rate of supersaturation development, reaction time and temperature, solution pH, and the presence or absence of additives. Supersaturation can be induced by several methods such as the mixing of two solutions containing the component solute ions, evaporation of the solvent, or control of the reaction temperature. The degree of supersaturation acts as a driving force to form a crystalline nucleus which then grows as a stable particle. Additionally, a precursor solid phase can be used in the synthesis reaction. In order to achieve the desired results, this precursor must have a higher solubility than the desired phase, and will thus act to supply solute for the particle synthesis. The formation of the particle surface (i.e. a phase boundary) creates a nucleation energy barrier which the supersaturation must exceed in order for precipitation to occur. The nucleation barrier and reaction conditions are related in the equation for nucleation frequency (9).

$$B = B_n \exp\left[\frac{-16\pi\sigma^3 \upsilon^2 b}{3k^3 T^3 (\ln S)^2}\right] = C \exp\left[\frac{-\Delta G^*}{kT}\right] \qquad (1)$$

where B is the nucleation rate, σ is the surface energy of the particle, ν is the molecular volume, k is Boltzman's constant, T is temperature in Kelvin, and b is a correction factor (less than 1) for the case of heterogeneous nucleation. S is the supersaturation or ratio of actual concentration (C) to the equilibrium solubility (C^*). The reaction variables are related to the change in Gibbs free energy ,$\Delta G°$, as shown in equation [1] where ΔG^* is the free energy nucleation barrier. B_n is a factor related to the number of sites available for nucleation, and can be increased by the use of seed materials. The Gibbs-Thomson equation relates the solubility of particles to their size using the supersaturation (9)

$$\ln\left(C/C^*\right) = \ln(S) = 2\sigma\upsilon / kTr_c \qquad (2)$$

where r_c is the critical nuclei size.

There are two major types of nucleation processes (8,9). Homogeneous nucleation involves the aggregation of component solute ions to achieve a stable ordered state without interaction with another surface. Homogeneous nucleation involves a balance between the decrease in free energy from the formation of the ordered, solid phase, and the generation of positive free energy in the form of the surface energy. This creates a critical size for nuclei formation. Nuclei are unstable below this size and will return to the solution, whereas nuclei above the critical size are stable and will undergo growth. Homogeneous nucleation will have the highest free energy barrier to nucleation.

Heterogeneous nucleation is much more common in solution synthesis, and occurs when nuclei formation is catalyzed by the presence of a surface that lowers the energy necessary to form a stable nucleus. Heterogeneous nucleation can result from solution contact with the reactor chamber, or by the deliberate addition of seed materials. In a seeded synthesis reaction, Turnbull (10) has identified several characteristics necessary to produce advantageous results. Seed particles must be homogeneously dispersed throughout the system, the crystalline lattice mismatch between seeds and product must be less than 20% (and preferably below 5%), and the number concentration of extrinsic seeds must exceed the concentration of intrinsic nucleation sites.

The use of seeds lowers the thermodynamic nucleation barrier to particle formation by creating an epitaxial surface for crystal growth, and by providing a reactive surface for the production of additional nuclei (11). By increasing nucleation frequency, the amount of solute available for growth is diminished and the particle size is decreased (12). Also, the use of seeds can change reaction kinetics by reducing the initial time to nucleate and therefore lower the reaction time. Alternatively, the reaction temperature can be lowered to achieve final product which is identical to the products resulting from unseeded synthesis reactions.

Previous work (13,14) has shown that the glycothermal synthesis process, a liquid phase precipitation at elevated temperatures under autogeneous pressure using a glycol as solvent, for α-Al_2O_3 is mediated by the formation of a precursor pseudo-boehmite phase (15). The pseudo-boehmite is thermodynamically unstable under the reaction conditions, and transforms to the stable α-Al_2O_3 (similar to the formation of alumina in the hydrothermal system (16)). Thus, the overall reaction is as follows.

$$2AlOOH \rightleftharpoons \alpha\text{-}Al_2O_3 + H_2O \quad (3)$$

LeChatelier's principle favors the generation of the product in a non-aqueous environment. In hydrothermal processing, the solvent is a product and the level of α-Al_2O_3 permitted in the reaction is inhibited. Hydrothermal processing produces alumina at 405°C and 34.5 MPa in a reaction time of days to weeks. In contrast, glycothermal processing of alumina proceeds at a minimum temperature of 270°C, 4.1 MPa and 12 hours.

Growth theories of particle morphology describe rate limiting mechanisms (8,17,18) derived from the steps of integrating a solute ion from solution into the crystal structure. The diffusion layer theory considers growth is limited by the slowest reaction step. These steps include the transport of material to the diffusion layer near the particle surface, diffusion through the boundary layer, incorporation into the crystal lattice, and (in melt growth) the removal of the heat of incorporation from the crystal surface to the solvent (or vapor). The boundary layer, δ, is a stationary solvent layer between the component concentration at the surface and the bulk concentration in the solvent. The thickness of the boundary layer is a function of temperature and hydrodynamics. The boundary layer is related to the mean linear rate of crystal growth, l, in the following equation.

$$\frac{1}{l} = \frac{1}{l_0 + ku} + \frac{1}{l_j} = \left(\frac{1}{k_d} + \frac{1}{k_i}\right) \times \frac{\rho_c}{\Delta c} \quad (4)$$

l_j, l_0, and k are constants and u is the relative velocity between the crystal and the solution. k_d is the diffusion coefficient of solution components divided by the boundary thickness (D / δ). k_i is the interface reaction rate constant for crystal growth.

Δc is the difference in concentration of components between the bulk (c) and the surface (c_k), and ρ_c is the density of the crystal.

In the case of $u \rightarrow \infty$ (high stirring rate), the growth rate becomes proportional to l_j, which is related to the interface reaction constant k_i.

$$\frac{1}{l} = \frac{1}{l_j} = \frac{\rho_c}{k_i \Delta c} \tag{5}$$

In the case of low stirring rate ($u = 0$), the diffusion of components to the surface becomes a factor in growth of a surface.

$$\frac{1}{l} = \frac{1}{l_0} + \frac{1}{l_j} = \frac{\rho_c}{k_d \Delta c} \tag{6}$$

Incorporation of components in a growing facet involves ion adsorption at the surface, dehydration of lattice ions, surface diffusion to favored sites, integration into the crystal, as well as the opposing process of ion dissolution. Different facet types will have different reaction rate constants, and the alteration of the thickness of the boundary layer can affect the growth rate to produce different growth morphologies.

The growth shape of a crystal depends on the relative growth velocities of the various faces, which determine the relative center-to-face distances (19). The equilibrium shape depends on the surface energies, whose relative values may be used in the Wulff theorem (20) to give the relative center-to-face distances. Therefore, the central distances may be measured to represent relative growth velocities (ideal growth shape) or surface energies (equilibrium shape). Based on crystallographic data, including space groups and lattice parameters, the computer program SHAPE (Dowty, E. and Richards, R.P. v4.0, 521 Hidden Valley Road, Kingsport, TN, 1991-1993.) was used to determine the face indices of the α–Al_2O_3 particles synthesized under different conditions and to generate the growth shape of the crystals as a function of relative growth rate.

Experimental

Seed preparation. α-hematite (α–Fe_2O_3) sol particles were synthesized as seed materials by the MacCallum method (21). 22.7 grams of $Fe(NO_3)_3 \cdot 9\ H_2O$ was dissolved in 100 ml of water and precipitated with 30 ml aqueous ammonia (28%) under stirring (all chemicals are reagent, Fisher Scientific, Fair Lawn, NJ.). The precipitate was filtered and redispersed in water four times. 12.5 ml glacial acetic acid (29%) was added and stirred gently for 18 hours. The sol was diluted to 400 ml and placed in a reflux apparatus, then heated with stirring and maintained at 80°C for 48 hours. The sol was then centrifuged at 15,000 rpm for 15 minutes. The suspension of α-hematite seeds was prepared by dispersing the centrifuged α-hematite in methanol. The synthesized particles were characterized by X-ray diffraction and bright field transmission electron microscopy.

The α-alumina seed dispersions were prepared by dispersing high-purity commercial α-alumina powder (high purity alumina AKP-53 (>99.99% purity), Sumitomo Chemical Company, Ltd., Japan.) in methanol and ultrasonicating in a Fisher ultrasonic bath to break up agglomerates. Large particles and agglomerates were removed by sedimentation and the fine particles were recovered. Figure 1 presents scanning electron micrographs and transmission electron micrographs of the seed particle morphologies.

Preparation of α-alumina particles in 1,4-butanediol solution. All samples were prepared from a high-purity commercial gibbsite powder having an average particle size of 0.2 μm (Hydral 710, Bayer hydrated alumina, Aluminum Company of America, Bauxite, AR). Ten grams of commercial gibbsite powder was dispersed in 100 ml methanol under rapid stirring. The methanol suspension of gibbsite was redispersed in 250 ml of 1,4-butanediol. After the suspension was stirred for two hours, measured amounts of seeds were added to the dispersed gibbsite suspension, giving various mixtures with seed concentrations from 10^9 to 10^{14} seeds/ml. The suspension was heated to 60° C for 12 hours under vigorous stirring to evaporate most of the residual methanol. The resulting suspension was placed in a 1000 ml stainless steel pressure vessel equipped with a magnetically stirred head (Parr Autoclave 4522, Parr Instrument Company, Moline, IL). The atmosphere inside the pressure vessel was flushed with nitrogen prior to heating the contents to the desired temperature, to minimize oxidation and decomposition of the 1,4-butanediol. The vessel was then heated to 300°C at a rate of 3°C/min. The reaction time was 12 hours.

Runs were aborted if reaction pressures exceeded 4.1 MPa. Reaction pressure is an indicator of solvent stability, with rising pressure resulting from solvent oxidation. Solvent oxidation not only alters the reaction medium, but can produce uncontrolled amounts of uncharacterized additives. This can inhibit morphological control. Uncontrolled oxidation leads to increased pressure that can cause equipment failure, and this concern must be monitored when using this technique. 1,4-butanediol has a pressure of 4.8 MPa at 370°C in contrast to the pressure of water (34.5 MPa) just below its critical temperature of 374.1°C. At 370°C, 1,4 butanediol will begin to oxidize over the course of a reaction, which limits dwell times to less than 5 hours.

After the glycothermal treatment, the vessel was cooled to room temperature, with excess pressure relieved via a pressure release valve. The reaction products were washed at least five times by repeated cycles of centrifugation and redispersion in isopropanol. After washing, the recovered powders were dried at 25°C in a desiccator for 48 hours. The dried, recovered powders were analyzed for phase composition using X-ray diffraction (X-ray Diffractometer APD 3720, Philips Electronics, Mahwah, NJ). The morphology of the synthesized particles was observed using scanning electron microscopy (Scanning Electron Microscope JSM 6400, JEOL, Boston, MA.). Particle size analysis of the synthesized α-alumina powders was performed using a centrifugal sedimentation technique (Horiba Capa-700, Horiba Ltd., Kyoto, Japan).

A

B

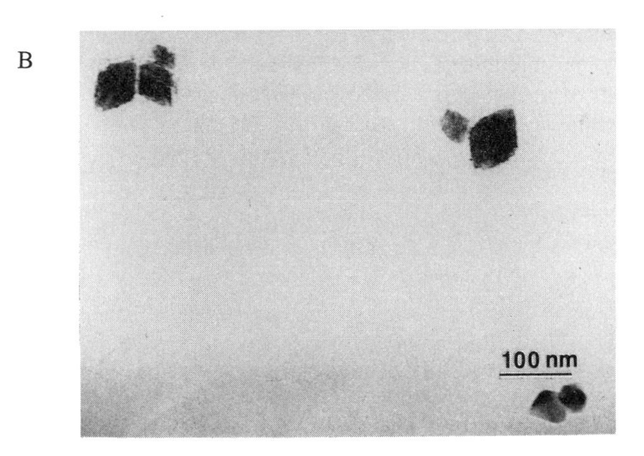

Figure 1. (A) Scanning electron micrograph of α-Al_2O_3 seed particles. (B) Transmission electron micrograph of α-Fe_2O_3 seed particles. (Reprinted with permission from ref. 23. Copyright 1997 The American Ceramic Society.)

Results and Discussion

Glycothermal Reaction. The glycothermal reaction proceeds via an intermediate pseudo-boehmite phase as identified by Inoue (15). The precursor gibbsite phase reacts to form the pseudo-boehmite phase within two hours, as shown in Figure 2. Residual pseudo-boehmite is present at six hours, and by twelve hours, the formation of α–Al_2O_3 is complete. The kinetics of the glycothermal synthesis reaction are currently under investigation, and are expected to vary with solids loading, stirring rate, and the presence of seeds. A higher solids loading should form a greater amount of pseudo-boehmite, and thereby slow the reaction. The use of seeds should increase the reaction rate according to the seed concentration.

Morphological Control. Previous work has demonstrated the morphological forms created by varying reaction conditions (13,14). Figure 3 presents scanning electron photomicrographs of α–Al_2O_3 polyhedra formed at low values of stirring rate in the reaction vessel. Figure 4 shows platelets formed at high stirring rates. Figure 5 shows SHAPE representations of growth particle shapes. Platelet morphologies result from high stirring rates and low solids loading, whereas the polyhedral shapes are formed at low stirring rates and high solids loading.

The effect of stirring rate and solids loading on morphology relates to the growth conditions of the synthesis reaction. At low stirring rates, the supersaturation in the bulk solution is not identical to the supersaturation level at the surface due to the boundary layer, and the growth is diffusion controlled. At high stirring rates, shear forces reduce the boundary layer to cause the supersaturation near growth planes to be identical to the bulk. Integration of solute components into the crystal is the rate controlling mechanism and surface controlled growth is active in the determination of particle morphology. As growth mechanisms are altered by control of reaction conditions, particle morphology can be tailored to the desired form.

Seeding Effects on Morphology. Figures 3 and 4 present scanning electron micrographs showing the effect of seed materials on particle morphology. Production of alumina at low stirring rate results in a polyhedra morphology, both as an unseeded product and at a seed concentration of 1.6×10^{12} α-Fe_2O_3 seeds/ml of 1,4-butanediol. Likewise, high stirring rates produce a characteristic platelet morphology in either the seeded or unseeded condition. It is experimentally evident that the use of seed particles does not alter morphological control in the glycothermal process. Morphology is related to reaction conditions such as stirring rate, solids loading and reaction time.

Size Control via Seeding. Figure 4 presents scanning electron micrographs of α–Al_2O_3 platelets formed with increasing number concentrations of α–Al_2O_3 seeds. It is evident that the characteristic face length (the basal plane of the alumina platelet) decreases with higher seed concentrations. The unseeded particles have a characteristic length in the range of 3 to 7 μm. The addition of 1.6×10^7 seeds/ml 1,4-butanediol decreases the characteristic length to 2 or 3 μm. Progressively increasing

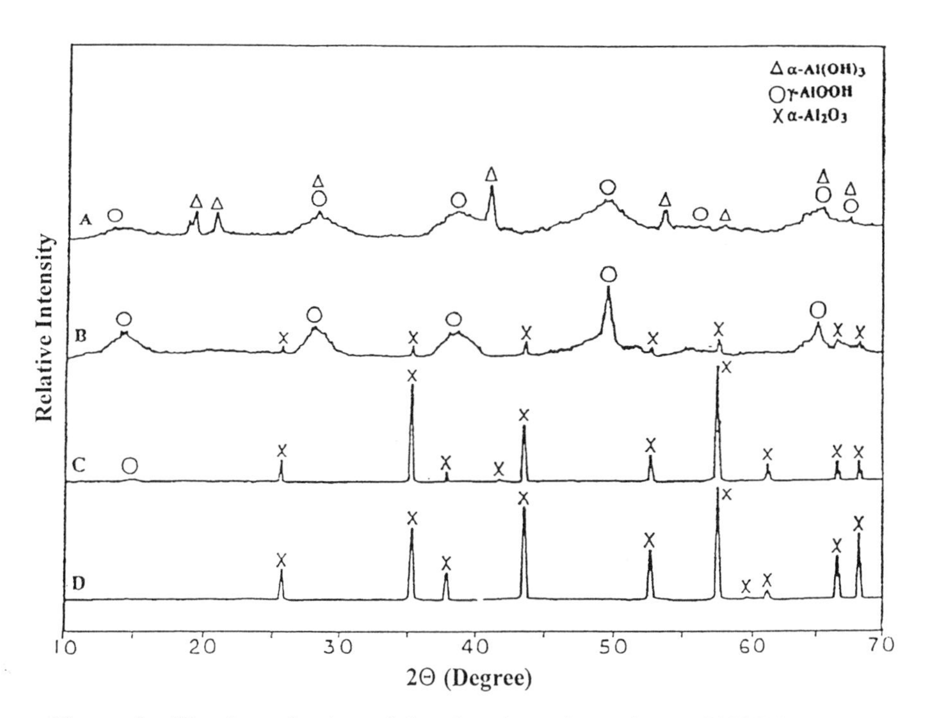

Figure 2. Kinetic evaluation of the glycothermal reaction at 300°C by phase identification via X-ray diffraction. (A) Precursor material. (B) 2 hours. (C) 6 hours. (D) 12 hours. (Reproduced with permission from ref. 13. Copyright 1996 The American Ceramic Society.)

A

B

Figure 3. (A) α-Al_2O_3 polyhedra particles formed at 36 rpm. (B) α-Al_2O_3 polyhedral particles formed at 36 rpm and $4{\times}10^{11}$ seeds/ml. (Reproduced with permission from ref. 23. Copyright 1997 The American Ceramic Society.)

the number concentration continues to reduce the particle size to values below 200 nm while maintaining morphological control. Increasing the number concentration beyond 4×10^{10} α–Al_2O_3 seed particles/ml creates aggregated particle clusters with loss of morphological control.

It is interesting to note that the particles of Figure 4D are smaller than the seed materials used to synthesize them. McArdle and Messing (12) have shown that the use of seed materials not only serves to create epitaxial growth sites, but also to provide a catalyst for the production of additional nuclei. At the high concentrations of seed materials employed for the smallest particles, the nucleation rate is high and lowers the supersaturation from the pseudo-boehmite phase. There is no longer a driving force for crystal growth due to supersaturation, but the free energy of the system can be lowered by the dissolution of non-equilibrium crystal facets. These non-equilibrium faces are dissolved in order for stable crystal growth to occur on equilibrium facets. The size of seed particles is therefore reduced to give a uniform particle size.

Figure 6 presents the particle characteristic face length as a function of the number concentration of seed material to compare the effectiveness of alumina seeds against hematite seeds. The particle size control is relatively linear against the log of concentration. The number concentration of hematite seeds required to achieve equivalent results to the alumina seeds is greater by approximately two orders of magnitude. Lattice mismatch in the hematite seeds is not a probable cause for the difference, as hematite seeds have been shown to act as epitaxial sites for seed growth (22).

Alternatively, the solubility of iron in 1,4-butanediol could play a role in the need for increased seed number concentrations. 1,4-butanediol is an organic analogue to water, and can support ionic species. α-alumina has been found to exhibit limited solubility at room temperature in 1,4-butanediol (10^{-4} M, J. H. Adair, University of Florida, unpublished data) and Fe^{3+} should also be supported in the 1,4-butanediol environment. The absolute value of the solubility of α-Fe_2O_3 at 300°C and 4.1 MPa are a subject of future study, but may be affected by the particle size of the seed materials according to the Gibbs-Thomson equation, which relates curvature effects to the solubility of small particles. Thus, the 60 nm α–Fe_2O_3 particles could exhibit significant solubility in the 1,4-butanediol solvent, and a higher seed number concentration would therefore be required to influence nucleation frequency.

Conclusions

In conclusion, the glycothermal synthesis technique has significant production advantages over the hydrothermal process for the synthesis of α–Al_2O_3. Use of glycols as a reaction solvent synthesizes α-Al_2O_3 at the lowest known temperatures (270°C) to the authors' knowledge. The particles formed are of uniform size and morphology The use of seed materials does not affect the control of particle morphology, and both α-alumina and α-hematite are effective seed materials for the control of particle size. Particle shape is a function of stirring dynamics and solids loading.

A

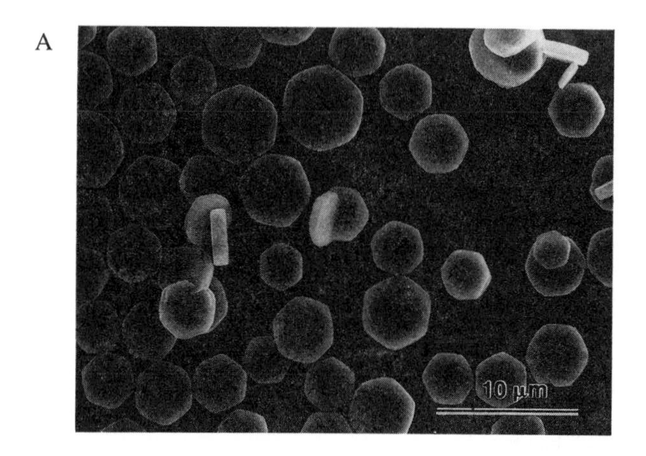

B

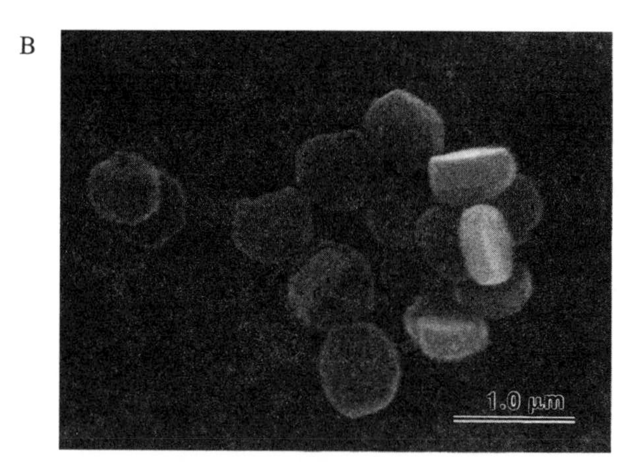

Figure 4. SEM photomicrographs of α-Al_2O_3 platelets synthesized by glycothermal treatment (300°C, 12 hr) of commercial gibbsite powders (Alcoa Hydral©) with various α-Al_2O_3 seed concentrations (#/ml): (A) without seeds, (B) 1.6×10^9, (C) 1.6×10^{10}, (D) 4×10^{10}. (Reproduced with permission from ref. 23. Copyright 1997 The American Ceramic Society.)

C

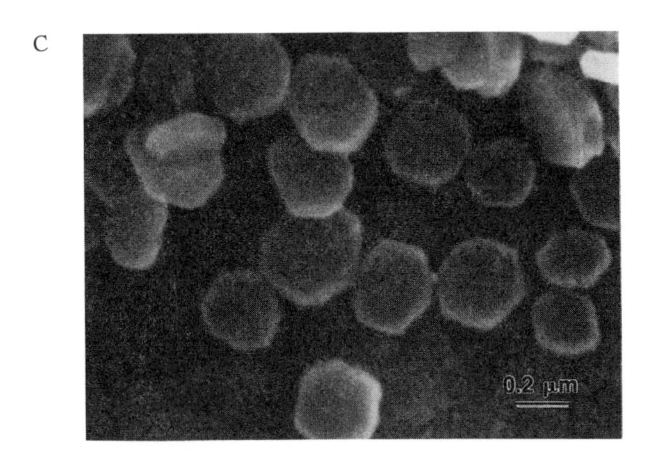

D

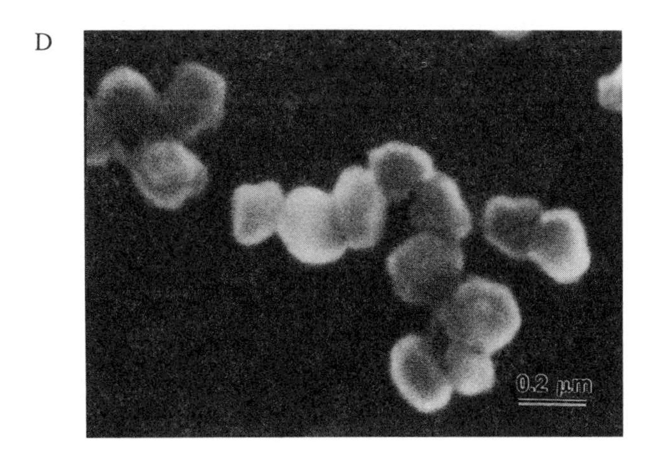

Figure 4. Continued.

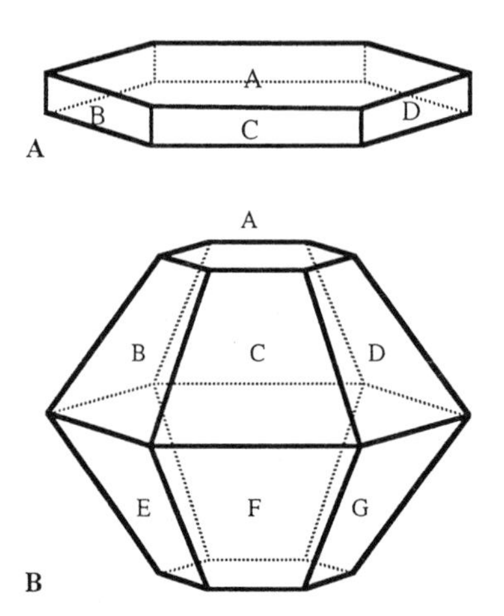

Figure 5. SHAPE representations of particle morphologies. (A) Platelet [A: (0001) B: $(2\bar{1}\bar{1}0)$ C: $(11\bar{2}0)$ D: $(\bar{1}2\bar{1}0)$] (B) Polyhedral [A: (0001) B: $(1\bar{1}02)$ C: $(10\bar{1}2)$ D: $(01\bar{1}2)$ E: $(1\bar{1}0\bar{2})$ F: $(10\bar{1}\bar{2})$ G: $(01\bar{1}\bar{2})$]. (Adapted with permission from ref. 23. Copyright 1996 The American Ceramic Society.)

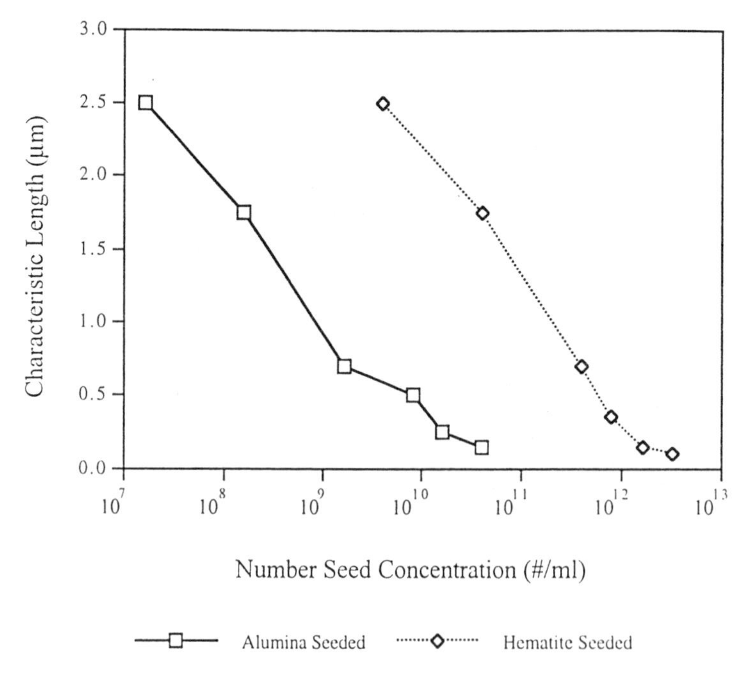

Figure 6. Characteristic platelet face length as a function of number concentration of $\alpha-Al_2O_3$ and $\alpha-Fe_2O_3$ seeds. (Reproduced with permission from ref. 23. Copyright 1997 The American Ceramic Society.)

Literature Cited

1. Johnson, D.W. in *Advances in Ceramics*; Messing, G.L., Mazdiyasni,K.S., McCauley, J.W., and Haber, R.A., Eds.; Ceramics Powder Science, The American Ceramic Society, Inc.: Columbus, OH, 1987,Vol. 21; pp. 3-20.
2. Kato, A. in *Materials Science Monographs*; Vincenzini, P., Ed.;,High Tech Ceramics, Part A; Elsevier Science Publishers B.V., Eds.: Amsterdam, 1987, Vol. 38; pp. 459-469.
3. Osseo-Asare, K.; Arriagada, F.J. in *Ceramic Transactions*; Messing, G.L.,Hirano, S.I., and Hausner, H., Eds.; Ceramic Powder Science III; The American Ceramic Society, Inc.; Westerville, OH, 1990, Vol. 12; pp. 3-16.
4. Chow, G.M.; Markowitz, M.A.; Singh,A. *J. of Matls.*, **1993**, *11,* pp. 62-65.
5. Benedetti, A.; Fagherazzi, G.; Pinna, F. *J. Am. Ceram. Soc.*, **1989**, *72*, pp. 467-469.
6. Hirano, S. *Am. Ceram. Soc. Bull.*,**1987,** *66*, pp. 1342-1344.
7. Dawson, W.J. *Am. Ceram. Soc. Bull.*, **1989,** *67,* pp.1673-1678.
8. Nancollas, G. H. in *Ceramic Transactions*; Messing, G. L. and Fuller,E. R. Jr., Eds.; Ceramic Powder Science II; The American Ceramic Society, Inc.: Westerville, Ohio,1987; Vol 1; pp. 8-22.
9. Randolph, A.D.; Larson, M. A. *Theory of Particulate Processes: Analysis and Techniques of Continuous Crystallization*; Academic Press, Inc.: San Diego, CA, 1988; pp. 109-134.
10. Turnbull, D.; Vonnegut, B. *Ind. Eng. Chem.*, **1952**, *44,* pp. 1292-1298.
11. McArdle, J.L.; Messing, G.L. *J. Am. Ceram. Soc.*, **1986**, *69*, pp. C98-C101.
12. McArdle, J.L.; Messing, G.L. *Adv. Ceram. Matls.*, **1988**, *3*, pp. 387-392.
13. Cho, S.B., Venigalla, S.; Adair, J.H. *J. Am. Ceram. Soc.*, **1996**, *79* , pp. 88-96.
14. Cho, S.B., Venigalla, S.; Adair, J.H. in *Ceramic Transactions*; Adair, J.H., Casey, J.A., Randall, C.A. and Venigalla, S., Eds.; Science, Technology, and Application of Colloidal Suspensions; American Ceramic Society: Westerville, OH, 1995, Vol. 54; pp. 139-150
15. Inoue, M., Tanino, H., Kondo, Y.; Inui, T. *J. Am. Ceram. Soc.*, **1989**, *72*, pp. 352-353.
16. Kennedy, G. *Am. J. Sci.*, **1959**, *257*, pp. 563-573.
17. *The Kinetics of Industrial Crystallization*; Nyvlt, J.; Sohnel, O.; Matuchova M.; ; Broul, M., Chemical Engineering Monographs 19, New York, New York, **1985**.
18. Boistelle, R.; Astier, J.P. *J. Crystal Growth,* **1988**, *90*, pp. 14-30.
19. Matijevic, E. in *Chemical Processing of Advanced Materials;* Hench, L.L. and West, J.K., Eds.; John Wiley and Sons, Inc.: New York, NY, 1992; pp. 513-527.
20. Wulff, G. *Z. Krist,*. **1901**, *34*, pp. 520-530.
21. MacCallum, R.B. U.S. Patent No. 3,267,041, August 16, 1966.
22. McArdle, J.L.; Messing, G.L. *J. Am. Ceram. Soc.*, **1989**, *72,* pp. 864-867.
23. Bell, N.S.; Cho, S.B.; Adair, J.H., *J. Am. Ceram. Soc.* (in press).

Chapter 13

Synthesis and Characterization of a Trimetallic Double-Alkoxide Precursor to Potassium Aluminosilicate

Ramasubramanian Narayanan[1] and Richard M. Laine[2,3]

Departments of [1]Chemistry and [2]Materials Science and Engineering and the Macromolecular Science and Engineering Center, University of Michigan, 930 North University, Ann Arbor, MI 48109–2136

A novel, soluble trimetallic double alkoxide precursor to potassium aluminosilicate ($KAlSiO_4$) has been synthesized by a one pot procedure and its structure characterized. The process termed the "oxide one pot synthesis" (OOPS) process uses simple, inexpensive chemicals, silica and aluminum hydroxide, to form a processable precursor in one simple step. The precursor is relatively stable to moisture and can be handled in air for long periods of time with minimal hydrolysis. On pyrolysis in air, to 1100° C, it produces phase-pure $KAlSiO_4$. The proposed precursor was characterized using thermogravimetric analysis, multinuclear NMR, and mass spectroscopy. The results suggest the precursor disproportionates in solution.

Better control over product properties and easier synthetic methods have always been challenging goals for chemists and ceramists. Chemical routes to ceramics (sol-gel or precursor) offer a significant advantage over traditional melt or solid-state reaction methods as they lower the processing temperature, lead to more homogeneous materials and allow greater control over the final microstructure of the resultant ceramic materials (*1,2*). Additionally, chemical approaches permit processing thin films and fibers. But the extant chemical routes cease to be simple and efficient, when used for the synthesis of multimetallic oxides.

While the sol-gel method offers a significant improvement over conventional solid state reactions, its utility for processing multimetallic oxides suffers from a major drawback. Rates of hydrolysis of metal alkoxides can differ by orders of magnitude (*3-5*). Differences in hydrolytic reactivities of different metal alkoxides often leads to preferential hydrolysis of one metal alkoxide over another, leading to segregation in the final ceramic material (*6*). Furthermore, sol-gel processing commonly involves the use of expensive, highly reactive metal alkoxides. Attempts to alter the hydrolytic reactivity of metal alkoxides by chemical modifications (e.g.

[3]Corresponding author

chelating ligands, formation of double alkoxides) before using them in multimetallic oxide syntheses have been successful (*6-8*). However this approach is not simple and still involves the use of expensive and hydrolytically unstable metal alkoxides.

The precursor method of ceramic processing is promising as it offers significant advantages such as control over chemical and phase purity of the resultant ceramic material, low processing temperatures, and processable precursors (*9*). Precursor syntheses are often easy and efficient. However, the processable precursors described in the literature are typically limited by the availability of salts of different metals and their solubilities in common solvents (*10*). To our knowledge, simple, inexpensive, general synthetic routes to processable multimetallic aluminosilicate precursors were unknown, until recently.

We recently reported the discovery of a general, low cost route to alkoxide precursors to aluminosilicates (spinel, mullite and cordierite), by direct reaction of any stoichiometric mixture of SiO_2, $Al(OH)_3$ and group I/II metal hydroxides with triethanolamine (TEA) in ethylene glycol (EG) solvent (*11*). This process, termed the 'oxide one pot synthesis' (OOPS) process provides stable, processable precursors to phase pure ceramic materials. A general reaction for the synthesis of OOPS derived precursors is given by:

$$xMOH + ySiO_2 + zAl(OH)_3 + (x+y+z)TEA \xrightarrow{200°C/x\text{'s } EG/5\text{-}10\ h/\ -H_2O} M_xSi_yAl_z\,(TEA)_{x+y+z} \quad (1)$$

Herein we describe the extension of this process to the synthesis of a trimetallic potassium aluminosilicate precursor and the delineation of the proposed precursor structure.

Experimental

Thermogravimetric Analyses (TGA). TGA of samples were performed using a 2950 Thermal Analysis Instrument (TA instruments, Inc., New Castle, DE). Samples (10-20 mg) were placed in a platinum pan and heated in flowing, dry air (60 cm^3/min), in "Hi-Res 4.0 mode" at 50° C/min. to 1000° C.

Nuclear Magnetic Resonance (NMR) Spectroscopy. NMR spectra were recorded using a Bruker Aspect 3000, AM-360 MHz spectrometer (Bruker Instruments Inc., Manning Park, Billerica, MA). $^{29}Si\{^1H\}$ NMR spectra were obtained with the spectrometer operating at 59.6 MHz and using a 32000 Hz spectral width, a relaxation delay of 15 s, a pulse width of 58°, and 32 K data points. The samples were dissolved in EG. C_6D_6 and tetramethylsilane, in a sealed inner tube, served as lock and reference materials respectively. ^{27}Al NMR spectra were obtained with the spectrometer operating at 93.8 MHz and using a 41000 Hz spectral width, a relaxation delay of 0.2 s and a pulse width of 13°. The samples were dissolved in EG. D_2O in a sealed inner tube served as the lock solvent and a 1M solution of

$AlCl_3$ in D_2O:H_2O (1:1) served as the external reference. ^{1}H and ^{13}C spectra were taken with the samples dissolved in $CDCl_3$.

Solid-state spectra were recorded at Laboratoire de la Matière Condensée on an MSL 400 Bruker spectrometer operating at 79.5 MHz for ^{29}Si. The spinning rate was 4 kHz. For ^{29}Si CPMAS experiments, contact times of 2 ms were used and 80 scans accumulated. A line broadening of 10 Hz was applied before Fourier transformation.

Mass Spectroscopy. Mass spectra of the samples were recorded at the University of Michigan, Department of Chemistry. FAB studies were conducted using a VG 70-70-E, a magnetic sector, double-focusing mass spectrometer made by VG Analytical, and the spectrometer is operated using the 11-250-J data collection software system supplied with the instrument.

The spectrometer was scanned from m/z 2800 to m/z 75, using an exponential-down magnet scan, and an external calibration against CsI salt clusters. The sample is dissolved in 3-nitrobenzylalcohol and deposited on the target, on the probe tip. Prominent matrix (3-nitrobenzylalcohol) peaks are subtracted out. Low intensity peaks were an average of 10 or more scans, averaged as continuum data. The continuum data is then smoothed and centroided by the data system. The FAB gun uses xenon gas, and is run at 1 mA current and 10 kV.

Syntheses. Silatrane glycol was synthesized as described in reference 11. Alumatrane was synthesized as described in reference 12. The OOPS synthesis of the KAS precursor was described in reference 13. The stepwise synthesis of the KAS precursor is described below.

Synthesis of KAS Precursor (stepwise). In a 500 mL Schlenk flask, silatrane glycol (7.22 g, 30.7 mmol), potassium hydroxide (2.00 g, 30.4 mmol) and ≈ 200 mL of EG were added and the mixture heated to distill off EG. The solution cleared on heating and was cooled after distilling off ≈ 30 mL of EG in ≈ 30 min. To that clear solution, alumatrane (4.47 g, 25.8 mmol) dissolved in ≈ 50 mL of EG was added. The resultant mixture was heated to distill off EG to concentrate the solution to ≈ 75 mL. The solution was then vacuum dried at 110° C for 6 h, and at 150° C for 10 h to yield a light brown solid. Yield: 12.8 g (93.8%). The TGA ceramic yield of the solid was 35.4 wt% (Theoretical ceramic yield calculated for the formation of $KAlSiO_4$ from the proposed precursor is 35.5 wt%).

Safety Considerations. Silatrane glycol, alumatrane and potassium hydroxide are corrosive solids and the solvent EG is an irritant. Therefore protective eyewear, clothing and gloves are recommended while handling these chemicals.

Results and Discussion

Reaction (2, path a) depicts the OOPS synthesis of the $KAlSiO_4$, (KAS) precursor. The precursor structure shown is the simplest conceivable, based on starting materials, solvents and previous studies (*12*). Upon heating in air to 1100° C, this precursor transforms to phase pure, fully crystalline $KAlSiO_4$, as described elsewhere (*13*). The exact structure of this precursor remains to be completely elucidated. Steps taken towards that goal are described here.

$Al(OH)_3 + SiO_2 + MOH + 2\ N(CH_2CH_2OH)_3$

Path a: X's $HOCH_2CH_2OH$, 200°C

Path b: X's $HOCH_2CH_2OH$, 200°C

(2)

To carefully delineate the exact components of the precursor, it was dissected into two constituent parts: silatrane glycol and alumatrane, which were synthesized as shown [Reactions (3) and (4)] in one step from SiO_2 and $Al(OH)_3$ respectively (*11,12*).

The products of reactions (3) and (4) were characterized by ^{1}H, ^{13}C, and ^{29}Si NMR, high resolution mass spectroscopy and elemental analysis (*11,12*). Reaction (2, path b) depicts the stepwise synthesis of the KAS precursor. Solutions of silatrane glycol and alumatrane in EG were combined along with KOH, and the mixture heated to distill off EG, resulting in KAS precursor in high yields. The order of addition of the three reactants does not have any effect on the product of the reaction.

Thermogravimetric Analyses. The TGA decomposition profile (Figure 1) of the KAS precursor prepared via reaction (2, path b), exhibits three mass loss regions. The first major mass loss occurs below 400° C and corresponds to ligand decomposition, the second (400° to 600° C) corresponds to oxidation of organic char produced during ligand decomposition, and the third one (800° to 950° C) to carbonate decomposition. Evidence for these proposed decomposition pathways are discussed elsewhere (*13*). It is also significant that the ceramic yield obtained for the reaction 2, path b precursor, exactly match that calculated. Note that the elemental analyses are rarely more accurate than ± 0.2 wt %. Thus the TGA supports the proposed precursor structure. However NMR and mass spectral data are less conclusive.

$$SiO_2 + N(CH_2CH_2OH)_3 \xrightarrow[\text{EG, 200 °C/ -H}_2\text{O}]{\text{10 mol \% TETA}} N(CH_2CH_2O)_3Si{-}OCH_2CH_2OH \quad (3)$$

$$Al(OH)_3 + N(CH_2CH_2OH)_3 \xrightarrow[\text{200°C/ -H}_2\text{O}]{HOCH_2CH_2OH} [Al(OCH_2CH_2)_3N]_4 \quad (4)$$

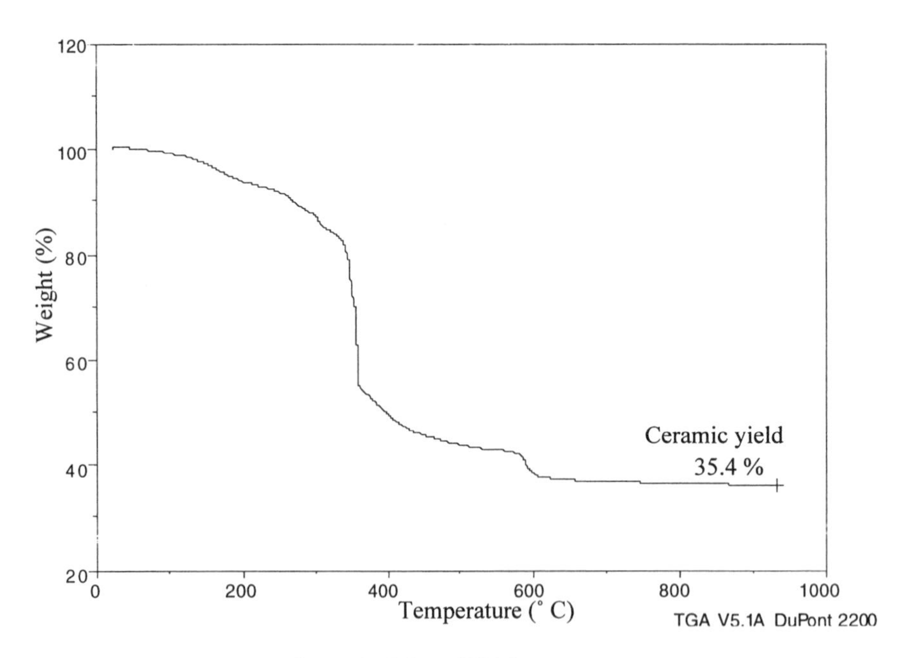

Figure 1: TGA of KAS precursor

NMR analyses

^{1}H and ^{13}C NMR. Table I provides the results for the KAS precursor. The peaks are too broad, because of coupling with quadrupolar ^{27}Al and ^{14}N nuclei, to permit detailed analysis of multiplicities in the ^{1}H NMR, and the intensities in the ^{13}C NMR. But chemical shifts and number of peaks are as expected for the proposed structure based on our studies on the spinel precursor structure (*12*).

^{27}Al NMR. The ^{27}Al NMR (Table I) is simple to interpret as it has only one peak at 69.2 ppm which indicates a tetracoordinate Al center, as expected for the proposed precursor structure (*14-16*). There is a definite change in ^{27}Al NMR signals on going from alumatrane to the precursor (Figure 2). Alumatrane has two very broad peaks, centered at 61.5 and 5.4 ppm, indicating tetracoordinate and hexacoordinate Al centers respectively, as expected for an asymmetric dimer in solution. The KAS precursor has only one narrow peak at 69.2 ppm, indicating a more symmetric, tetracoordinate Al center.

To study the events associated with precursor formation, reaction pathway 'a' in Scheme 1 was run and ^{27}Al NMRs were taken for the intermediate and the product.

Table I. NMR data for KAS precursor

Nucleus	Chemical shifts (ppm)
^{1}H	2.6 (broad), 2.9 (m), 3.6 (broad), 3.8 (t)
^{13}C	51.2, 53.2, 57.5, 63.6
^{29}Si	-96.7 and -104.7
^{27}Al	69.2
^{29}Si CPMAS	-95.9

The two broad peaks in alumatrane (A) change to one narrower peak at 61.7 ppm (B), after addition of potassium ethylene glycolate, indicating the formation of a symmetrical tetracoordinate Al center, which remains tetracoordinate (albeit with a different environment) after the addition of silatrane glycol (69.2 ppm) (C). It should be noted that the order of addition shown in Scheme 1, path 'a' is different from that in experimental section. But as mentioned above, the precursors formed either way are identical according to mass spectral and TGA analyses.

^{29}Si NMR. Table I and Figure 3, show that the ^{29}Si NMR of the KAS precursor exhibits two peaks: a large one at -96.7 and a smaller one at -104.7 ppm, corresponding to different pentacoordinate Si centers. The peak at -96.7 is that typical of silatranes (*11*) where the fifth ligand is the TEA nitrogen. In contrast, the peak at -104.7 ppm is typical of pentacoordinate, anionic silicates (*17*).

To rationalize the presence of pentacoordinate anionic Si, and to follow the events during precursor formation by ^{29}Si NMR, the reaction shown in Scheme 1, path "b" was run, and ^{29}Si NMRs of the intermediate and the product were taken.

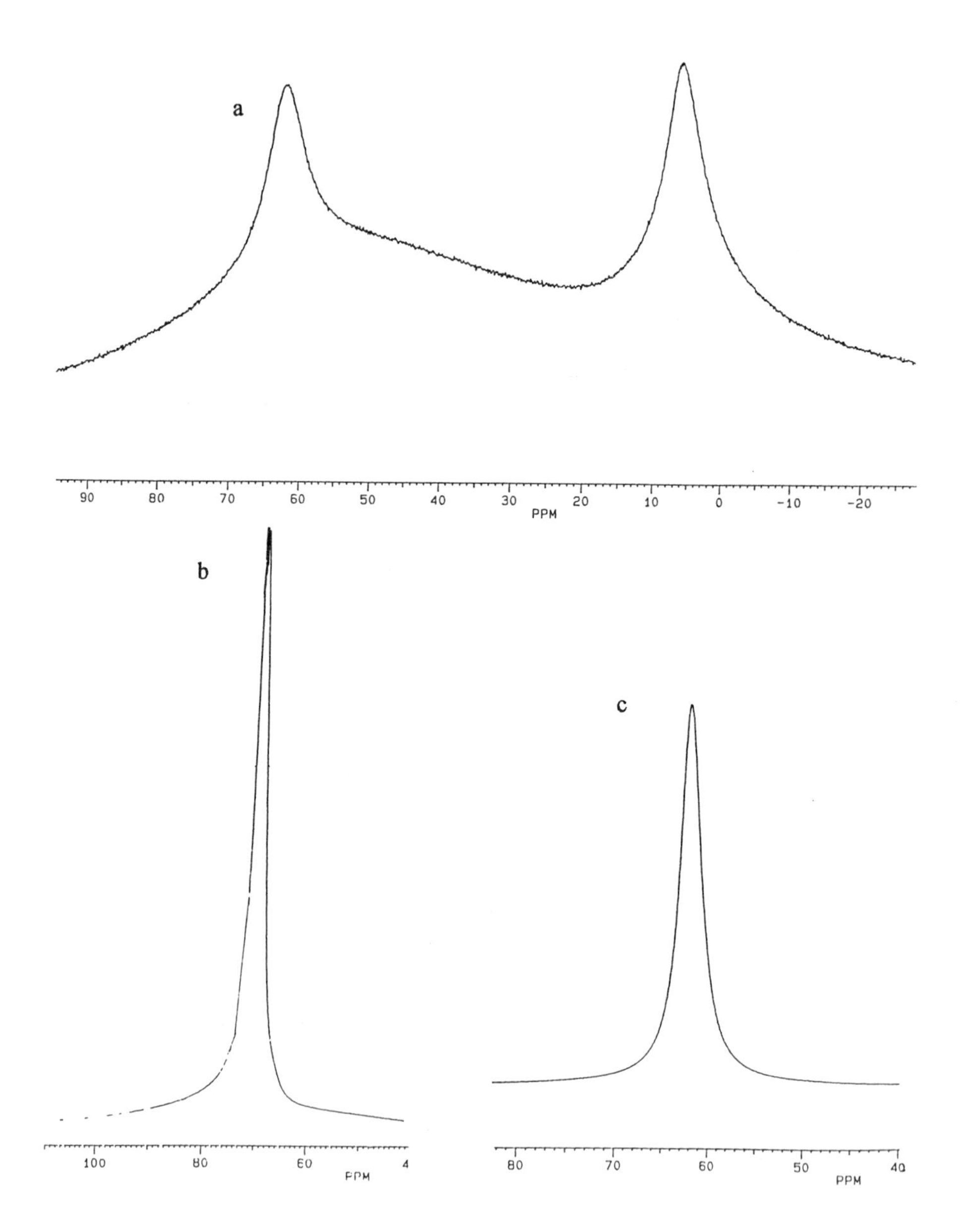

Figure 2. Change in Al environment during precursor formation
a) Alumatrane b) Species B in Scheme 1 c) KAS precursor

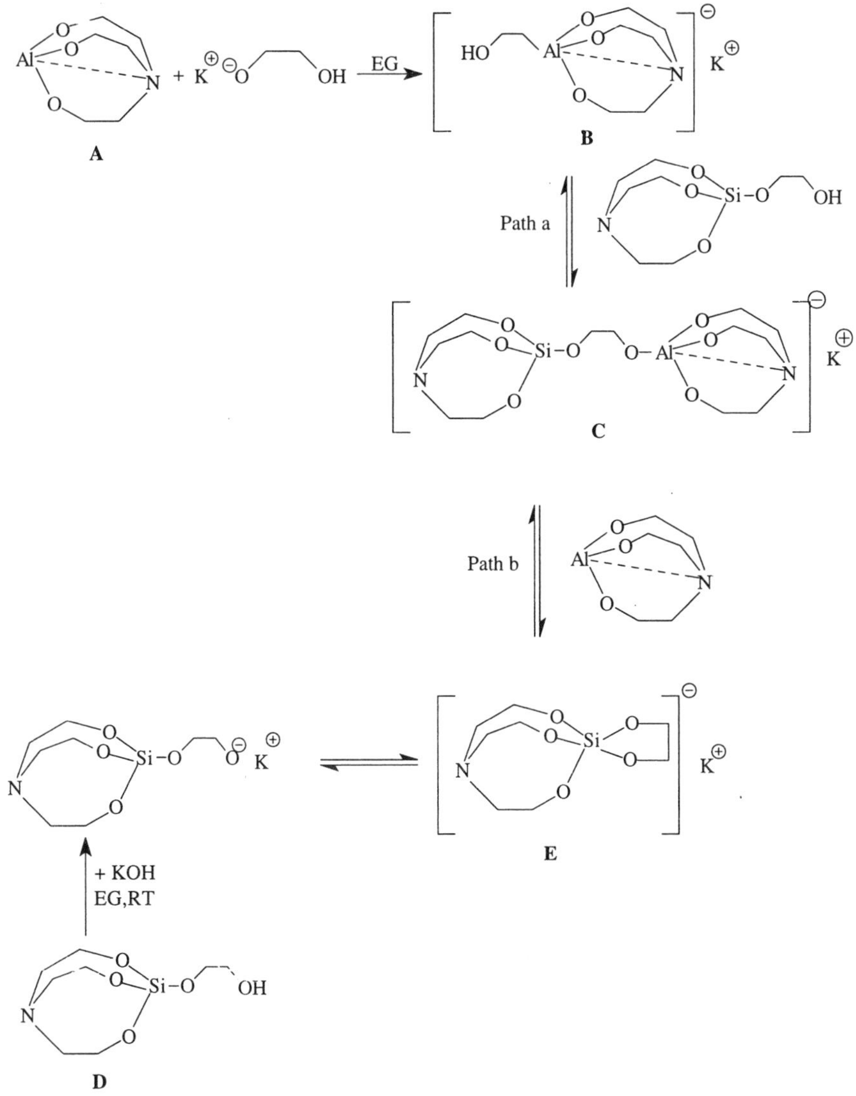

Species	^{27}Al NMR Chemical shifts (ppm)	Species	^{29}Si NMR Chemical shifts (ppm)
A	61.5 and 5.4	**D**	-95.6
B	61.7	**E**	-105.5
C	69.2	**C**	-96.7 and -104.7(solution) -95.9 (solid state)

Scheme 1. Precursor formation -proposed pathways

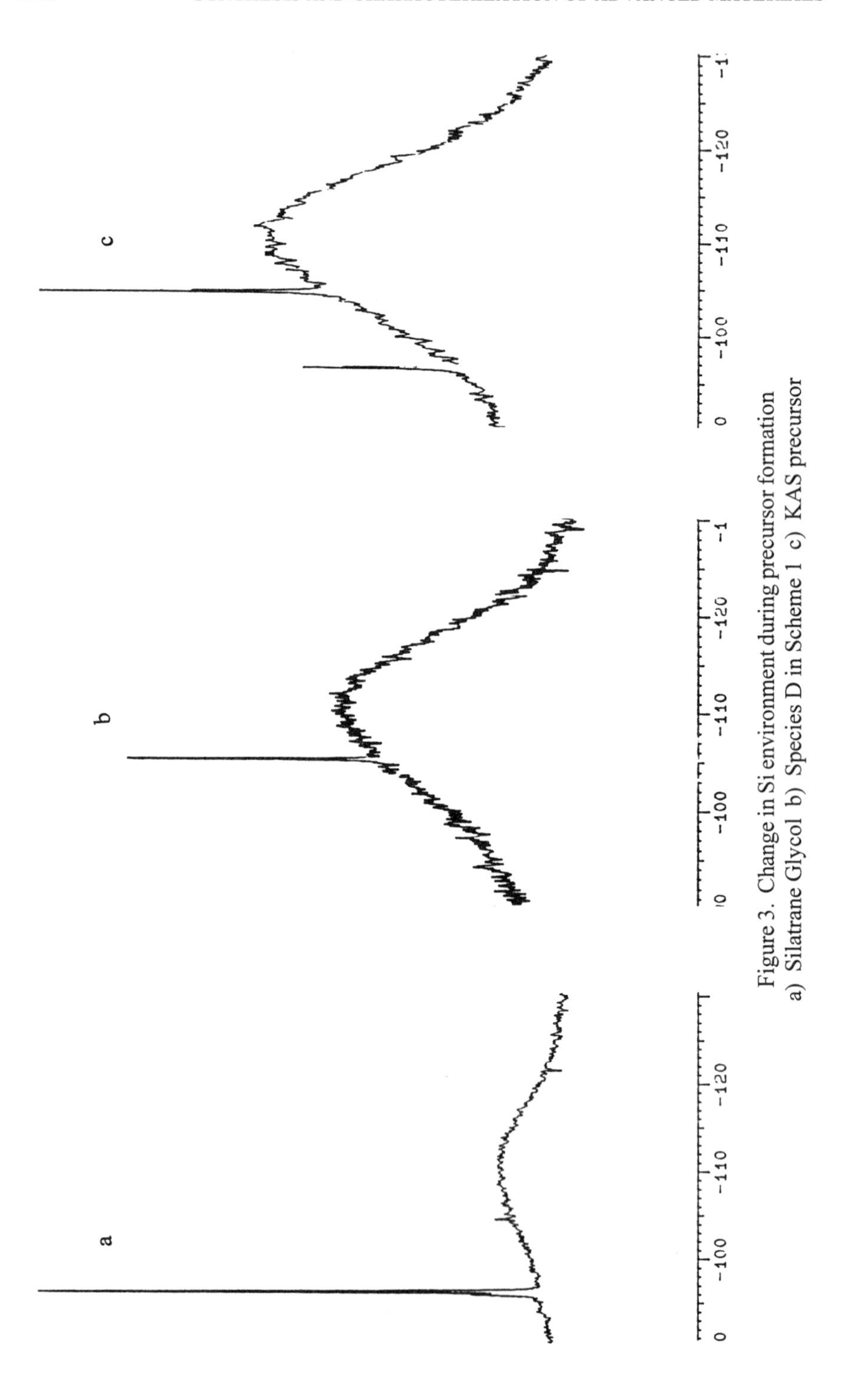

Figure 3. Change in Si environment during precursor formation
a) Silatrane Glycol b) Species D in Scheme 1 c) KAS precursor

The potassium salt of silatrane glycol exhibits a peak at -105.5 ppm, indicating a pentacoordinate anionic Si (E in Scheme 1) as opposed to the -95.6 ppm peak of silatrane glycol (*11*). The KAS precursor formed by the subsequent addition of alumatrane to the salt, exhibits two peaks (C in Scheme 1). However, the solid state ^{29}Si NMR of the KAS precursor, shows only one peak, at -95.9 ppm, in accordance with the proposed precursor structure. Therefore, the two peaks seen in solution NMR, arise probably as a result of equilibration via path 'b'. The intermediates in Scheme 1 (species B and E) were characterized both by multinuclear NMR and FAB mass spectroscopy. The proposed equilibration process is corroborated by FAB mass analysis as follows.

Mass Spectral Analysis of KAS Precursor. Fast atom bombardment (FAB) is the only technique that can be used for salts like the KAS precursor. FAB requires that the sample be dispersed in a non-interactive matrix for analysis. Unfortunately, all three typical matrix materials, viz. TEA, p-nitrobenzylalcohol (NBA) and glycerol, interact/react with the sample. This problem was encountered during analysis of the spinel precursor (*12*). NBA was found to interact least with the spinel precursor and was used as the matrix in those studies (*12*). On this basis, an NBA matrix was used with the KAS precursor. EG can also be used as matrix but its high vapor pressure compared to other matrix materials, leads to much lower intensities for ion fragments. Results of FAB analyses done in both EG and NBA are shown in Table II. At the outset, nominal mass analysis by positive ion FAB were in accord with expectations. The molecular ion peak expected from the proposed precursor structure was seen, along with peaks resulting from interaction with the NBA matrix (Table II).

Table II. FAB mass spectral analysis of KAS precursor

Fragmentassigned	m/z	Int. in NBA matrix	Int. in EG matrix
TEA-Al + H^+	174	37.5	1.7
K^+ $^-O-CH_2-C_6H_4-NO_2$ + H^+	192	78.1	---
TEA-Al-EG-Si-TEA K + H^+	447	100.0	5.7

Corresponding negative ion FABs do not provide any useful information as matrix interactions predominate obscuring all other peaks. Unfortunately, high resolution mass spectral analysis did not confirm the proposed structure. As shown in Table III, the exact mass analysis of the molecular ion peak corresponds to that of the disilicate structure with a potassium counterion, rather than the proposed protonated trimetallic double alkoxide structure.

The disilicate species cannot form given the stoichiometry of the precursor (confirmed by TGA) and it is also significant that the corresponding dialuminate species was not seen in the mass spectrum. FAB mass spectral analyses in EG, of intermediate E in Scheme 1, path 'b', substantiates the proposed structure of the intermediate, and does not exhibit a fragment corresponding to the disilicate species, thus precluding the formation of the latter as part of the equilibration process--it

Table III. High resolution mass spectral analysis of molecular ion peak of KAS precursor by positive ion FAB in NBA matrix

Fragment	Mol. Weight (g/mol) calculated
TEA-Al-EG-Si-TEA K + H^+	447.1146
TEA-Si-EG-Si-TEA K	447.1021
High Res. Mass Spectral analysis of Precursor	447.1021

must arise from the proposed KAS structure. FAB mass spectrum of intermediate B in Scheme 1, path "a" also confirms the proposed structure and has a similar fragmentation pattern to that of the KAS precursor. It is likely that the matrix interaction is predominant and promotes the disproportionation equilibria suggested in Scheme 1, thereby giving anomalous mass spectral results. While it is clear that the mass spectroscopy shows the presence of disilicate species, the source of that species is questionable. It probably arises from disproportionation in solution and/or the matrix interaction with NBA.

Conclusions

The OOPS process appears to offer access to well defined chemical intermediates that provide the atomic mixing desirable in precursor processing of ceramics. It also appears to provide a viable alternative to sol-gel and existent chemical methods for processing multimetallic oxides. The as- synthesized KAS precursor can be handled in air for practical purposes and is soluble in common organic solvents, which makes it a suitable candidate for processing into desired shapes and sizes. The corresponding Li and Na aluminosilicate precursors were also synthesized and they behave similarly to the KAS precursor (Narayanan, R., Laine, R.M., unpublished results). Thus phase pure lithium and sodium aluminosilicates can also be obtained. The structural characterization is complicated by the probable disproportionation in solution and we are currently pursuing methods (solid state NMR, non/less interactive matrix) that would prove the trimetallic double alkoxide structure of the precursor unambiguously.

Acknowledgments

We thank the Army Research Office for support of this work through ASSERT Grant No. DOD-G-DAAL03-92-G-0053. We would also like to thank Federal Aviation Administration for partial support of this work through Contract No. 95-G-026. We also thank Dr. Florence Babonneau of the Laboratoire de Chimie Matière Condensée, Université Pierre et Marie Curie, for efforts to characterize the precursor by solid state NMR.

Literature Cited

1. Brinker, C. J.; Scherer, G. W. *Sol-gel Science: The physics and chemistry of Sol-Gel processing*, Academic press: Boston, MA (1990).

2. Kansal, P; Laine, R. M. *J. Am. Ceram. Soc.* **1995**, *78*, 529-38.
3. Huling, J. C.; Messing, G. L. *J. Non-Cryst. Solids*, **1992**, *147-148*, 213-21.
4. Wang, Y.; Thomson, W. J. *J. Mater. Res.*, **1995**, *10*, 912-7.
5. Laine, R. M. in *Sol-Gel Processing of Glasses*; Mackenzie, J. D.; Ulrich, D. R., Ed.; SPIE Proc., vol 1328, pp 16.
6. Livage, J; Henry, M.; Sanchez, C. *Prog. Solid State Chem.* **1988**, *18*, 259-341.
7. Babonneau, F.; Bonhomme-Coury, L.; Livage, J. *J. Non-Cryst. Solids*, **1990**, *121*, 153-57.
8. Thomas, I. M. in *Sol-Gel Technology for Thin Films, Fibers, Preforms, Electronics, and Specialty Shapes*; Klein, L., Ed.; Noyes Publ: Park Ridge, N.J., 1988, pp. 2-15.
9. McColm, I. J.; Clark, N.J., *High Performance ceramics*, Chapman and Hall: NewYork, NY (1988).
10. Eror, N. G.; Anderson, H. U. In *Mat. Res. Soc. Symp. Proc.*; Brinker, C. J.; Clark, D. E and Ulrich, D. R., Ed.; Better Ceramics Through Chemistry II, Materials Research Society, Pittsburgh, PA, vol 73, pp 571-77.
11. Laine, R. M.; Treadwell, D. R.; Mueller, B. L.; Bickmore, C. R.; Waldner, K. F.; Hinklin, T., *J. Mater. Chem.*, **1996**, *6*, 1441-3.
12. Waldner, K. F.; Laine, R. M.; Dhumrongvaraporn, S.; Tayaniphan, S.; Narayanan, R., *Chem. Mater.*, **1996**, *8*, 2850-7.
13. Baranwal, R.; Laine, R. M., *J. Am. Ceram. Soc.*, in press.
14. Kriz, O.; Casensky, B.; Lycka, A.; Fusek, J.; Hermanek, S., *J. Magn. Res.*, **1984,** 60, 375-381.
15. Gerardin, C.; Sundaresan, S.; Benziger, J.; Navrotsky, A.; *Chem. Mater.* **1994**, *6*, 160-70.
16. Akitt, J. W., Multinuclear studies of aluminum compounds. In: Progress in NMR spectroscopy, Pergamon press, NY, 1989, vol 21, pp 1- 149.
17. Blohowiak, K. Y.; Treadwell, D. R.; Mueller, B. L.; Hoppe, M. L.; Jouppi, S.; Kansal, P.; Chew, K. W.; Scotto, C. L. S.; Babonneau, F.; Kampf, J.; Laine, R. M., *Chem. Mater.* **1994**, *6*, 2177-92.

Chapter 14

Synthesis of β″-Alumina Polymer Precursor and Ultrafine β″-Alumina Composition Powders

David R. Treadwell[1], Anthony C. Sutorik[1], Siew Siang Neo[1], Richard M. Laine[1], and Robert C. Svedberg[2]

[1]Departments of Chemistry and Materials Science and Engineering, University of Michigan, 2300 Hayward Avenue, Ann Arbor, MI 48109–2136
[2]Advanced Modular Power Systems, Inc., 4667 Freedom Avenue, Ann Arbor, MI 48108

β″-Alumina, an important high-temperature solid electrolyte, was synthesized from inexpensive starting materials via a polymeric, metalloorganic precursor. The precursor, made by reacting NaOH, LiOH, $Al(OH)_3$, and triethanolamine (TEA) in ethylene glycol (EG) while distilling off product water, is a glassy thermoplastic at room temperature, and dissolves in polar solvents, e.g. ethanol. Spheroidal, particles of <100 nm diameter were formed by flame-spray pyrolysis of ethanolic solutions of the precursor. X-ray powder diffraction showed the as-formed powders to be an intermediate phase, possibly m-alumina, which transforms to β″-alumina on heating above 1200 °C.

Alkali metal thermo-electric converters (AMTEC) are high-efficiency (≥35%) cells for the production of electricity from heat(*1*). A key component of this cell, as well as of Na/S batteries, is a nonporous, β″-alumina ceramic electrolyte that conducts Na^+ ions when either system is operating (≈ 800 °C for AMTEC, ≈ 400 °C for Na/S). Despite the high performance ratings for both power sources, their widespread use is limited by a very high cost of production. The manufacture of high quality β″-alumina tubes contributes to this high cost.

We recently described two discoveries that offer solutions to this problem. First, a new, simple synthesis of polymeric precursors to β″-alumina made directly from NaOH, alumina and triethanolamine, TEA (*2*). Second, we learned to form a wide variety of ultrafine aluminosilicate ceramic powders by flame spray pyrolysis processing of these polymers. The goal of the current research program is to demonstrate that it is possible to produce reasonable quality β″-alumina shapes using these low cost precursors and powders.

Precursor Design

We recently discovered (*2-10*) several very simple, synthetic routes to alkoxysilanes and alanes directly from SiO_2 and $Al(OH)_3$. This work is summarized below, the references, especially reference 2, provide complete details.

Alkoxysilanes. SiO_2 in a variety of forms will react on heating in excess ethylene glycol (EG) with a base (Group I/II metal oxide or hydroxide), to give anionic, pentacoordinated monomers or dimers with group I metals (*3-5*):

$$2SiO_2 + 2KOH + 5HOCH_2CH_2OH \xrightarrow[-4H_2O]{198°C} 2K[(OCH_2CH_2O)_2Si\text{-}OCH_2CH_2OH]$$

$$\updownarrow \; 130°C/vacuum \;\; -EG$$

$$K_2[(OCH_2CH_2O)_2Si\text{-}OCH_2CH_2O\text{-}Si(OCH_2CH_2O)_2]$$

With group II metals, depending on the initial M:SiO_2 stoichiometry, either hexacoordinate, dianionic silicates (*5-7*) or monoanionic, pentacoordinated silicates can be isolated (*11*).

Neutral Alkoxysilanes from SiO_2.. Replacing KOH with catalytic amounts (e.g. 2-10 mol %) of triethylenetetramine [$NH_2(CH_2CH_2NH)_3H$, TETA, bp ≈270 °C] promotes reaction to form a clear solution with one ^{29}Si NMR peak at -82 ppm, indicative of a neutral monomer or simple oligomer:

$$SiO_2 + x\text{'s } EG \xrightarrow[2 \text{ mol \% TETA}]{200°C/-H_2O} Si(OCH_2CH_2OH)_4 \text{ and/or } [Si(OCH_2CH_2OH)_2\text{-}O\text{-}CH_2CH_2]_n \qquad (1)$$

Further heating leads to a crosslinked polymer, $Si(eg)_2$, which will redissolve in EG to give a tractable oligomer that can be processed into thin films, fibers and cast shapes. $Si(eg)_2$ is a precursor to many other materials through ligand exchange processes (*4*).

Neutral Alkoxyaluminanes. Attempts to dissolve $Al(OH)_3$ in EG with catalytic amounts of TETA were unsuccessful; however reaction with TEA gives a well defined material, $(TEAAl)_4$:

$$Al(OH)_3 + N(CH_2CH_2OH)_3 \xrightarrow{200°C/EG/-H_2O} [N(CH_2CH_2O)_3Al]_4 \qquad (2)$$

The solid compound is tetrameric and has been well characterized (*14*). $(TEA\text{-}Al)_4$ is highly soluble in common organic solvents. It is also relatively moisture stable. Most importantly, $(TEAAl)_4$ can be made directly from the common aluminum ore, bauxite [$Al(OH)_3$, mostly gibbsite]. Bauxite (9 m^2/g), reacts (200 °C, 24 h) to give 40 wt. % dissolution.

Based on the above discoveries, we explored the possibility of preparing a variety of aluminosilicate polymer precursors.

OOPS Alkoxyaluminosilicates. We find that by combining the simple reactions shown above, it is possible to prepare numerous aluminosilicate alkoxide precursors by direct reaction of any mixture of SiO_2, $Al(OH)_3$ and group I/II metal hydroxides/oxides with EG and triethanolamine (TEA) (*4*). These precursors offer good-to-excellent hydrolytic stability. This process is termed the "oxide one pot synthesis" (OOPS) process.

Using OOPS processing, precursors to spinel, mullite, and cordierite are easily made. One kg of spinel precursor can be made (6 h) by reacting MgO, $Al(OH)_3$ with TEA in EG. The structure is a double alkoxide with a $HMgAl_2(TEA)_3$ stoichiometry (*12-16*). A mullite precursor can be prepared by reaction of a 2:6 mixture of SiO_2 and $Al(OH)_3$ (*14-17*). A cordierite precursor was also synthesized by direct reaction of MgO, SiO_2 and Al_2O_3 (*17*). Based on the OOPS method, a low cost synthesis of β″-alumina was developed as described below.

Synthesis of a β″-alumina precursor. The target precursor system required, is an 11:1 ratio of Na:Al. An OOPS precursor formulated as:

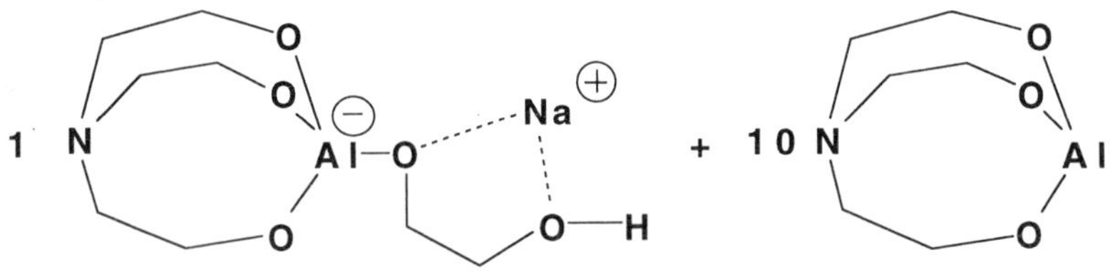

1 N(CH₂CH₂O)₃Al⁻–OCH₂CH₂OH···Na⁺ + 10 N(CH₂CH₂O)₃Al

was synthesized. This was found to be a successful precursor and thus, new precursors were prepared that incorporate Li and Mg ions as stabilizers.

Experimental

Synthetic Procedures: Reagents. All materials were used as received from standard commercial sources. β″-alumina precursor powder and an authentic β″-alumina sample were obtained from Ceramatec. All materials are handled in air except where noted. Ethylene glycol was recycled by redistilling twice, and so the supply used was either redistilled or freshly received.

Synthesis of Polymeric Precursor $Na_{1.67}Al_{10.67}Li_{0.33}(TEA)_{10.67}\cdot x(EG)$. In a 2 L round bottom flask, 199.9 g (2197 mmol) of $Al(OH)_3\cdot 0.72H_2O$, 326.7 g (2190 mmol) triethanolamine (TEA) and 1 L ethylene glycol (EG) are mixed with constant stirring (magnetic). The reaction is heated to ≈ 200 °C to distill off some EG and byproduct water. After approximately 2 h, the reaction turns clear, indicating that the reaction to from the $[(TEA)(Al)]_4$ complex is complete. Then, 13.71 g (343 mmol) of NaOH and 2.84 g (68 mmol) of $LiOH{\cdot}H_2O$ are added to the solution, and the reaction is refluxed for 1 h. The reaction is then distilled under N_2, first to remove by-product water, then excess EG, until the magnetic stir bar stops stirring. Care was taken to prevent over-heating of the product. If the product is intended for preparation of ultra-fine powders in the next step, ethanol is cautiously and slowly added to prepare the flame-spray solution. Otherwise, the remaining EG is removed by vacuum distillation at 180 °C for 4 h. On cooling, a glassy yellow solid is recovered from the flask, and may be stored in air. The ratio of Na/Al/Li = 1.67/10.67/0.33 was chosen as it is the composition of commercially available Ceramatec β″-alumina ceramic tubes. Yields are essentially quantitative.

Synthesis of Flame Spray Pyrolyzed Nanopowder, $Na_{1.67}Al_{10.67}Li_{0.33}O_{17}$. A more detailed description of the general procedure for nanopowder synthesis was published (*18*). The apparatus used is illustrated in Figure 1. Care has been taken to add engineering safety controls to prevent explosion, owing to the obvious concerns of creating an atomized mixture of fuel in oxygen. Smoke, heat and CO detectors and manual kill switches will instantly shut off the feeds, and appropriate shielding adds a measure of safety. A 10-20 wt. % ceramic (as precursor)-ethanol solution of the precursor is fed into the device's ignition chamber. The solution is atomized to an aerosol and ignited via natural gas/oxygen pilot torches. Combustion occurs at temperatures ≈ 2000 °C. The precursor is instantly combusted to an ultrafine oxide

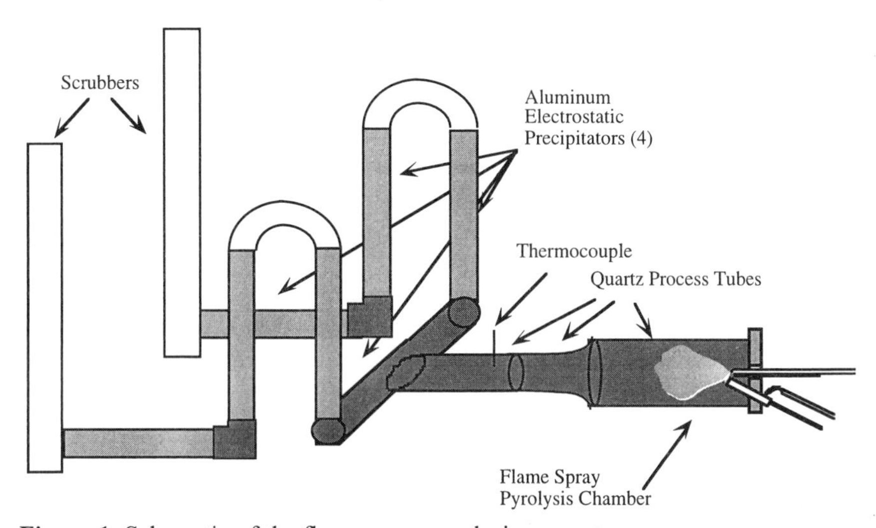

Figure 1. Schematic of the flame spray pyrolysis apparatus.

powder. The powders are collected down stream in a series of well-insulated and shielded electrostatic precipitation (ESP) tube-wire collectors held at 10 kV.

Instrumental Methods: TGA, DTA, XRD and DRIFTS were performed using standard methods, as described elsewhere (*4*).

Results and Discussion

The efforts to date can be divided into progress on Precursor Synthesis and Characterization, and on powder processing as presented below.

Precursor Synthesis and Characterization: The OOPS process permits tailoring of compositions just by choosing initial metal ratios. Thus, the initial metal ratio of Na/Al/Li = 1.67/10.67/0.33 is expected to be retained in the resulting ceramic product. TEA is high boiling (b.p. = 335 °C) and coordinates strongly to Al^{3+}; thus, barring thermal decomposition during the reaction, the stoichiometry of the final solution should remain constant. Consequently, a working formula of $Na_{1.67}Al_{10.67}Li_{0.33}(TEA)_{10.67}(EG)$ was assumed initially.

The polymer TGA profile is shown in Figure 2. The first major mass loss begins at ca. 310 °C with the steepest change occurring at 365 °C. This mass loss is associated with organic ligand decomposition. The major fraction of the decomposition products are gaseous and are lost. Some smaller fraction chars. Hence, a second major mass loss is seen between ≈ 365 °C and 600 °C, as this char oxidizes. A final ceramic yield of 29.3 % is observed. If the initial formulation of the precursor's composition were correct, a ceramic yield of 30.8 % would be expected. The observed lower yield indicates the presence of excess EG, and a final formulation of $Na_{1.67}Al_{10.67}Li_{0.33}(TEA)_{10.67}(EG)\cdot 1.6\ EG$ can be calculated.

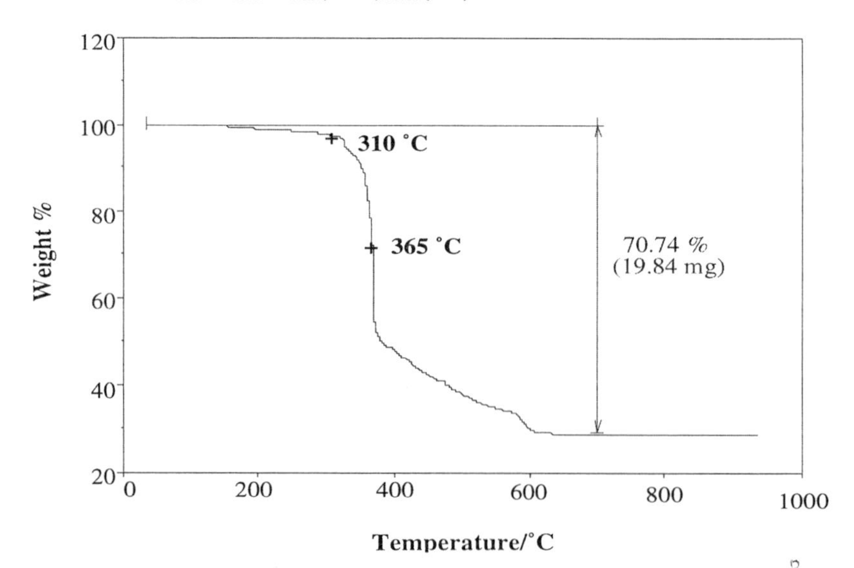

Figure 2. TGA of the precursor $Na_{1.67}Al_{10.67}Li_{0.33}(TEA)_{10.67}(EG)\cdot 1.6\ EG$.

The IR spectrum of the polymer is shown in Figure 3. Present are several peaks characteristic of the organic component: νC-H bond stretching at 2862 cm^{-1} and a collection of νC-O and νC-N bands at 899, 1110, 1271, 1364, and 1451 cm^{-1}. The strong peak at ≈ 671 cm^{-1} is likely due to νAl-O bond vibrations. The band centered at 3550 cm^{-1} is associated with hydrogen bonded νO-H.

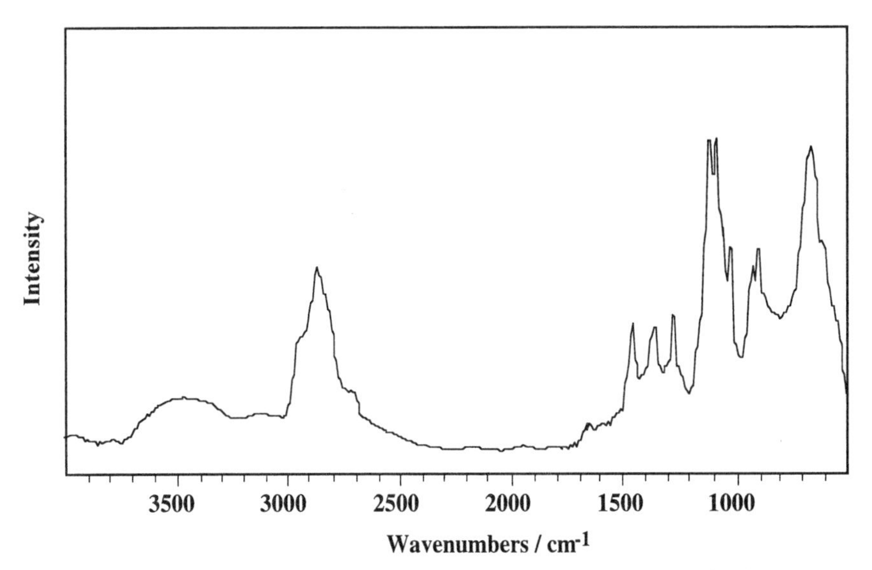

Figure 3: DRIFT spectrum of $Na_{1.67}Al_{10.67}Li_{0.33}(TEA)_{10.67}(EG)\cdot 1.6$ EG vacuum dried (180 °C/4 h).

Nanopowder Characterization: For combustion in the flame spray apparatus, the polymer was dissolved in anhydrous ethanol. Typically, a 20 wt. % solution is prepared as it gives the appropriate viscosity for use in the apparatus. The polymer is stable in ethanol for limited periods only, and a gel precipitates out if the solution stands too long. In 1 h of operation, 2 L of solution can be combusted, providing 50-100 g of collected nanopowder.

The XRD powder pattern of the as-synthesized nanopowder (Figure 4a) exhibits significant crystallinity. Most of the peaks can be assigned to β″-alumina. The peaks at 18.7° (4.75 Å), 45.9° (1.98 Å), and 66.9° 2θ (1.40 Å) are disproportionately intense relative to the other β″ peaks and may indicate the presence of γ-Al_2O_3 or other transition alumina phases which share these particular reflections. No peaks for other likely phases, e.g. Na_2O, Li_2O, were found.

Some phase separation is implied, although it is also possible that, because of the unique synthetic approach employed in forming the nanopowder, a metastable phase, m-alumina, related to β″-alumina was synthesized. No peaks are present that suggest the presence of the β phase, which has distinguishing diffraction peaks at 20.0° (4.45 Å), 21.8° (4.07 Å), and 33.5° 2θ (2.68 Å). Because flame spray pyrolysis involves rapid heating and cooling, it appears that m-alumina is the kinetically stable product. Much of the previous synthetic exploration in this system relied heavily on traditional solid state syntheses (*20*). Clearly, the flame spray approach provides a unique alternative for forming kinetically stable phases.

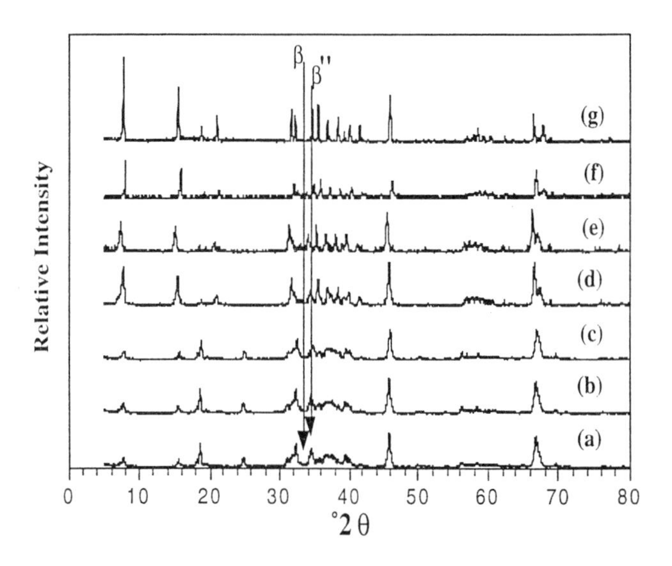

Figure 4: XRD of the nanopowder after various heat treatments: (a) as-prepared, (b) 800 °C for 1 h, (c) 1000 °C for 1 h, (d) 1200 °C for 1 h, (e) 1400 °C for 1 h, (f) 1600 °C followed immediately by cool down. Spectrum (g) is that of 100% β″-alumina crushed from a Ceramatec sample. The position of the characteristic peak from β (33.5°) and β″ (34.5°) are highlighted.

The XRDs of the powder samples remain unchanged after heating at 800 ° or 1000 °C (1 h). Only after 1 h at 1200 °C does a change in the pattern occur as a traditional β″ pattern appears (e.g. Figure 4g, β″ sample from Ceramatec). All unassigned peaks disappear. Thus, the observed pattern can unambiguously be assigned to phase pure β″-alumina. The absence of extraneous peaks at any point along the heating curve suggests that the "as-processed" powder may indeed be a novel metastable form of β″.

A TEM micrograph of the as-prepared nanopowders is given in Figure 5. As seen in the synthesis of other oxides (*26*), the particle size ranges from 10 to 150 nm. The particle shapes are spherical and with limited faceting. The structure of the β/β″-aluminas is layered (*29*), and if left to grow naturally, crystallites of these phases would likely have plate-like morphologies.

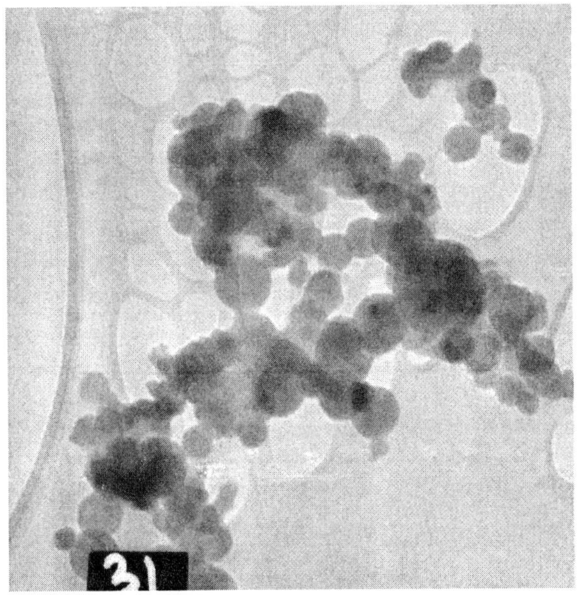

Figure 5: TEM micrograph of the as-prepared nanopowder. Top edge of scale box is 170 nm.

TGA (Figure 6) and DTA studies of the nanopowder were run to develop a picture of its response to thermal processing. In Figure 6 a very small total mass loss (2.4 wt. %) is observed. Two mass loss events occur; the first centered at 100 °C, where removal of surface bound moisture would be expected, and the second at 800°C is typical for the loss of CO_2.[5] A DTA experiment was performed as well, but no thermal events were observed from 25-1400 °C. Given that crystallization is observed ≥ 1200 °C in the bulk heating experiments, the absence of thermal events in the DTA indicates either a transformation of very low energy or one which is driven by the high surface energy already present in the nanopowders.

IR spectra of the nanopowder were taken of the as-formed material and after treatment at 1200 °C (Figure 7b and 7d, respectively). The untreated powders possess a broad and overlapping manifold of strong peaks from 500 to 1050 cm^{-1}. After heating to 1200 °C, where crystallization of β″ is clear, the IR now reveals more well-defined peaks at 620, 670, 740, and 800 cm^{-1}. These compare favorably (although not exactly) with those observed when the Ceramatec precursor powder is heated to 1200 °C (Figure 7c).

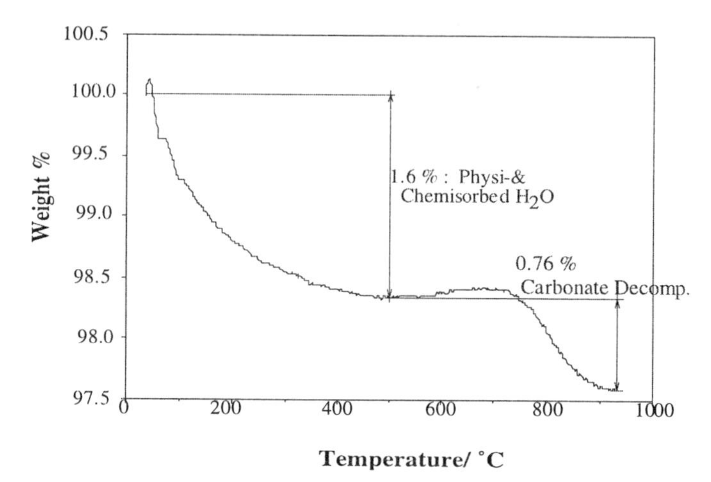

Figure 6: TGA plot of the as-prepared nanopowder.

Nanopowder Stability. The question of nanopowder stability with respect to reproducible formation of β″-alumina is important. Many of the processing steps explored so far have not been executed with rigorous regard to purity, i.e. preparations are done in air with undistilled solvents. Still, the highly pure β″ is consistently reproduced. Nanopowders have even been left exposed to the atmosphere for days at a time, prior to sintering, with no change in the results.

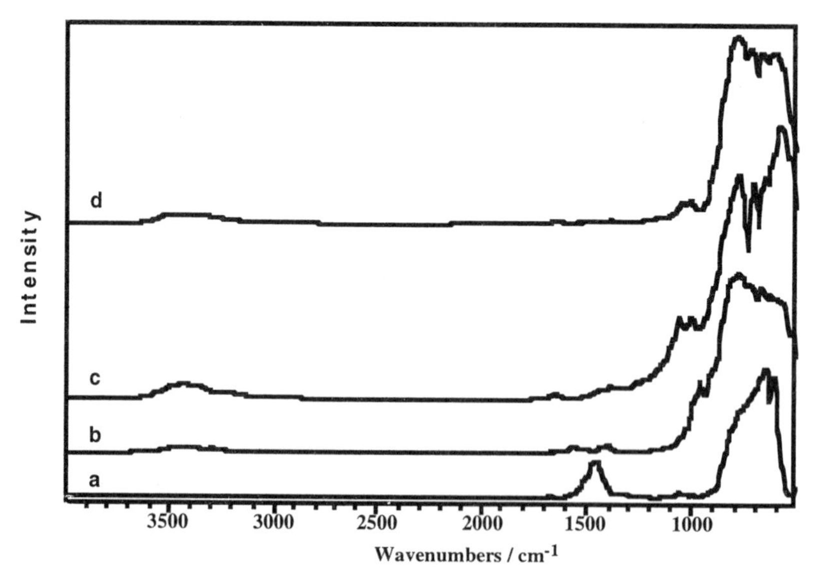

Figure 7: DRIFTs-IR spectra of (a) β″-alumina precursor powder as received from Ceramatec, (b) as-prepared nanopowder, (c) Ceramatec powder heated to 1200 °C, and (d) nanopowder heated to 1200 °C.

Conclusions

We have demonstrated that we could synthesize a bimetallic metalloorganic precursor to β″-alumina from NaOH, LiOH, $Al(OH)_3$ and triethanolamine. On flame spray pyrolysis, it gave ultrafine powders, <100 nm, of a metastable form of alumina, that on heating to >1200 °C converts to pure β″-alumina.

Acknowledgments

This research was funded through a cooperative agreement with Advanced Modular Power Systems, Inc., Ann Arbor, MI 48108 under a NASA STTR grant—NAS3-27776. The authors wish to thank the editor and the reviewers for valuable suggestions.

Literature Cited

1. a. Krummer, J.T.; Weber, N. U.S. Patent 3,404,036, 1968. b. Weber, N. *Energy Conservation*, **1974**, *14*, 1. c. Cole, T. *Science*, **1983**, *221*, 915.
2. Laine, R.M.; Treadwell, D.R.; Bickmore, C.R.; Waldner, K.F.; Hinklin, T., *J. Mater. Chem.*, **1996**, *6*, 1441-1443.
3. Laine, R.M.; Blohowiak, K.Y.; Robinson, T.R.; Hoppe, M.L; Nardi, P.; Kampf, J. Uhm, J. *Nature* **1991**, *353*, 642-644.
4. Blohowiak, K.Y.; Treadwell, D.R.; Mueller, B.L.; Hoppe, M.L.; Jouppi, S.; Kansal, P.; Chew, K.W.; Scotto, C.L.S.; Babonneau, F.; Kampf, J.; Laine, R.M. *Chem. Mater.* **1994**, *6*, 2177-2192.
5. Hoppe, M.L.; Laine, R.M.; Kampf, J.; Gordon, M.S.; Burggraf, L.W. *Angew. Chem. Int. Ed., Engl.* **1993**, *32*, 287-9.
6. Kansal, P.; Laine, R.M. *J. Am. Ceram. Soc.*, **1994**, *77*, 875-882.
7. Kansal, P.; Laine, R.M. *J. Am. Ceram. Soc.*, **1995**, ***78***, 529-38.
8. Laine, R.M.; Youngdahl, K.A.; Nardi, P. U.S. Patent 5,099,052 March 24, 1992.
9. Laine, R.M.; Youngdahl, K.A. U.S. Pat. No. 5,216,155, June 1993.
10. Laine, R.M.; Mueller, B.L.; Hinklin, T. U.S. Patent No.5,418,298 May 23, 1995.
11. a. Stone, R.L.; Tieman, T.D. *Soc. Min. Eng. Trans.*, **1964**, (June), 217-22 and references therein. b. Tieman, T.D. Soc. Min. Eng. Trans., **1964**, (Sept.), 258-9.
12. Laine, R.M.; Treadwell, D.R.; Mueller, B.L.; Bickmore, C.R.; Waldner, K.F.; Hinklin, T. *J. Mater. Chem.* in press.
13. Laine, R.M.; Mueller, B.L.; Hinklin, T.; Treadwell, D.R.; Dhumrongvaraporn, S.; Jiraporn, M. submitted for publication.
14. Pinkas, J.; Verkade, J. *Inorg. Chem.* **1993**, *32*, 2711-16 and references therein.
15. Narayanan, R.; Laine, R.M. unpublished work.
16. Waldner, K.F.; Laine, R.M.; Dhumrongvaraporn, S.; Tayaniphan, S.; R. Narayanan *Chem. Mater.*, 1996, 8, 2850-57.
17. Kansal, P. Dissertation, Jan. 1996, University of Michigan.

18. Bickmore, C. R.; Waldner, K. F.; Treadwell, D. R.; Laine, R. M. *J. Am.Ceram. Soc.*, **1996**, *79*, 1419-23.
19. Hodge, J. D. *Ceram. Bull.*, **1983**, *62*, 244-8.
20. (a) Johnson, D. W.; Granstaff, S. M., Rhodes, W. W. *Am. Ceram. Soc. Bull.*, **1979**, *58*, 849-52. (b) Miller, M. L.; McEntire, B. J.; Miller, G. R.; Gordon, R. S. *Am. Ceram. Soc. Bull.*, **1979**, *58*, 522-24. (c) Foster, L. M.; Scardefield, J. E. UK Patent 1,507,426, 1977. (d) Eddy, D. S.; Rhodes, J. F. US Patent 4,052,538, 1977. (e) Morgan, P. E. D. US Patent 4,339,511, 1982. (f) Duncan, J. H.; Barrow, P.; Brown, P. Y. *Proc. Br. Ceram. Soc.*, **1989**, *41*, 231. (g) Van Zyl, A.; Duncan, G. K.; Barrow, P.; Thackeray, M. US Patent 4,946,664, 1990. (h) Van Zyl, A.; Thackeray, M. M.; Duncan, G. K.; Kingon, A. I.; Heckroodt, R. O. *Mater. Res. Soc. Bull.*, **1993**, *28*, 145.

Pyrolysis and Combustion Synthesis and Characterization

Chapter 15

High-Rate Production of High-Purity, Nonagglomerated Oxide Nanopowders in Flames

Y.-J. Chen[1], N. G. Glumac[1], G. Skandan[2], and B. H. Kear[1]

[1]College of Engineering, Rutgers University, Piscataway, NJ 08855–0909
[2]Nanopowder Enterprises Inc., A Division of SMI, Piscataway, NJ 08854–3908

High rate (~50 g/hr) synthesis of nanoparticles of oxides such as SiO_2, TiO_2 and Al_2O_3 has been demonstrated in a reduced pressure flat flame reactor. The new process, called Combustion Flame - Chemical Vapor Condensation (CF-CVC), is a modification of the original CVC process which involves pyrolysis of chemical precursors in the gas phase. Careful selection of flow parameters has resulted in powders which are only loosely agglomerated, significantly enhancing their usefulness in commercial applications. The nanopowders have a narrow particle size distribution with a mean particle size controllable between 5 and 50 nm. The powder characteristics, post treatments and commercial applications of these powders will be discussed. Furthermore, chemical modeling and laser diagnostics have been applied to these flames, and the ability to measure flame features, including temperature profiles and radical species concentrations, has been demonstrated, and will be discussed in terms of model development.

Nanostructured materials have recently received much attention due to the dramatic changes in material properties as grain or particle sizes approach the nanometer scale, effectively placing a greater fraction of molecules at or near grain boundaries (1). Physical, optical, and magnetic properties can all be enhanced in such materials. For example increases in hardness in nanostructured WC-Co materials, as compared to conventional micrograined WC-Co, have been demonstrated (2). In addition, nanostructured powders of metal oxides have been compacted into optically transparent disks (3). The applications areas for nanomaterials are widespread and include: catalysts and catalyst-supports, thermal barrier coatings, abrasives, phosphors for displays, filters, batteries, and many others.

In most of these applications, powder with particle sizes in the nanometer range serves as the starting material. This powder is either compacted into the desired shape and sintered, dispersed in solution, or sprayed onto a surface as a film. In all cases, the exploitation of these applications opportunities requires a source for large quantities of non-agglomerated nanopowders. Non-agglomeration is essential in advanced materials applications for several reasons. For example, the sintering temperature required to achieve maximum densification is reduced as the degree of agglomeration decreases. This lowered sintering temperature then prevents the decomposition of constituents in the powder. Also, for materials which are processed by a colloidal suspension method, non-agglomeration helps to insure that the powders will be dispersed uniformly in solution.

Currently all techniques used to produce non-agglomerated nanopowders have severe limitations. Liquid phase synthesis has been used to create a variety of nanopowders with small particle sizes and negligible agglomeration. However, the particles are often coated with residue from the synthesis process, and thus high purity is difficult to attain, most often requiring extensive post-processing of the powders. Both Inert Gas Condensation (IGC) and Chemical Vapor Condensation (CVC) (4) have been used to produce non-agglomerated nanopowders in small batch quantities. In IGC, an evaporative source produces a metal vapor in a low pressure, inert gas environment. The vapor is quenched in the gas phase, causing condensation of the metal into nanoparticles. In CVC, the evaporative source is replaced by a hot wall tubular reactor through which a precursor species flows. The precursor pyrolyzes in the tube, and the desired powder species condenses upon exiting the tube. In both cases, synthesized powders are deposited thermophoretically onto a cold surface. However, the scalability of either technique to an industrial sized reactor is questionable.

Building on our experience with IGC and CVC, we have replaced the heat source by a flame in the Combustion Flame - Chemical Vapor Condensation (CF-CVC) technique. This technique offers several advantages over previous methods and has the potential to continuously generate non-agglomerated powders at high rates typical for industrial processes. These advantages have been exploited in other research and commercial flame synthesis processes for the production of diamond, carbon black, other particulates, and a variety of thin films, but not to date for the large scale production of nanoscale powders.

The setup for CF-CVC is similar to that used for large area low pressure combustion synthesis of diamond (5). A schematic of the flame geometry is shown in Figure 1. A flat flame burner stabilizes a disk shaped flame at low pressure above a substrate held at constant temperature. The burner design insures a uniform flow velocity and temperature at the burner face. Under these conditions, provided the gap distance is small compared to the burner diameter, the flow admits a one-dimensional similarity solution (6) which dictates that the thermodynamic variables and chemical species concentrations will only vary in the axial direction, i.e. perpendicular to the burner and substrate. Thus, there is a high degree of radial uniformity within the flame, insuring that conditions for surface deposition and/or particle production will be similar throughout the gap region.

A precursor vapor is mixed in with the premixed flame gases. The precursor species pyrolyzes in the flame as the temperature rises in the preheat zone. After complete pyrolysis, the appropriate monomers are formed in the gas-phase, and these condense to form clusters as the temperature cools closer to the substrate. Clusters grow and agglomerate to form nanoparticles further downstream, and these particles are collected on a cooled substrate. The low pressure insures a minimum of particle-particle collisions which can lead to agglomeration. In addition, the inherent chemical and thermal uniformity of the flow environment leads to powders with a very narrow size distribution.

Also enhancing the usefulness of the combustion environment is the fact that flames have been the subject of many detailed studies, and much is known about the chemical structure. Indeed, fairly accurate models incorporating detailed gas-phase chemistry are directly applicable to the deposition environment to predict local flame chemistry. In addition, many diagnostic probes are available for characterizing flames to validate model predictions.

The scalability of flame synthesis has been demonstrated in other reactors. In general, the process is scaled up by simply increasing the burner and substrate diameters and the corresponding flow rates, or by using burner arrays. The CF-CVC process is currently being scaled for use by Nanopowder Enterprises, Inc. (Piscataway, NJ).

Experimental

A schematic of the experimental setup is shown in Figure 2. The flat flame burner and substrate reside inside a water-cooled vacuum chamber which is maintained at low pressure (5-50 mbar) by a vacuum pump and throttle valve. Gases are fed from bottles through calibrated rotameters and fully mixed before entering the chamber. A precursor species such as hexamethyldisilazane (HMDS) is placed in a bubbler through which a small amount of carrier gas (e.g. helium or nitrogen) is passed, entraining precursor vapor. The vapor and carrier gas are then mixed with the premixed flame gases before entering the burner.

Samples are collected from the substrate and analyzed for size and size distribution using the TEM and SEM, and for impurity content using an Inductively-Coupled Plasma (ICP) technique.

The vacuum chamber sits on an optical table and has 4 ports for optical access. A Nd:YAG-pumped dye laser is available for probing the deposition flames. The setup for diagnostics measurements is shown in Figure 3. The arrangement is optimized for collection of either absorbed or scattered signal, allowing for a variety of diagnostics to be performed including fluorescence, absorption, emission, Rayleigh scattering, and particle scattering. Since the probed environment involves deposition, maintaining clean optical surfaces is essential. This is accomplished by purging the laser inlet and collection windows with a flow of nitrogen.

Safety

The primary safety concerns in CF-CVC are explosion and powder handling. Explosion hazards are minimized by a reactor design with blow-out ports and

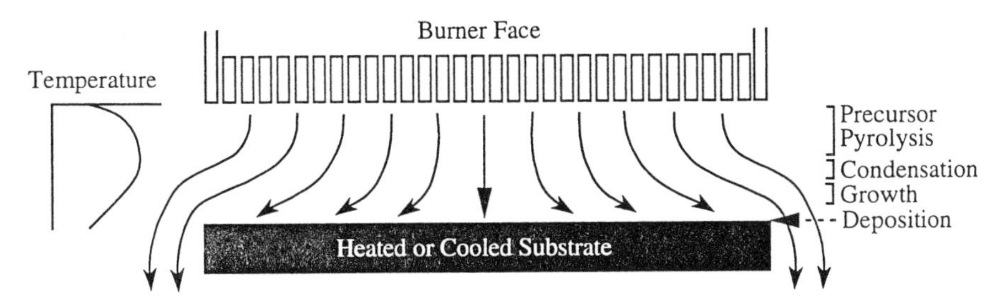

Figure 1. A schematic of the CF-CVC process.

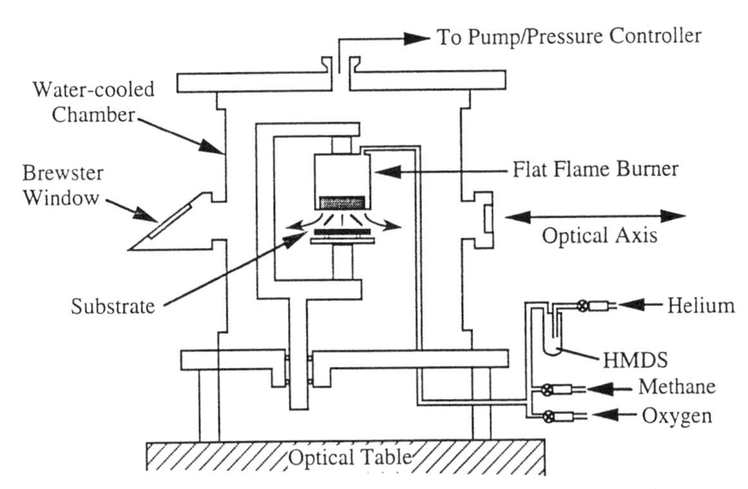

Figure 2. The experimental setup for the synthesis experiments.

closed-loop pressure control which keeps the chamber pressure at a level sufficiently low such that the post-explosion pressure is still below atmospheric. Ignition of the flame is performed in an inert atmosphere, and, if ignition does not occur within 10 seconds after the fuel and oxidizer flows have been started, flow to the chamber is stopped, and the chamber pumped down and refilled with nitrogen.

Oxide nanopowders can cause skin irritations upon contact. Thus, all handling of powders is performed with gloved hands, and all personnel wear goggles and breathing filters during powder collection and cleanup.

Results: Synthesis and Characterization

We have scaled the CVC process to produce test quantities (up to 50 g/hr) of nanophase oxide powders with particles on average less than 20 nm in diameter. This ten-fold increase in production rate was made possible by substituting the hot wall reactor by the flat flame combustor. A short and uniform residence time across the entire surface of the burner at a high temperature, ensures the formation of uniformly sized particles. Since the pyrolysis and subsequent condensation into nanoparticles occurs in a reduced pressure environment, the powder is typically non-agglomerated.

Figure 4 shows production rate (g/hr) of SiO_2 as a function of precursor flow rate (moles/min) for the 6 cm diameter burner. Optimizing the flame parameters, combined with increasing the burner diameter, should lead to higher production rates.

Figure 5 shows a TEM micrograph of as-synthesized SiO_2 powder. As shown in the micrograph, the particles touch each other, and occasionally form necks, but they are not cemented aggregates. The non-agglomeration is borne out by the fact that these powders form stable suspensions in water or alcohol. The surface area of SiO_2 powders, measured by single point nitrogen adsorption using the Brunauer, Emmett, and Teller (BET) method, was found to be in the range 275 - 320 m^2/g. Table I shows the impurity content in the powder, determined by an ICP technique. For elements present in amounts too small to detect, the detection limit of the instrument is given in the amount column. It should be noted that even though the burner is made of copper, the concentration of the copper was very low. This was possible because of the unique design of the burner which allows it to be cooled to the desired temperature. If the burner is too cold, the precursor will condense, and if the burner is too hot, the precursor will react with copper and the reactants could get entrained in the vapor phase. From the data, it can be concluded that we do not introduce any process related contaminants. The high purity of the powder makes it suitable for use in applications such as chemical mechanical polishing of semiconductor wafers.

The same process was also used to synthesize gram quantities of titania and alumina. Figure 6 shows TEM micrograph of as-synthesized TiO_2 powder and Figure 7 shows a photograph of a transparent as-compacted Al_2O_3 pellet. The high degree of transparency shows that the pores are too small ($< 1/20$ th the wavelength of light) to scatter light. In contrast, agglomerated nanopowders are

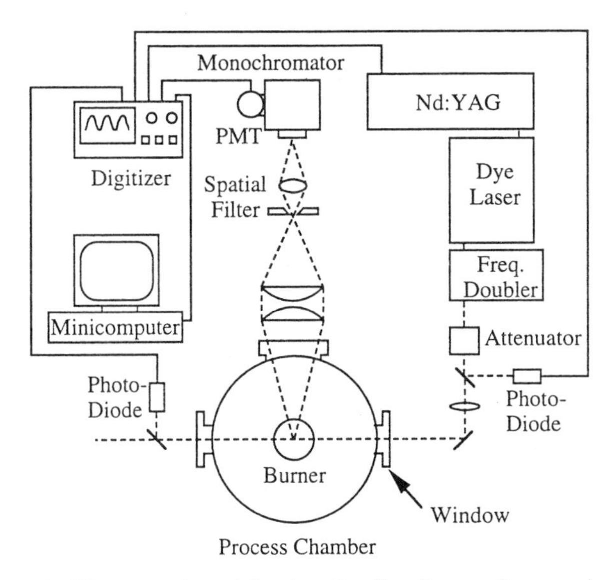

Figure 3. The experimental setup for the diagnostic experiments.

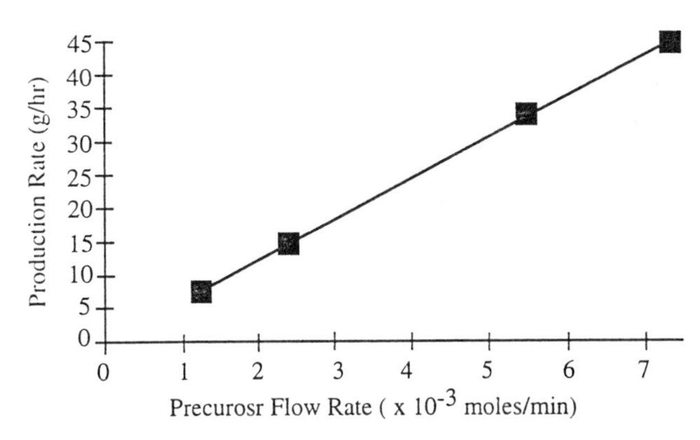

Figure 4. Production rate of SiO_2 as a function of precursor concentration.

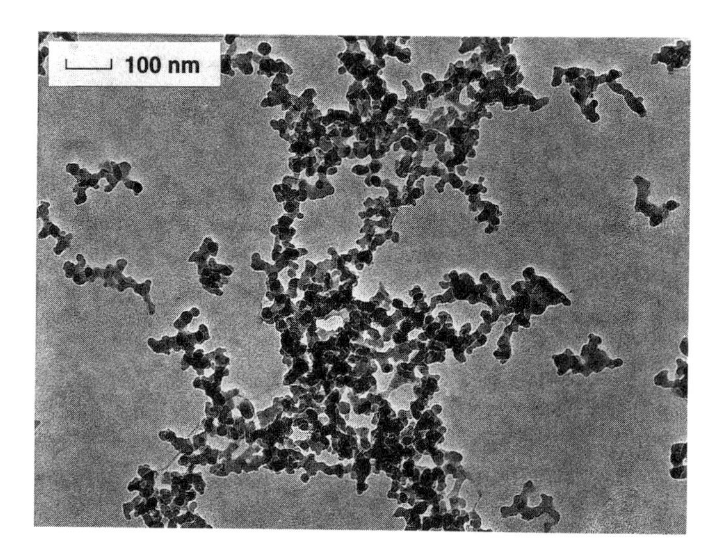

Figure 5. TEM micrograph of as-synthesized SiO_2 powder.

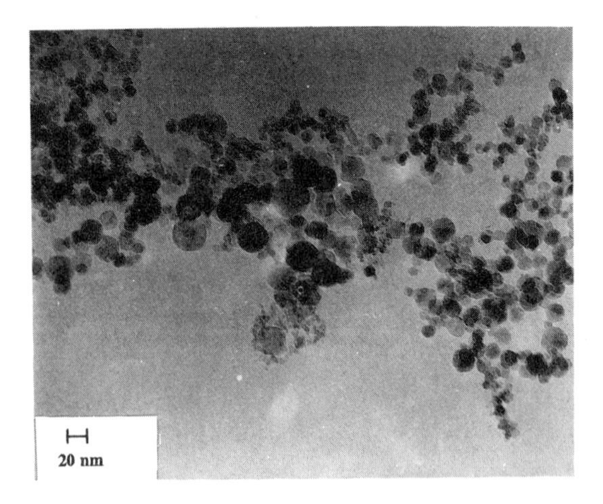

Figure 6. TEM micrograph of as-synthesized TiO_2 powder.

Table I: Chemical Composition of the As-synthesized powder

Element	*Amount (ppm)*	*Element*	*Amount (ppm)*
Calcium	<10	Manganese	< 1
Cadmium	<1	Molybdenum	< 10
Chromium	<2	Nickel	<2
Cobalt	<2	Potassium	63
Copper	11	Sodium	<10
Iron	4.9	Vanadium	<10
Lead	<5	Zinc	<2
Magnesium	<10	Zirconium	<16

opaque since the interagglomerate pores are too large and scatter light. The as-synthesized TiO_2 powder consists of a phase mixture of rutile and anatase, which on annealing at temperature above 800°C, transforms to rutile. This sequence of phase transformation is shown for a series of x-ray diffraction measurements in Figure 8.

Results: Modeling and Diagnostics

One goal of the present study is to develop a process model which can predict particle sizes, size distributions, and production rates for a specified precursor given the process parameter inputs (e.g. pressure, fuel equivalence ratio, flow rates, etc.). Such a model would inevitably involve complex gas phase chemistry which would be capable of predicting the chemical and thermal environment through which the precursor gas passes. Considerable progress has been made with combustion chemical mechanisms, especially for methane and hydrogen fuels. These full mechanisms can readily be applied to one-dimensional flows such as those in our study, and the numerical solution remains computationally tractable.

Two problems arise, however, when applying these models to the deposition environment. In general, the 1-D models cannot predict the temperature profile with a high degree of accuracy in unstrained flames. This is typically because the models do not account for radiative losses or radial conduction. In addition, temperature predictions in hydrogen/oxygen flames are especially poor due to the rapid radial diffusion of H and H_2. In stagnation-point flames with high burner diameter to gap ratios, the models have been demonstrated to predict temperatures much better since axial conduction and chemical heat release are the dominant terms which control the temperature profile. Inability to predict the temperature profile will lead to inaccurate chemical species concentrations even if the kinetics in the model are correct. In these flames, the diameter to gap ratio is only about 1.5, and so there is much concern as to whether the temperature profile is predictable, especially when hydrogen is used as a fuel.

The second problem concerns the effect of the precursor addition on the flame chemical structure. Past studies have shown that in some environments,

Figure 7. Photograph of a transparent Al_2O_3 compact placed over the letters N,L and U. As-synthesized powders were compacted into a pellet at a pressure of 500 MPa.

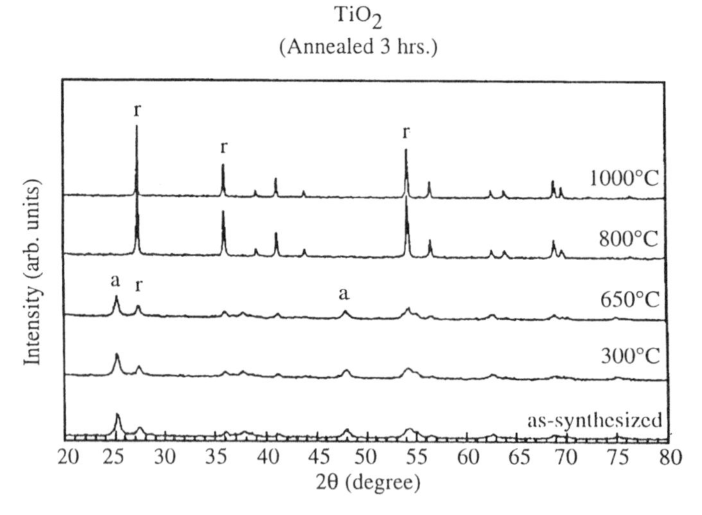

Figure 8. Evolution of the rutile phase during annealing. The as synthesized powder is a 30:70 mixture of anatase to rutile. The metastable anatase phase is stablized at small grain/particle sizes. On annealing the grain/particle size increases accompanied by transformation to the stable rutile phase.

the addition of HMDS does not affect the OH number density profile in the flame, and thus probably does not alter the fundamental chemical structure of the flame. This enables the process model to decouple the flame kinetics from the precursor pyrolysis/condensation kinetics, which is a valuable simplification since the kinetics of species like HMDS in a reactive environment are not well known. The decoupling allows the flame structure to first be calculated without the precursor, then the time/temperature/chemical environment history of an 'inert' mass of fluid can be inferred. It is this history that is then applied to a simplified precursor kinetic model to predict particle sizes and production rates. Establishing whether or not this decoupling is allowable was a goal of the initial diagnostics effort.

To probe the chemical and thermal environment of these deposition flames, we chose to use fluorescence of OH which provides a strong signal in most lean flames and can be used to determine temperature. To eliminate problems due to a strong particle scattering signal, we pumped the (1-0) band of the $A\Sigma \leftarrow X\Pi$ transition near 282 nm and collected fluorescence in the (1-1) band near 315 nm. Since the deposition runs are short, signal strengths must be large in order to collect the requisite data with a minimum of signal averaging. Therefore, we operate in the partially saturated regime for fluorescence. This allows us to readily obtain relative OH number density profiles, though conversion to absolute number densities is made more difficult. For the temperature measurements, we collect fluorescence from 2-4 lines to create a Boltzmann plot from which we estimate the local temperature. We choose lines with very similar B coefficients in order to minimize the effects of partial saturation. Thus, the data can be processed assuming linear or fully saturated conditions, and the difference is less than 20 K. Typical 2σ uncertainties for the flame temperature measurements range from 50-150 K. For the OH number density profiles, we choose to excite ground states which have population fractions which do not vary much over the expected ranges of temperatures in the flames. This minimizes the temperature corrections required to the obtained profiles.

Figure 9 shows a temperature profile obtained in a hydrogen/oxygen flame, and the predictions of the 1-D model. The 1-D model is as much as 200 K too high at the centerline, though the shape of the profile is very similar. In methane/oxygen flames the agreement has been slightly better, but still of the order of 100 K off the predictions. This suggests that a more comprehensive modeling approach must be taken in order to accurately predict the flame temperature profile.

The effect of the precursor appears to depend, as expected, on the precursor flow rate. Figures 10 and 11 show OH number density profiles before (circles) and after (triangles) precursor addition for two different flames, an H_2/O_2 flame with a very low precursor flow rate and a CH_4/O_2 flame with a precursor flow rate typical of production conditions. The low flow condition shows a very small difference between the profiles with and without precursor addition, while the average flow case shows a much more dramatic difference both in shape and magnitude of the OH profile. Thus, the precursor under normal operating con-

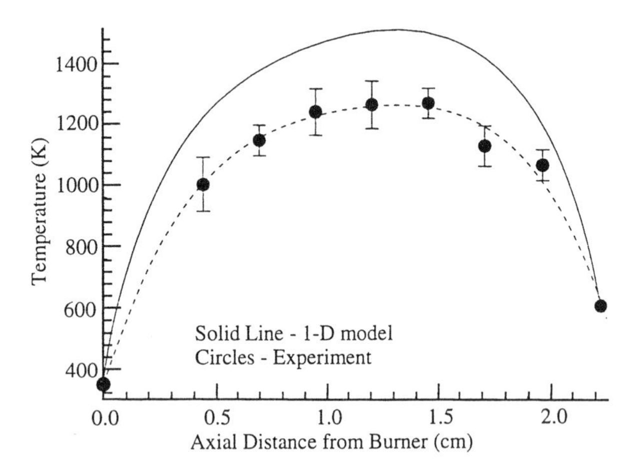

Figure 9. The experimental and 1-D model predictions of the temperature profile in a hydrogen/oxygen flame. The substrate is at 2.22 cm.

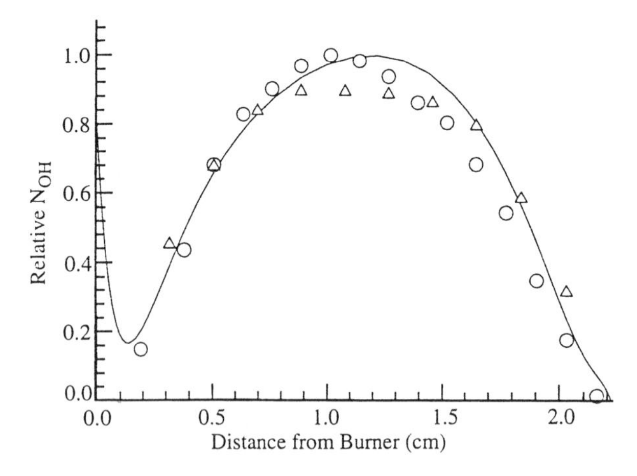

Figure 10. The relative OH number density profiles in a hydrogen/oxygen flame with (triangles) and without (circles) precursor addition for the case of a very low precursor flow rate. The line is the prediction of the 1-D model, normalized at the peak N_{OH} value.

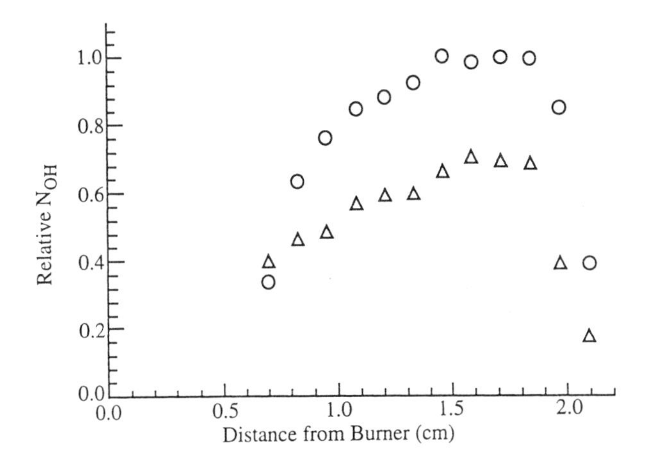

Figure 11. The relative OH number density profiles in a methane/oxygen flame with (triangles) and without (circles) precursor addition for the case of a typical precursor flow rate used in production conditions.

ditions, is significantly affecting the flame structure, and cannot be neglected in the chemical calculations.

Conclusions

We have demonstrated high rate synthesis of several oxide nanopowders using CF-CVC. The powders exhibit virtually no agglomeration and are of high purity. The technique is readily scalable to production of non-agglomerated oxide nanopowders in commercial quantities. Initial studies aimed at characterizing the synthesis environment have demonstrated the ability to measure temperature and radical concentrations in these deposition environments. However, the simplified model predictions are currently only qualitatively correct. There is evidence for substantial coupling of the precursor decomposition chemistry and the flame chemistry at higher precursor flow rates, which must be addressed in future modeling efforts.

Literature Cited

1. Gleiter, H. *Prog. Mat. Sci.* **1989**, *33*, 223.
2. Kear, B. H.; McCandlish, L. E. *Nanostructured Materials.* **1993**, *3*, 19.
3. Skandan, G. *Nanostructured Materials.* **1995**, *5*, 111.
4. Chang, W.; Skandan, G.; Hahn, H.; Danforth, S. C.; and Kear, B. H. *Nanostructured Materials.* **1994**, *4*, 345.
5. Glumac, N. G.; Goodwin, D. G. *Materials Letters.* **1993**, *18*, 119.
6. Evans, G.; Greif, R. *Trans. ASME.* **1987**, *109*, 928.

Chapter 16

In-Situ Particle Size and Shape Analysis During Flame Synthesis of Nanosize Powders

S. Farquharson[1], S. Charpenay[1], M. B. DiTaranto[1], P. A. Rosenthal[1], W. Zhu[2], and S. E. Pratsinis[2]

[1]Advanced Fuel Research, Inc., 87 Church Street, East Hartford, CT 06108
[2]Department of Chemical Engineering, University of Cincinnati, Cincinnati, OH 45221–0171

The unique chemical, electromagnetic and mechanical properties of nanosize particles have generated a demand for a supply of uniform and well characterized particles to develop new applications. Although flame synthesis has been successfully scaled up to commercial production of micron and sub-micron particulates, control of nano-particle size, shape, phase composition, and aggregate size is difficult because flame temperature, residence time and precursor loading all influence these parameters. A Fourier transform infrared spectrometer was interfaced to a diffusion flame reactor to measure these parameters during the production of TiO_2 and SiO_2 by oxidation of $TiCl_4$ and $SiCl_4$, respectively. Rayleigh scattering theory was used to generate spectra to match measured infrared absorbance spectra based on particle size, shape and number density. The approach successfully monitored changes in particle size and shape as a function of process conditions. Theoretical spectra matched to in-situ measurements identified shapes, ranging from spheres to needles. These shapes were confirmed by post-process transmission electron microscopy.

Recently, nanosize particles (diameters below 100 nm) have demonstrated enhanced properties in a number of applications. For example, ceramic layers formed from nano-particles demonstrate improved adhesion, ductility, and mechanical strength (*1*). Changes in chemical, physical and mechanical properties compared to bulk materials, such as a lower melting point, are attributed to the relative number of atoms or molecules on the surface of the particle becoming comparable to that inside the particle (*1*). The predominant methods of preparing these particles are based on aerosol processes, such as flames (*2*), tube furnaces (*3*), gas-condensations (*4*), thermal plasmas (*5*), etc., designed to provide sufficient temperature to promote gas-to-particle conversion. However, until now, only flame processes have been scaled up to produce commercial quantities of ceramic particulates, such as titania (TiO_2) and silica (SiO_2), albeit with typical particle diameters of 200 nm and above. The challenge is to design aerosol flame reactors capable of producing particles well below this size, indeed less than 100 nm in diameter. Furthermore, control of the particle shape, size distribution and, in some cases, crystalline phase is often important for various applications. For example, rutile TiO_2 has a significantly higher index of refraction than anatase TiO_2, and is preferred in pigments where high opacity is desired (a $2 billion/year business worldwide (*6*)). Although extensive studies have shown that flame temperature, residence time and precursor loading are key variables in determining the final particle size and phase composition (*7-13*), control of the particle properties remains difficult. This is exemplified in Figure 1, where different reactant mixing configurations were used to promote different size TiO_2 particles, but also profoundly influenced the shape and possibly the phase composition of the particles.

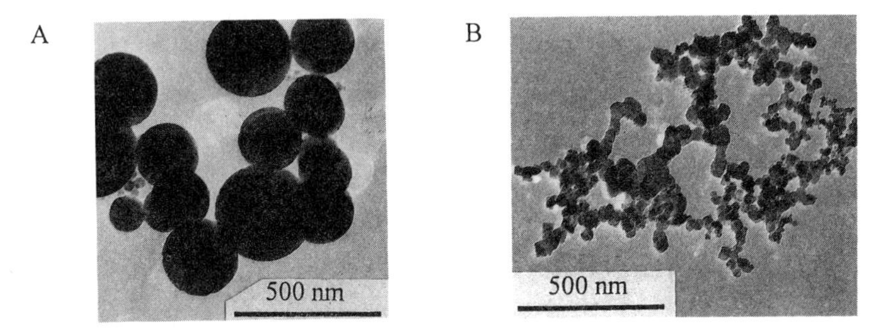

Figure 1. Transmission electron micrographs of titania formed by different reactant mixing configurations, A and B. See Reference 14 for details.

We have been investigating the influence of various reactor design parameters in the flame synthesis of titania and silica produced from the oxidation of $TiCl_4$ and $SiCl_4$, respectively (*14,15*). A diffusion flame reactor design is used since it provides a stable flame over a wide range of operating conditions (*16*). Nucleation sites form directly from the chemical product, e.g., TiO_2, and grow by Brownian coagulation and sintering (*17*). In the present study, we describe the development and use of infrared spectroscopy to perform in-situ measurements of particle size, shape and distribution within the aerosol flame. The goal is to better understand particle formation and growth, and ultimately control the flame reactor and produce nanoscale particles with the desired properties.

Previously, we demonstrated the ability of infrared spectroscopy to measure particle size in streams of ash, coal, and silica (*18*). The size determinations are based on Mie theory, which provides a general relationship between particle size and the wavelength dependence of infrared light scattering. Here we focus on the determination of particle shape, in addition to size, and report in-situ measurements of both titania and silica particles formed in a diffusion flame reactor.

Infrared Light Scattering by Particles

The analytical approach is based on the general relationship between the measured infrared absorbance (A_v) and the absorption of gases and extinction of particles, which is given by (*19*):

$$A_v = -\log \tau = [k_v C_{onc} L + \Sigma_i(N_i A_i F_{ext,i,v} L)]/2.303 \quad (1)$$

where τ is the transmittance, k_v is the absorptivity for the gas (cm^2/mole) at infrared frequency v (cm^{-1}), C_{onc} is the concentration of the gas (mol/cm^3), N_i is the number density of the ith particle (number/cm^3), A_i is the projected area of the ith particle (cm^2), $F_{ext,i}$ is the extinction efficiency of the ith particle at a given collection angle (dimensionless), and L is the pathlength through the sample stream. The first term represents gas absorption while the second term represents particle absorption and scattering. The parameters k_v and $F_{ext,i,v}$ are usually independent of C_{onc} and N_i respectively, except when C_{onc} and N_i are very large. The major difference between the two terms is that gas absorption bands are a few hundred wavenumbers wide and, when sufficient spectral resolution is used, are often accompanied by sharp bands due to the distribution of the absorption to rotational energy levels. In contrast, particle absorption bands do not contain rotational fine structure and the scattering term results in a broad feature that varies smoothly across the whole spectrum. The combination of the gases and particles produces a spectrum with gas and particle absorption peaks superimposed upon a baseline shift due to particle scattering. For example, Figure 2 shows gas phase infrared absorptions due to H_2O, CO, CO_2, and CH_4 on top of a tilted baseline due to TiO_2 infrared particle extinction (note: absorption by TiO_2 solid phase particles at 765 cm^{-1}). The absorption and scattering features are easily separated.

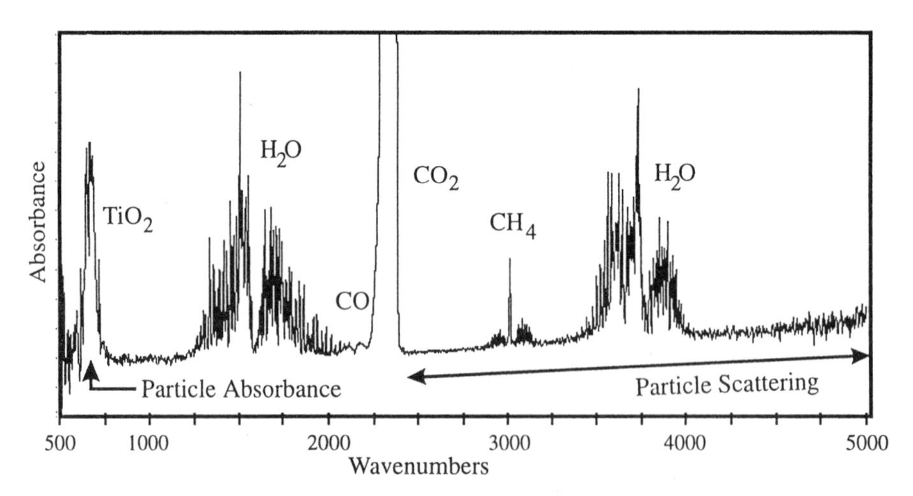

Figure 2. Infrared spectrum of TiO_2 in an aerosol flame.

Mie theory gives a general description of light scattering by particles, and allows prediction of the extinction spectrum for a collection of particles given the particle sizes and their optical constants (*20*). Assuming no gaseous species are present (i.e. $C_{onc} = 0$), Equation 1 can be rewritten in terms of the extinction cross-section, $C_{ext,i}$ (the extinction efficiency per cm^2 of particle surface area) and n_i, the number of particles per cm^2 of infrared beam.

$$A_v = \Sigma_i(n_i C_{ext,i,v})/2.303 \qquad (2)$$

The equation for Mie scattering comes directly from solutions to Maxwell's equations applied to a spherical particle and the medium surrounding it (*20*). The extinction cross-section can further be defined as a combination of absorption and scattering, such that:

$$C_{ext} = C_{abs} + C_{sca} \qquad (3)$$

The contribution of the two effects is apparent in the infrared spectrum of TiO_2 particles (Figure 3), where the low frequency (< 1000 cm^{-1}) is dominated by the infrared absorption at ~ 760 cm^{-1}, and the high frequency is dominated by scattering resulting in a sloping baseline.

Although Mie theory provides an excellent description of the scattering (high wavenumber) term, the description of the absorption term is limited to spheres. This is unfortunate, since it is clear from Figure 1 that nanoscale TiO_2 particles may be better described as ellipsoids. Furthermore, for particles smaller than the incident infrared radiation wavelength, the electric field component of the infrared radiation induces a polarization charge at the surface of the particle. These surface modes result in a strong shape dependency in the infrared absorption (*21*). This shape dependency of the absorption can be addressed by Rayleigh scattering theory, which is a limited subset of Mie theory and applies when the particles are much smaller than the infrared wavelength, specifically when $|m_v| 2\pi r v < 1$, where $m = n + ik$, m is the complex index of refraction, n is the refractive index, k is the absorptive index, r is the particle radius, and ν is spectral frequency. This condition is met for the TiO_2 particles examined here, i.e., $v < 1000$ cm^{-1} ($\lambda > 10$ μm, and $2\pi r < 10$ μm).

Rayleigh theory defines the extinction of light for spherical particles in terms of their polarizability, α_v(*20*):

$$C_{abs} = 6\pi v n_m V \mathrm{Im}\alpha_v \qquad (4)$$

$$C_{sca} = 24\pi^3(v n_m)^4 V^2 |\alpha_v|^2 \qquad (5)$$

where $$\alpha_v = (\varepsilon_v - \varepsilon_m)/(\varepsilon_v + 2\varepsilon_m) \qquad (6)$$

and $$\varepsilon = m^2 = (n + ik)^2 \qquad (7)$$

n_m is the refractive index for the matrix, V is the particle volume ($4/3\pi r^3$), α_v is the polarizability of a sphere, and ε_v and ε_m are the dielectric constants for the particle and the matrix, respectively. Im signifies the imaginary component of the complex expression.

The crystalline structure of rutile TiO_2 is tetragonal and is optically anisotropic, and consequently two sets of optical constants are required, one parallel (m_{para}) and one perpendicular (m_{perp}) to the crystal c-axis. The literature values of n and k are used to compute the dielectric function as 1/3 m_{para} and 2/3 m_{perp} for TiO_2 in accordance with Equation 7 (*22*). Once calculated, the spectra along with the dielectric function of the medium ($\varepsilon_m = 1$ for air, $\varepsilon_m = n_{KBr}^2 = 2.33$ for KBr) are entered into the above equations as needed.

Figure 3 shows the calculated extinction spectra for several sizes of TiO_2 particles. The absorptive index, k is represented by an absorbance band at 730 cm^{-1}, since it is proportional to the concentration of TiO_2 present in the infrared beam according to $k = k_v/(4\pi\nu)$ and Equation 1. While the refractive index, n yields a broad scattering contribution throughout the mid and near infrared. The principal difference between Rayleigh and Mie scattering theory, is that the latter accounts for internal optical reflections, which become apparent as the particle size approaches infrared wavelengths (see Figure 3B). This results in intensity maxima and minima due to constructive and destructive reflective interferences.

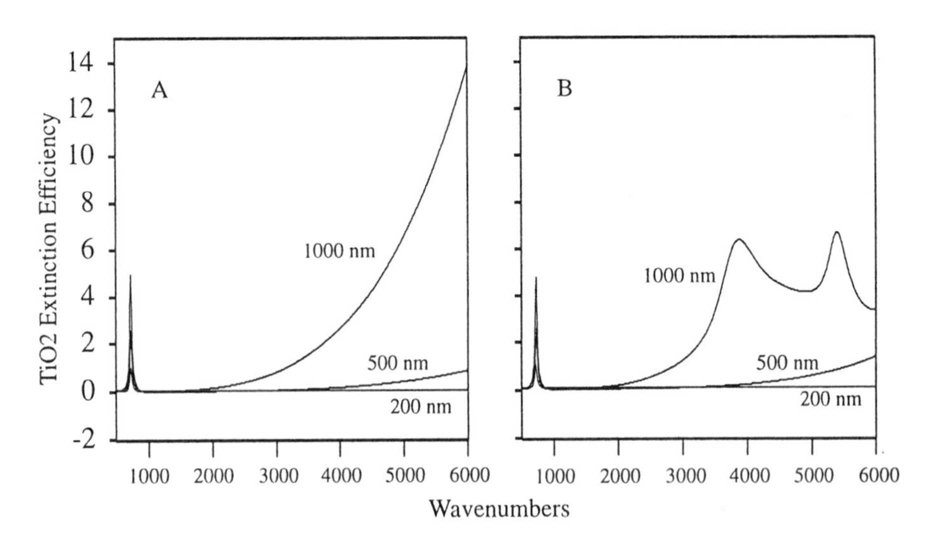

Figure 3. Calculated extinction spectra for 200, 500 and 1000 nm (1 μm) diameter TiO_2 particles using A) Rayleigh and B) Mie scattering theory.

Rayleigh theory may be extended to encompass ellipsoids (including spheres, discs, and needles), cubes, or a continuous distribution of ellipsoids. The average polarizability can be rewritten to include a geometric shape parameter, L_j

$$<\alpha_{vj}> = 1/3\Sigma_j\,[(\varepsilon_v - \varepsilon_m)/(\varepsilon_m + L_j(\varepsilon_v - \varepsilon_m))] \quad (8)$$

where $$L_1 = abc/2\int[(a^2 + q)f(q)]^{-1}dq\,,\ (b^2 \text{ for } L_2,\ b^3 \text{ for } L_3) \quad (9)$$

and $$f(q) = [(q + a^2)(q + b^2)(q + c^2)]^{1/2} \quad (10)$$

The principal dimensions of the ellipsoid are a, b, and c, and the integral of q is from zero to infinity. L_j can have values from 0 to 1, with the constraint, $\Sigma L_j = 1$. For example, for a sphere, $a = b = c$, and $L_1 = L_2 = L_3 = 1/3$, for a disk, $a = b >> c$, and $L_1 = L_2$ (as $c \Rightarrow 0$, $L_1, L_2 \Rightarrow 0$, $L_3 \Rightarrow 1$), and for a needle, $a >> b = c$, and $L_2 = L_3$ (as $b, c \Rightarrow 0$, $L_1 \Rightarrow 0$, $L_2, L_3 \Rightarrow 1/2$). A detailed discussion of the geometric factor is provided in reference 20. The effect of particle shape on the TiO_2 absorption band is shown in Figure 4A, in which the infrared absorption (Equation 2) is calculated using 100 nm particles in Equation 3 and assuming $C_{abs} >> C_{sca}$. Although an aerosol flame would ideally generate particles that consist of identical spheres or ellipsoids, a distribution of shapes is more likely. For this reason, a continuous distribution of ellipsoids is introduced for L_j in Equation 8 composed of a Gaussian distribution about L_j with a standard deviation σ. Figure 4B shows absorption spectra for a normal distribution of ellipsoids about 100 nm radius TiO_2 spheres with a standard deviation in geometric shape of 0.02, 0.05, 0.1 and 0.2.

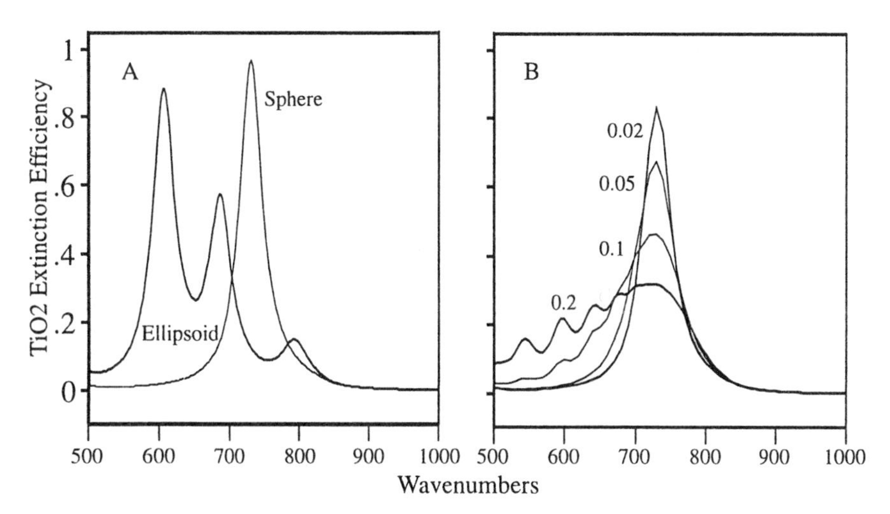

Figure 4. TiO_2 absorption band for 100 nm spheres and ellipsoids. A) Distinct values of L_j yield three distinct absorption maxima. B) Increasing the standard deviation, $\sigma = 0.02$, 0.05, 0.1 and 0.2. broadens the absorption and shifts it to lower wavenumbers. Scattering = 0 in this spectral region.

Experimental

Salt Pellets. Initial testing of these software programs were conducted by casting KBr salt pellets containing either TiO_2 or SiO_2 particles of known size. The average grain size, d_p, was determined from the particle density, ρ_p and the powder specific surface area, A, measured by nitrogen adsorption (Gemini 2360, Micrometrics) according to $d_p = 6/(\rho_p A)$, the BET equation (Brunauer, Emmett, and Teller) (*23*). The average density of anatase (3.84 g/cm^3) and rutile (4.174 g/cm^3) was used for the TiO_2 calculations, whereas 2.2 g/cm^3 was used for SiO_2. In both cases, the error is approximately 10% of the calculated diameter. The particle diameters or d_p were: 1) 19 nm TiO_2, 2) 144 nm TiO_2, 3) 13 nm SiO_2 and 4) 85 nm SiO_2. The pellets concentrations typically were 1 mg of sample per 300 mg of KBr. Transmission electron micrographs (TEM) were obtained on a Philips EM 400 microscope operating at 100 kV or a CM 20 operating at 200 kV.

Diffusion Flame Reactor. The diffusion flame reactor used to study the formation of titania and silica particles consists of a series of three concentric quartz tubes, similar to that described by Allendorf *et al.* (*24*) (see Figure 5(a)). A detailed description of the apparatus is given by Pratsinis *et al.* (*15*). The central tube was 5 mm in diameter and the spacing between successive tubes was 1 mm. The vertically positioned furnace can be heated to 1450 °C. Methane (Wright Brothers, 99%) was used as fuel to produce a "hydrocarbon assisted" flame, while air or pure oxygen (Wright, 99.9%) was used as oxidant. The chemical vapor precursor used was generated by passing argon through a bubbler (b) containing either liquid $TiCl_4$ or $SiCl_4$ (Aldrich, 99.9%). The argon (Wright Brothers, 99.8%) and all reactant flow rates were metered by rotameters (c) (Linde), and the temperatures were measured by thermocouples T2 and T3 (Omega) or thermometer T1 (d). The maximum temperature of each flame was measured by placing a 0.015" Pt-Rh R-type thermocouple (Omega) in the flame tip in the absence of the precursor. The precursor loading was controlled by carefully controlling the bubbler temperature using an electrically heated mantel. For example, at 65 °C, an argon flow rate of 0.25 L/min resulted in 1×10^{-3} mole/min TiO_2.

Numerous flame configurations can be obtained by selecting which gas (fuel, oxidant or precursor) or gas combination passes through which quartz tube. For the present study two flame configurations were used. For TiO_2 production, the Ar, $TiCl_4$ precursor vapor, and CH_4 fuel passed through the central tube of the burner (a), and air as oxidant passed through the outer tube (flame configuration 1). For SiO_2 production, the Ar and $SiCl_4$ precursor vapor passed through the central tube of the burner (a), the CH_4 fuel passed through the middle tube, and the oxidant (air or O_2) passed through the outer tube (flame configuration 2). The fuel-to-oxidant ratio was varied from lean to rich by holding the fuel flow rate constant at 0.4 L/min and changing the air (or O_2) flow rate to 2.5, 3.8 or 5.5 L/min. The precursor mantel temperature was set to 65 °C for TiO_2 and 25 °C for SiO_2 to regulate the precursor loading. The argon flow rate was maintained at 0.25 L/min.

The temperature of the burner and the manifold are kept 200 °C above the bubbler temperature to prevent condensation of the precursor in the lines.

Particles synthesized in the flame were collected on glass fiber filters with the pore size of 0.2 μm (Gelman Scientific; 143 mm) placed inside a stainless steel open-faced filter holder (e). The particle collection unit was placed 120 mm above the tip of the burner in all experiments. Since the luminous part of the flame was in the range of 20-90 mm, a distance of 120 mm between the collector and the flame sufficed to quench particle sintering since the temperature drops very quickly downstream of the flame front. For example, the measured temperature at the filter is less than 300 °C, while at the flame front it can be as high as 1600 °C. Sampling was facilitated by a vacuum pump (g) and, for safe operation, the corrosive by-products such as Cl_2 and HCl were removed from the exhaust by passing them through a 1M NaOH aqueous solution (f) before being released to the fume hood.

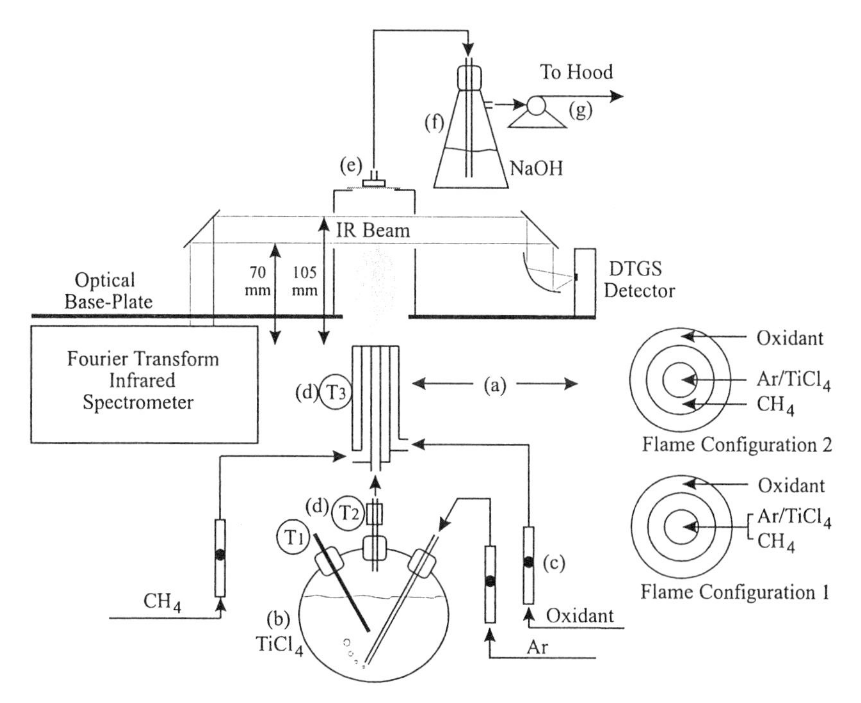

Figure 5. Schematic of diffusion flame reactor, flame configurations 1 and 2, and FT-IR interface.

FT-IR Optical Interface. The optical interface employed an aluminum base-plate attached to a Bomem MB-155 interferometer (see Figure 5). The base-plate extended over the burner and supported a deuterated triglycine sulfate (DTGS) detector. A hole was located in the middle of the base-plate to pass the flame. Two mirrors were mounted on pitch and yaw stages to transmit the infrared beam from the interferometer through the flame to an off-axis parabolic mirror, which focused the beam on the detector. The top of the burner was 1" below the base-plate to position the bottom of the 1.25" diameter infrared beam at the tip of the flame. Previous studies indicated that the chemical reaction and particle growth by sintering were complete at this position, however, the extent of agglomeration was unknown.

An experimental set (usually 10 spectra) consisted of 200 averaged scans per spectrum at 4 cm^{-1}. The absorptions, based on peak height, were used to compare changes in relative concentration due to experimental conditions (e.g. ratio of fuel to oxidant). Since methane is overlapped by HCl, the HCl band at 2727 cm^{-1} (equivalent in intensity to the HCl band at 3015 cm^{-1}) was subtracted from the methane band at 3015 cm^{-1}. TiO_2 and SiO_2 absorptions were measured at 765 and 1130 cm^{-1}, respectively.

Results and Discussion

Salt Pellet Measurements. Initial method validation was performed by casting KBr pellets containing particles of known size and recording infrared transmission spectra. Figure 6 shows the measured infrared spectrum for TiO_2 particles in KBr. The figure includes the best match to the height of the TiO_2 absorption and the curvature of the base-line calculated according to Rayleigh theory. The full spectrum match defines both a unique particle diameter and a unique concentration factor. Indeed, the program predicts a diameter of 154 nm for the TiO_2 sample, a close match to the measured diameter of 144 nm by nitrogen adsorption. The accuracy is demonstrated in an expanded view of the high wavenumber range, which includes spectra predicted for ± 6 nm. The program also fit the TiO_2 absorption with a narrow distribution of nearly spherical ellipsoids (L_1, L_2, $L_3 = 1/3$, $\sigma = 0.075$). The mismatch between the measured and calculated band intensity maxima at 660 cm^{-1} is due to complete absorption of the infrared radiation at this wavenumber by the sample (saturation), while the calculated intensity is scaled by the concentration factor. This factor represents the number of particles in the infrared beam n_i, and can be used along with the calculated particle radius and the density of TiO_2 to determine the mass of TiO_2 in the beam. Fitting the measured absorption band along the sides yields the diameter given and a mass of 0.34 mg based on $m = (4/3)\pi r^3 \rho_p n_i$, with n_i equal to 4.46×10^{10} particles in the beam. The expected mass is 0.58 mg based on the amount of material used to prepare the sample. The difference is likely due to the error in fitting the attenuated TiO_2 infrared band.

In-Situ Flame Measurements. The reactor is designed such that a number of parameters can be changed and investigated. This includes flame configuration, oxidant type, oxidant-to-fuel ratio, relative precursor concentration, and overall flow rates. Here we investigate oxidant type, air or pure oxygen, and oxidant-to-fuel ratio as a function of increasing oxidant flow rate. The latter is known to affect particle size, while the former is known to affect particle shape (*14,15*). Figure 7 shows infrared spectra recorded during the production of TiO_2 particles in which

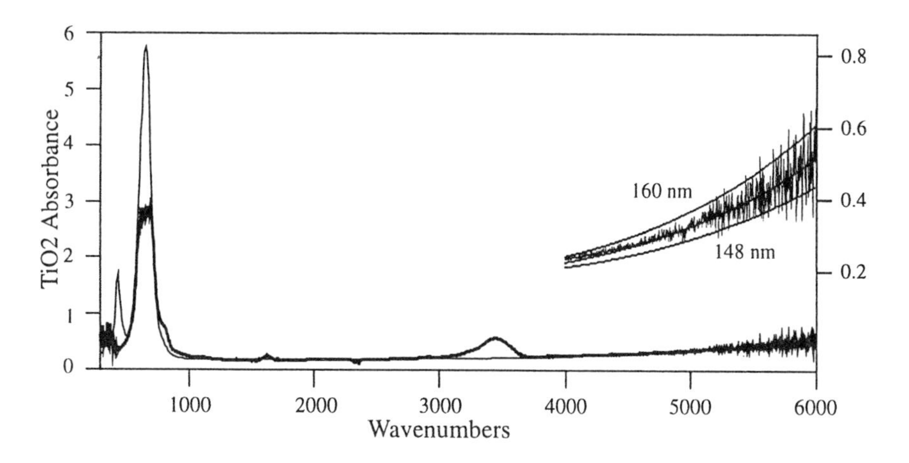

Figure 6. Measured and calculated infrared absorbance spectra for TiO_2 particles in KBr pellet. Diameter measured: 144 nm, calculated: 154 ($\pm$6 nm, expanded view).

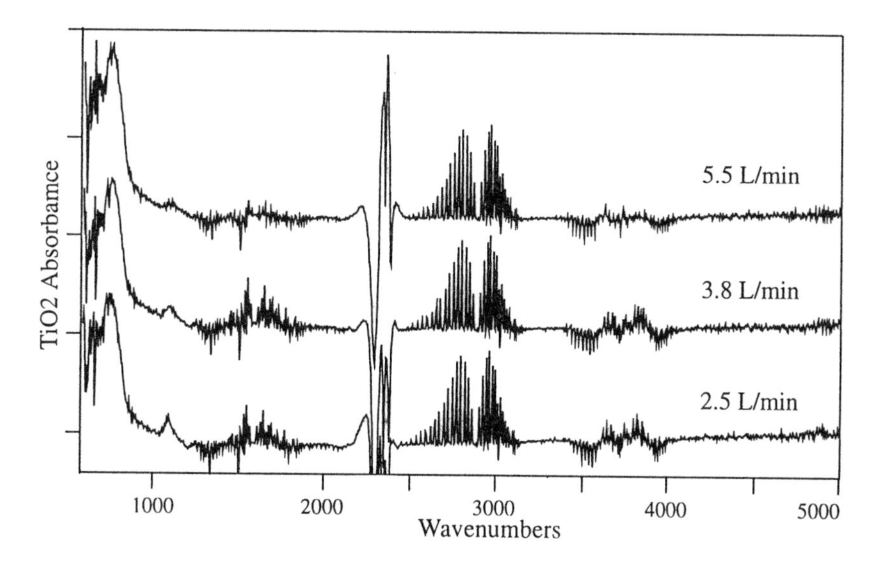

Figure 7. Measured infrared absorbance spectra for TiO_2 particles in a diffusion flame reactor for air flow of 5.5, 3.8, and 2.5 L/min. (See Figure 8 for y scale.)

the oxidant (air) flow rate was increased, while the fuel (methane) flow rate was held constant. The spectra are referenced to the flame burning fuel in the absence of $TiCl_4$ (air: 5.5, methane 0.4 L/min). This results in both positive and negative peaks, indicative of more or less of a chemical compound with respect to the reference flame. This is understood in terms of the oxidation and hydrolysis reactions of $TiCl_4$ (*25*). In the absence of $TiCl_4$, the hydrocarbon combustion reaction is:

$$CH_4 + 2O_2 \rightarrow 2H_2O + CO_2 \quad [1]$$

resulting in H_2O and CO_2 bands in the reference spectrum, as well as any unburned CH_4. In the presence of $TiCl_4$, the oxidation reaction is:

$$TiCl_4 + O_2 \rightarrow TiO_2 + 2Cl_2 \quad [2]$$

and the hydrolysis reaction is (where H_2O is provided by Reaction 1):

$$TiCl_4 + 2H_2O \rightarrow TiO_2 + 4HCl \quad [3]$$

These reactions result in positive TiO_2 and HCl (fine structure between 2500 and 3000 cm^{-1}) spectral bands, but negative CH_4 bands (band at 3015 cm^{-1}, consumed in Reaction 1). For the present set of flow conditions, increasing the oxidant flow rate has virtually no effect on these concentrations, while the CO_2 concentration slightly increases (excess produced in Reaction 1) and the H_2O concentration slightly decreases (consumed in Reaction 3). These data suggest that this flame configuration, which premixes the CH_4 and $TiCl_4$, tends to regulate the oxidation and hydrolysis chemistry.

Again, the measured spectra were matched with theoretical spectra based on Rayleigh scattering theory using a unique particle diameter, concentration and shape factor. The base-line was fit at spectral regions where the gas phase absorbencies are absent, i.e. at 2000 cm^{-1} and above 4200 cm^{-1}. For example, Figure 8 shows the best fit for an oxidant flow rate of 3.8 L/min. The predicted particle diameter is 152 nm, with a significant distribution about spheres ($\sigma = 0.2$, see Inset of Figure 8). Although the specific surface area was not measured for particles produced at this precursor temperature of 65 °C, particle diameters of 110 and 150 nm have been determined for precursor temperatures of 50 and 80 °C, respectively. All other conditions were identical. A TEM of particles produced at the precursor temperature of 50 °C, yielded an average diameter of 100 nm (statistical average of 40 measured particles, see Figure 8B). Both post-process size measurements indicate that the diameters calculated from in-situ infrared measurements are reasonably accurate, especially considering the departure from perfect spheres.

The infrared data indicate that the particle diameter above the flame decreases with increasing oxidant flow, while the number of particles increase, and overall mass increases slightly. This data is consistent with particle size determinations by specific surface area for an identical set of flow rates using flame configuration 1 (see Table I). Both data sets suggest that the increase in flow rate reduces the extent of sintering and consequently smaller particles are produced.

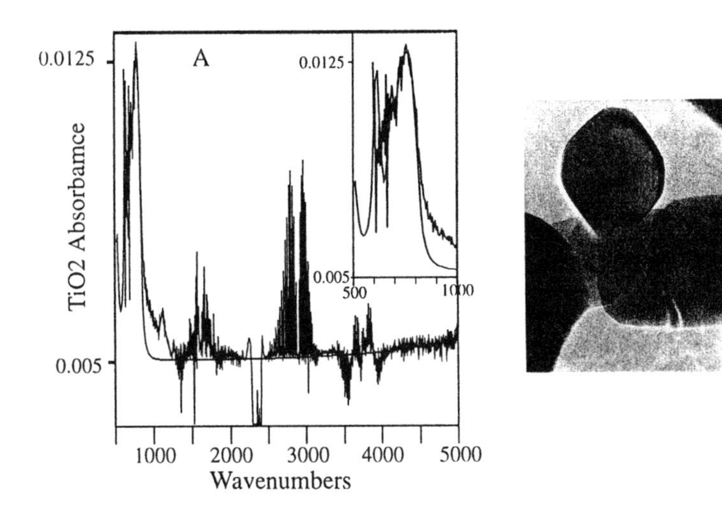

Figure 8. A) Theoretical fit to middle spectrum in Figure 7 corresponding to 2.9×10^7 particles/cm^2 of 152 ± 8 nm diameter, with L_1, L_2, L_3 = 1/3, and σ = 0.2 ±0.02. B) TEM of particles produced using the same flame configuration and flow rates, but a 50 °C precursor temperature.

Table I. Parameters Determined for Three Oxidant Flow Conditions for TiO_2

Air Flow Rate (L/min)	Particle Diameter (nm) BET*	Particle Diameter (nm) Infrared	Number of Particles n_i ($\times 10^7$)	Particle Mass (mg/cm^2) ($\times 10^{-6}$)
2.5	136	200	1.1	1.92
3.8	88	152	2.9	2.23
5.5	75	138	4.3	2.47

* BET data for flame configuration 2.

The wavenumber at which the TiO_2 absorption band occurs in the measured infrared spectra (see Figure 7) is shifted from the wavenumber predicted by theory. The absorption maximum, which is usually coincident with the Frohlich frequency (ν_F, the lowest-order surface mode) (*20*), can be calculated from the transverse optical mode of the material (ν_{TO}), as long as the dielectric function is known for the material at zero (ε_0) and infinite frequency (ε_{inf}), and the dielectric function for the matrix (ε_m), according to (*20*):

$$\nu_F^2 = \nu_{TO}^2[(\varepsilon_0 + 2\varepsilon_m)/(\varepsilon_{inf} + 2\varepsilon_m)] \quad (11)$$

For TiO_2, $\nu_{TO} = 214\ cm^{-1}$, $\varepsilon_0 = 10^2$ at 10 cm^{-1}, $\varepsilon_{inf} = (2.566)^2$ at 10,000 cm^{-1}, and ε_m = 1.0 for air. This predicts an absorption at 732 cm^{-1} for air, and 660 cm^{-1} for KBr (ε_m = 2.33). Although we observe the absorption band at the correct frequency for TiO_2 in KBr (see Figure 6), the absorption occurs at 765 cm^{-1} for TiO_2 in the flame reactor. Several possible changes in the properties of the particle could explain this discrepancy. We believe two explanations are reasonable. First, the high temperature of the particles within the flame may result in a change in the optical constants. Indeed research by Farquharson *et al.* has shown significant changes in the reflectance spectra of a thin layer of SiO_2 on Si at similar elevated temperatures (*26*). Here the TiO_2 particles are at temperatures greater than 1200 °C. Unfortunately, temperature data at present are unavailable. A second possibility is that, in the flame, the TiO_2 particles are not pure rutile, but may also contain a significant percent of the anatase crystalline form. Indeed, x-ray diffraction measurements (Siemens, D500) of TiO_2 formed under identical conditions yield approximately 85% anatase by weight (*14*)! The optical constants for anatase are also unavailable. However, it is known that anatase has a more open structure and has a refractive index less than rutile in the visible (*27*). A value of $\varepsilon_{inf} = (2.37)^2$ yields the desired wavenumber shift. In this regard, a change in optical constants due to either temperature or crystalline form could explain the observed spectral shift in the TiO_2 absorption band. Other possible changes in particle properties, such as porosity (*18*) and TiO_xCl_y intermediate formation (*28*, *29*), were considered, but these properties predict a shift in the TiO_2 band to lower, not higher, wavenumbers.

Figure 9 shows infrared spectra recorded during the production of SiO_2 in which pure oxygen or air was used as oxidant. Again the spectra are referenced to the flame burning fuel in the absence of $SiCl_4$ (oxygen or air: 3.8, methane 0.4 L/min). The spectrum obtained for pure oxygen was fit with a spectrum calculated for SiO_2 particles with a diameter of 185 $\pm$ 5 nm with a narrow distribution about spheres (L_1, L_2, L_3 = 1/3, σ = 0.16 $\pm$ 0.02), and a concentration factor, n_i corresponding to 2.9×10^7 particles per cm^2 of infrared beam.

The most significant change was observed when the oxidant was changed from oxygen to air. The use of air produces a doublet for the SiO_2 absorption peak. Similar to the TiO_2, this doublet may be due to a change in the optical constants as a function of temperature or crystal structure. (X-ray diffraction measurements indicate a highly amorphous content (*14*)). Surprisingly, the doublet can be fit if a unique needle shape is assumed for the particles (L_2, L_3 = 0.495, and σ = 0.07). The uniqueness of the shape is demonstrated by the detailed comparison of variations in L and σ, shown in Figure 10. A simple relationship between the geometric factor and the principle dimensions of the needle can be derived from Equations 9 and 10, in terms of the eccentricity, e (*20*):

$$L_1 = [(1-e^2)/e^2][(1/2e)\ln[(1+e)/(1-e)] - 1], \qquad \text{where } e^2 = 1 - b^2/a^2 \quad (12)$$

The fit in Figure 10 corresponds to $L_1 = 0.01$, and Equation 12 yields a = 20b = 20c (a cigar shape), which gives a = 1253, and b = c = 63 ± 4 nm, and a concentration factor of 4.8×10^6.

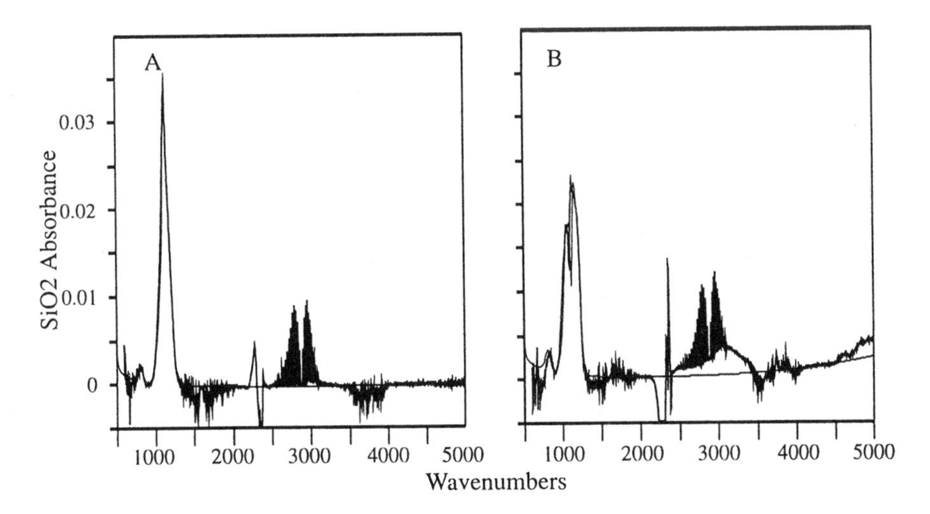

Figure 9. Measured and calculated infrared absorbance spectra for SiO_2 particles in a diffusion flame reactor using A) oxygen or B) air as oxidant. Rayleigh theory successfully fits both measured spectra using A) spheres for oxygen and B) needles for air as oxidant.

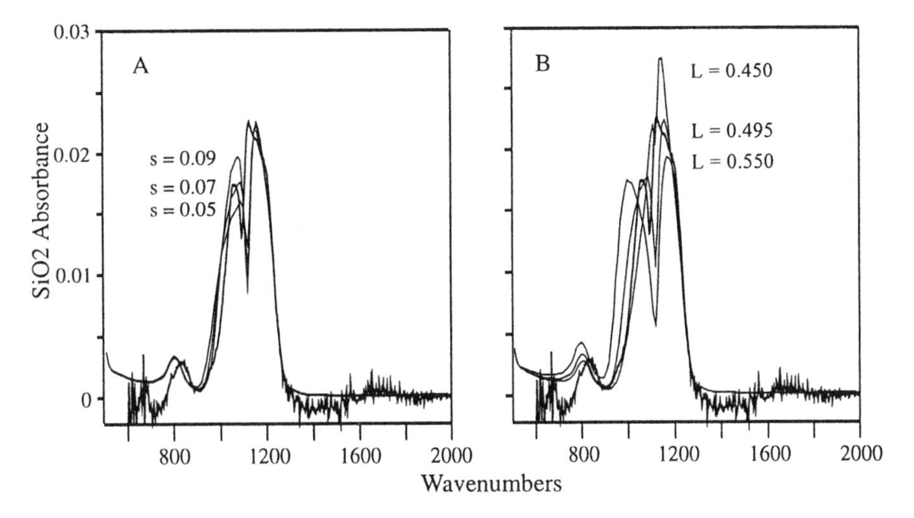

Figure 10. Comparison of calculated values for A) σ (s) and B) L with measured spectra.

The cigar shape is reasonable in view of TEMs measured for TiO_2 particles produced in air versus oxygen as the oxidant, as seen in Figure 11 (*14*). The TEMs suggest that air yields faceted particles that are fused together forming strings or ribbons, while oxygen produces larger, spherical particles. This is likely due to the higher flame temperature (by approximately 500 °C) obtained with pure oxygen, which promotes greater sintering and less agglomeration.

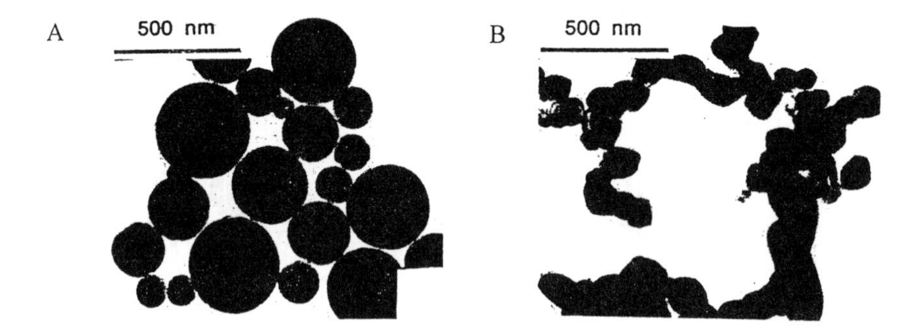

Figure 11. Transmission electron micrographs for TiO_2 produced using A) oxygen and B) air as oxidant. Reproduced with permission from Reference 14. Copyright 1996 ACS.

Conclusions

The ability of infrared spectroscopy to measure particle size and shape in a diffusion flame reactor was successfully demonstrated. Rayleigh scattering theory provided precise size and shape predictions for TiO_2 and SiO_2 particles as long as the particles were below 500 nm in diameter and reasonably mono-dispersed. These conditions are relevant for production of nanoscale particles. Accurate size determinations were obtained for particles contained in KBr pellets. In-situ infrared particle size determinations also compared favorably to post-process nitrogen adsorption and transmission electron microscopy size determinations, especially when shape is considered. A splitting of the SiO_2 absorption peak occurs when the oxidant is changed from pure oxygen to air, which is indicative of a change in particle shape from spheres to needles. TEM measurements support this conclusion. An anomalous shift in the TiO_2 absorption band from 732 to 765 cm^{-1} was observed in the infrared spectra measured in the flame. It is likely that this shift is due to a change in the optical constants for TiO_2 as a function of temperature or crystalline composition. X-ray diffraction measurements of particles produced under identical conditions showed that these particles are primarily anatase.

It can be concluded that infrared spectroscopy provides valuable information on the following properties: 1) particle size and shape; 2) reaction efficiency, as determined from infrared absorption bands corresponding to reactant and product concentrations; and 3) temperature (gas and solid phase). The latter measurements are the subject of a separate report (*30*). Thus, infrared spectroscopy could help elucidate the formation of nanosize particles, as well as provide a method for process monitoring and control. This technique could prove particularly useful to investigations on the effect of dopants, such as Si^{4+} and Al^{3+}, designed to enhance the phase transformation of anatase to rutile (*31*), or electric fields applied in the flame to control particle size (*32,33*).

Acknowledgments

The support of this research by the National Science Foundation, grant number DEB9211764 is gratefully acknowledged.

Literature Cited

1 Ichinose, N., Ozaki, Y., and Kashu, S., *Superfine Particle Technology*, Springer Verlag, London, 1992.
2 Zachariah, M.R.; and Huzarewicz, S., *J. Mater. Res.*, **1991**, *6*, 264.
3 Akhtar, M.K.; Pratsinis, S.E.; and Mastrangelo, S.V.R., *J. Am. Ceram. Soc.*, **1992**, *75*, 3408.
4 Sanders, W-A.; Sercel, P.C.; Atwater, H.A.; and Flagan, R.C., *Appl. Phys. Lett.*, **1992**, *600*, 950.
5 Vissokov, G.P.; Stefanov, B.I.; Gerasimov, N.T.; Oliver, D.H.; Enikov, R.Z.; Vrantchev, A.I.; Balabanova, E.G.; Pirgov, P.S., *J. Mater. Sci.*, **1988**, *23*, 2415.
6 Stamatakis, P.; Natalie, C.A.; Palmer, B.R.; and Yuill, W.A., *Aerosol Sci. Technol.*, **1991**, *14*, 316.
7 Ulrich, G.D., *Comb. Sci. Tech.*, **1971**, *4*, 47.
8 Ulrich, G.D.; Milnes, B.A.; and Subramanian, N.S., *Comb. Sci. Tech.*, **1976**, *14*, 243.
9 Ulrich, G.D.; and Subramanian, N.S., *Comb. Sci. Tech.*, **1977**, *17*, 119.
10 Ulrich.G.D.; and Riehl, J.W., *J. Colloid Inter. Sci.*, **1982**, *87*, 257.
11 Zachariah, M.R.; Chin, D.; Semerjian, H.G.; and Katz, J.L., *Combustion and Flame*, **1989**, *78*, 287.
12 Hung, C-H.; and Katz, J.L., *J. Mater. Res.*, **1992**, *7*, 1861.
13 Formenti, M.; Juillet, F.; Meriaudeau, P.; Teichner, S.J.; and Vergnon, P., *J. Colloid Inter. Sci.*, **1972**, *39*, 79.
14 Zhu,W.; and Pratsinis, S.E., in *Nanotechnology*, ACS Symposium Series No. 622, 1996, 64.
15 Pratsinis, S.E.; Zhu, W.; and Vemury, S. *Powder Technology*, **1996**, *86*, 87.
16 Fotou, G.P.; Vermury, S.; and Pratsinis, S.E., *Chem. Eng. Sci.*, **1994**, *49*, 4939.
17 Kusters, K.A., and Pratsinis, S.E., *Powder Technology*, **1995**, *82*, 79.
18 Solomon, P.R.; Carangelo, R.M.; Best, P.E.; Markham, J.R.; and Hamblen, D.G., *Fuel*, **1987**, *66*, 897.

19 Morrison, P.W., Jr.; Solomon, P.R.; Serio, M.A.; Carangelo, R.M.; and Markham J.R., *Sensors*, **1991**, *8*, 32.

20 Bohren, C.F. and Huffman, D.R., *Absorption and Scattering of Light by Small Particles*, John Wiley and Sons, New York, 1983.

21 Ocana, M.; Fornes, V., Garcia Ramos, J.V., and Serna, C.J., *J. Solid State Chem.*, **1988**, *75*, 364.

22 Jasperse, et al., in *Handbook of Optical Constants for Solids*, Palik, E.D., Ed., Acad. Press, 1985.

23 Brunauer, S.; Emmett, P.; and Teller, E., *JACS*, **1938**, *60*, 309.

24 Allendolf, M.D.; Bautista, J.J.; and Potkay, E. *J. Appl. Phys.*, **1989**, *66*, 5046.

25 Akhtar, M.K.; Vemury, S.; and Pratsinis, S.E., *AIChe Journal*, **1994**, *40*, 1183.

26 Farquharson, S., et.al. "Rapid Thermal Annealing Process with Real-Time Monitoring for SiO_2 Layer Formation," Final Report, DOD Army Grant No. DAAH04-94-C-0041, 1995.

27 Shannon, R.D.; and Pask, J.A., *Amer. Minerologist,* **1964**, *49*, 1707.

28 Pratsinis, S.E.; Bal, H.; Biswas, P.; Frenklach, M.; and Mastrangelo, S. *J. Amer. Ceram. Soc.*, **1990**, *73*, 2158.

29 Person, W.B.; and Maier, W.B., *J. Chem. Phys.*, **1978**, *69*, 297.

30 Morrison, P.W., Jr.; Rhaghavan, R.; Timpone, A.J.; Artelt, C.P.; and Pratsinis, S.E., *Chem. Mater.*, **1996** submitted.

31 Vemury, S. and Pratsinis, S.E., *J. Amer. Ceram. Soc.,* **1995**, *78*, 2984.

32 Vemury, S. and Pratsinis, S.E., *Appl. Phys. Lett.*, **1995**, *66*, 3275.

33 Vemury, S. and Pratsinis, S.E., *J. Aerosol Sci.*, **1996**, *27*, 951.

Chapter 17

Reactants Transport in Combustion Synthesis of Ceramics

O. E. Kashireninov

Institute of Structural Macrokinetics, Russian Academy of Sciences, 142432 Chernogolovka, Moscow Region, Russia

Diffusion in solids does not ensure the experimentally observed velocity of combustion wave propagation in the systems which are traditionally considered as gasless and burned in the mode of solid flames (gasless solid-state combustion). The phenomenology of indirect interactions, the thermochemistry and dynamics of the gas-phase carriers formation, as well as their participation in the reactants transport are studied in the systems Mo-B and Ta-C. The distributions of the main species in the gas phase of the combustion wave are measured *in situ* with the use of a dynamic mass-spectrometry (DMS) technique which allows for high temporal and spatial resolution. The detailed chemical pathways of the processes were established. It was shown that the actual mechanism of combustion in the systems under study is neither solid state nor gasless and the reactions are fully accomplished in a narrow front.

For the last 30 years, the combustion synthesis of oxygenless ceramics has been drawing attention, first and foremost because of its evident technological advantages, i.e., simplicity of the equipment, high productivity and low consumption of energy. In many cases, the combustion synthesis results in compounds, phases and materials that are impossible to be synthesized by other methods. These investigations were pioneered by A. Merzhanov and I. Borovinskaya (*1*). In later work they, together with their co-workers, created the essentials of the thermal theory of powder mixtures combustion and developed a well known method and various modifications of self-propagating high-temperature synthesis (SHS).

SHS is based on layer-by-layer combustion of a metal powder mixture with another reactant (reactants). Metal-reactant compositions can vary but have to meet the condition of adequate heat release to sustain the overall process. It has to be noted that processes of this kind have been known for many years (*2*), but up to the end of the 1960's they were not used to synthesize materials of complex chemical composition.

It is evident that SHS, by its chemical nature, is an exothermic redox reaction occurring in the frontal mode. Metals (Ti, Zr, Cr, Mo, W and many others) are reductants, and carbon (soot, graphite), nitrogen, halogens and others - oxidants. The elements with amphoteric properties (B, Al, Si) depending on the second reactant can be reductants (nitrides, carbides) as well as oxidants (borides, aluminides, silicides). Various theoretical and experimental aspects (including technological ones) of SHS have been considered in many summarizing publications (see, for example (*3,4*) where references were made on all main reviews published before). The subsequent works have considerably extended the number of system types and resulted in the more appropriate classification as a type of combustion synthesis.

It is not essential that, for combustion synthesis, the reactants are elements and initially solid (at least one of them). Refractory carbides and nitrides with complex chemical compositions (metal halogenides, organometallics, etc.) are also formed in combustion of gaseous systems. The special features of combustion synthesis in the gas phase were analyzed in the recent review by Brezinsky (*5*). The synthesis in a gaseous system proceeds at a stationary combustion front with moving flows of reactants and product. Consequently, these processes differ in essence from combustion of condensed systems by the conditions of nucleation and growth of the new phase, which manifest themselves in the product morphologies.

The products of combustion synthesis can be sinters, melts or powders (ultrafine included) of carbides, nitrides, borides, silicides, halogenides and other compounds, solid solutions or mixed phases, depending on the parameters and conditions of the process. There are many preparative and technological procedures for producing various compounds, materials and items in the combustion mode. These are described in the scientific and patent literature and their consideration goes beyond the framework of the present paper. The same has to be said about the numerous theoretical and experimental results concerning various aspects of combustion, that are also considered in the above-mentioned reviews.

Process Types in Combustion Synthesis

The processes of combustion synthesis differ considerably in the state of the substance in the reaction front (solids, melts, gases). Correspondingly, the transport processes responsible for chemical transformations in the combustion front are different as well. On the basis of the classifications presented in (*3*), four main types of processes can be recognized.

In combustion synthesis in gas systems (*gas+gas*), the reactions occur in counter-flow or co-flow diffusion flames ($SiH_4+NH_3 \rightarrow Si_3N_4$ (*6*), $Si(CH_3)+C_2H_4+O_2 \rightarrow SiC$ (*7*), $TiCl_4+N_2+Na \rightarrow TiN$ (*8*), etc.). Here, the process is controlled by the diffusion in gases. If the characteristic time of diffusion, τ_D, is lower than the characteristic time of chemical reaction, τ_R, then a process occurs in the kinetic mode. The conversion (η) for *gas+gas* systems may be as great as 100%.

The interaction of gases with porous samples formed of metal powders ($Ta+N_2 \rightarrow TaN$ (*9*), $Ti+N_2 \rightarrow TiN$ (*10*), etc.) is often limited by the diffusion of the gases in pores. As a rule, for the system *gas+solid* $\tau_D > \tau_R$. In combination with thermal dissociation of the products, this can lead to conversion lowering, with $\eta = 20\text{-}70\%$.

When the combustion temperature, T_f, for a mixture of initially solid powders is higher than the melting point, T_m, for at least one of the reactants or intermediate phase (eutectics) formed in the reaction, the interaction in the system *liquid+solid* takes place (e.g., Ti+C→TiC (*11*), Ti+B→TiB (*12*), etc.). The process is controlled by diffusion in the melt in this case ($\tau_D > \tau_R$). Estimates show that, for these systems, at the front width L~1mm and a combustion velocity u = 1-10 mm/s, the reactant residence time in the front (τ_r) at T_f = 1500-3000 K is several orders of magnitude higher than τ_D, resulting in close to complete conversion.

If, for the set of reactants and intermediate products (substances and phases), $T_m > T_f$, and the partial pressure of reactants and products (P_i) at T_f is much lower than the ambient pressure (P_0), it is believed that solid-state combustion takes place and the so called "solid flame" is observed. Following common logic, one can suppose that combustion synthesis in the system *solid+solid* has to proceed in the mode of reactive diffusion. However, the τ_D value (at the diffusion coefficient in solids $D_S \sim 10^{-9}$ cm^2/s and the distance equal to the particle size of the metal powder $r \sim 10$ μm) can be estimated as $\tau_D = r^2/D_S \sim 10^3$ s. This value considerably exceeds the maximum residence time for particles in the front, $\tau_r = L/u \sim 10^{-1}$ s and seems to be in conflict with the conversions $\eta > 70\%$ usually observed for these systems. Estimation of the minimum effective diffusion coefficient (D_{eff}) required for the process occurring in a combustion mode (*13*) gives $D_{eff} = u^2r^2/6\gamma\, a_{eff} \sim 10^{-8}$-$10^{-6}$ cm^2/s at the combustion rate u = 0.1 cm/s, coefficient of thermal auto-acceleration of the reaction $\gamma \sim 0.1$ and effective thermal diffusivity $a_{eff} \approx$ 0.55. Judging from these evaluations, the process cannot be frontal in the reactive diffusion mode. The possibilities of defect generation with over-equilibrium concentrations due to very fast heating and acceleration of diffusion along the grain boundaries do not remedy this contradiction that is repeatedly mentioned elsewhere (see, e.g. (*3,4*)).

In order to explain and theoretically prove the phenomenon of solid flames, the model for wide combustion zones was formulated (*14,15*). According to this model, the degree of chemical conversion in the front was supposed to be insignificant (as opposed to the classic theory by Zel'dovich and Frank-Kamenetsky) and the final product formation was assumed to occur in a wide zone behind the front through reactive diffusion. Alternative mechanisms of chemical reactions were not considered.

In this connection, it should be noted that, taking into account the basics of transport phenomena, chemistry and the results of a large body of research in this area, the above mentioned criterion of gasless combustion $P_i(T) << P_0$ (*16*) calls for serious argumentation. However, despite having the data on the physics of combustion, phase formation and the preparative-structural aspects of the combustion synthesis (interrelation of product characteristics with conditions of combustion), information about its detailed chemical mechanism(s) is still lacking. At the same time, the chemical mechanism defines in many respects other process characteristics and it is hardly (if ever) possible to control the synthesis without this understanding. The situation is made more complicated due to the fact that the synthesis and phase formation quite often coincide spatially and temporally n the combustion wave. Therefore, the sequence of phase transformations is usually termed as the chemical mechanism and the question how this or that phase of the reaction product was formed remains open. The direct dependence of the chemical mechanism on the system composition and conditions of the reaction (combustion) as well as the variety of these mechanisms are evident. In order to determine the chemical mechanisms for the systems *solid+solid*, one must, first of all, define the types and parameters of the key process, i.e., reactant transport.

The study of the detailed chemical mechanism of solid-state combustion, also called gasless combustion (*17*) and solid flames (*18*), was the goal of present work. The combustion processes of the systems Mo-B and Ta-C are considered to be classic examples of this phenomenon and were chosen as the subjects of this study.

Experimental

Reactants. The size classified powders of Mo (1-6 μm particle size), Ta (1-4 μm), B (amorphous, < 0.1-2.0 μm) and carbon black were used in the experiments. The admixed oxygen contents were 5, 6, 1, and ≅ 1 at.%, respectively. The relative densities of the pressed samples were 0.6-0.7.

Contactless Interaction. As far as diffusion in solids does not obviously provide the necessary reactant transport in the combustion wave, it was interesting to study the behavior of the Mo-B and Ta-C systems in the absence of direct metal-reactant contact. For this purpose, metal wires with diameter 1 mm and length 2-3 cm were placed into the channel in the pressed reactant (boron) or into the tube (graphite) with internal diameters of 2-5 mm, so that the clearance between the metal and reactant surfaces was not less than 0.5 mm. The fastening of the wire ends also excluded metal-reactant contact. The cross-section of the assembly with a tantalum wire inside a graphite tube is shown in Figure 1. The assemblies were heated in a vacuum or a controlled atmosphere in a vacuum furnace (Mo-B) or by Joule heat released on the wire as a filament of the electrothermograph (Ta-C). The temperature in the furnace was controlled with an accuracy of ±10 K. Time was measured from the moment of insertion of the tube into the furnace (or the current in the wire was turned on) until it was removed from it (or the electrothermograph was turned off).

Gas Phase Composition. The dynamic mass-spectrometry technique (DMS) with molecular beam sampling was used to study directly the high-temperature gases formed in the combustion front (*19*). The installation diagram and location of the sampling cone in relation to the burning sample are shown in Figures 2a and 2b. The ion currents were recorded with a time interval of 1-5 ms at a spatial resolution of the combustion wave not worse than 0.2 mm (usually 0.1 mm). The pressure sensitivity in the sampling point was not less than 0.1 Pa. The evaporation of zinc was used to define the position of the combustion wave in relation to the sampling cone. Zn powder (99 mass %) with particle size < 80 μm in the amount of 0.02-0.03 mass % was added to the samples when prepared. According to the data available (*20*), the temperature in the combustion wave on the sample surface differs from that in the volume by not more than 100 K. Therefore, the composition of gases on the surface and in the sample body can be considered to be identical. The time dependence of ion currents allows one to obtain the spatial structure of gas release from the combustion wave propagated with constant velocity u, assuming

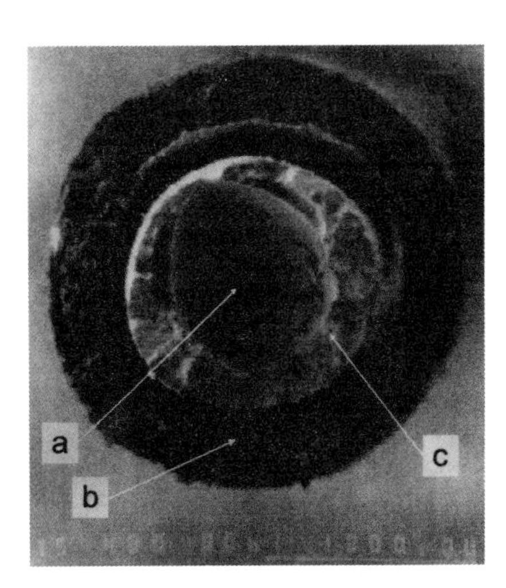

Figure 1. Cross-section of an assembly for the study of contactless interaction in a metal- reactant system: a - metallic wire, b - reactant, c - clearance.

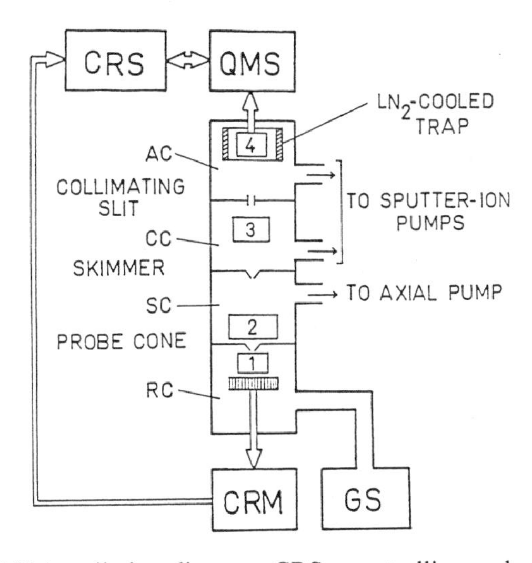

Figure 2a). DMS installation diagram: CRS - controlling and recording system, QMS - quadrupole mass-filter MS-7303, CRM - combustion rate measurement, GS - gas supply and exhaust system, AC - analyzer chamber, CC - collimator chamber, SC - skimmer chamber, RC - reaction chamber, 1-burning tablet, 2-adjustment system, 3-beam chopper, 4-QMS ion source.

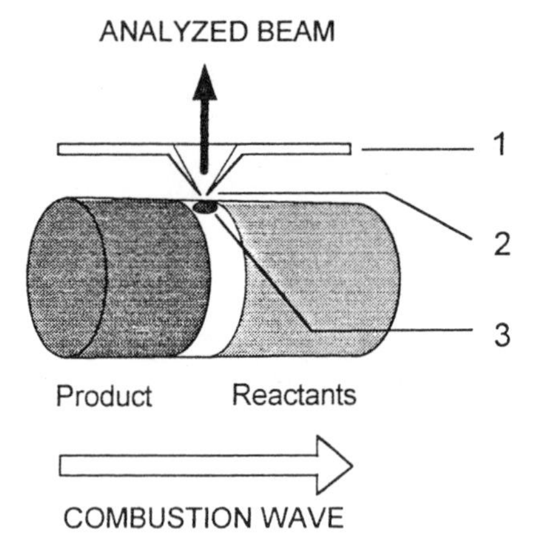

Figure 2b). Gas sampling from the surface of the burning tablet: 1-aluminum probe cone; 2-alumina cone tip, 30-70 μm; 3-sampling spot, 0.1-0.2 mm.

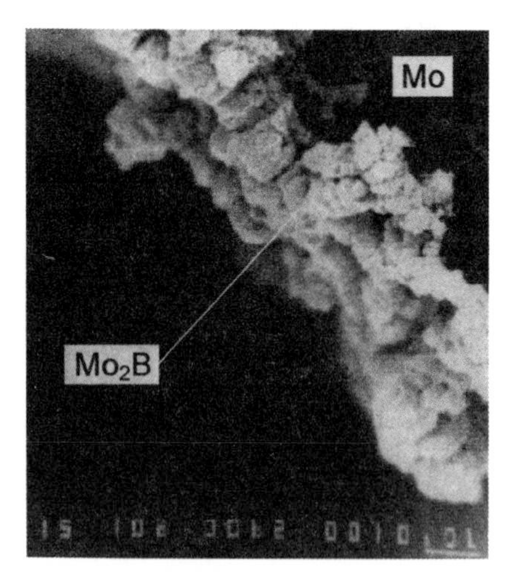

Figure 3. Mo_2B layer on molybdenum ($T = 1600$ K, $P = 20$ kPa (Ar), $t = 150$ s).

that $\lambda = u \cdot \Delta t$, where λ is the distance along the front propagation and Δt is the time interval from the beginning of the process $(\Delta t = t_i - t_0)$.

In order to obtain reliable data for the Ta-C system, some of its peculiarities need to be taken into account. Usually, the combustion of this system is accompanied by a very intensive gas release and strong flying apart of the sample particles. This resulted in uncontrollable clogging of the probe cone orifice. To avoid this, the samples were preheated in argon at 870 K for 20 min. The ion current of carbon monoxide was recorded for $^{12}C^{18}O^+$ with m/e = 30 in order to eliminate the influence of the background mass N_2 (m/e = 28). Hence, the natural isotopes ratio $^{18}O/^{16}O = 2\times10^{-3}$ (*21*) has to be taken into account in the evaluation of the concentration ratio for CO and CO_2 in the gas phase of the combustion front.

Results and Discussion

The study of the Mo-B and Ta-C systems without direct contact of reactants shows that formation of MoB and TaC is under way both in vacuum and in argon at high temperatures close to that for the combustion front (Figures 3-5). During prolonged runs (5 min and more) some sections of the molybdenum wire react almost completely, whereas others remain unreacted. MoB is formed only on the metal surface, and TaC only on the carbon surface. The gas-phase transfer is the only mode of reactants transport in these experiments. Moreover, one can suppose that the transport is unidirectional in the Mo-B system and that a two-way process occurs in the Ta-C system.

It should be noted that the possibility of the formation of an active gas phase during combustion of the Mo-B and Ta-C systems was pointed out earlier (*22,23*). However, the gas release in these systems was considered as a secondary process that "has nothing in common with the fundamentals of chemical interaction between the reactants" (*24*).

If gas-phase transport is assumed possible in the systems under study, the volatile oxides are the most probable carriers. The role of carbon oxides in graphite evaporation is well known, as well as the role of B_2O_3 vapors in boron gasification (see, e.g. (*25*)). The partial pressure over molybdenum and tantalum oxides at these temperatures is rather distinct and reaches, for example, 2.5 Pa for MoO_3 at 1800 K and 80 Pa for TaO_2 at 2700 K (*26*). Thus, the formation of oxide vapors in the combustion wave is not in doubt. At the same time, the problem of its composition and participation in the formation of the final product still needs to be solved.

Gas Phase in the System Mo-B. The temporal and spatial changes of ion currents for the gas-phase species in the combustion front of the system Mo-B are shown in Figure 6. It should be noted that only B^+, BO^+, BO_2^+, $B_2O_2^+$, and $B_2O_3^+$ ions are observed in the mass-spectrum for this system. The ion currents of molybdenum oxides are not detected even with the sensitive limits. It was shown (*27*) that the processes of dissociative ionization play a considerable role in B^+ and BO^+ formation in a mass-spectrometric study of the equilibrium vapor over the system B-B_2O_3, but the parameters of these processes remain unknown. Therefore, only the analyses of BO_2, B_2O_2, and B_2O_3 behaviors, which are the only precursors of corresponding ions, should be considered as correct.

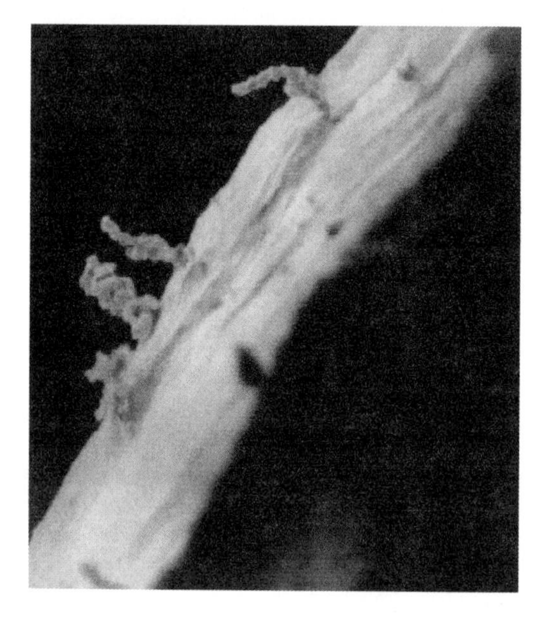

Figure 4. Ta_2C crystals on tantalum (T = 3200 K, P~10^{-2} Pa, t = 300 s).

Figure 5. TaC crystals on graphite (T = 3200 K, P~10^{-2} Pa, t = 300 s).

Before the indication of the combustion front, which moves with a velocity $u = 6$ mm/s to the probe ($t << t_0$), only the background masses are observed. On the distance about 1 mm ahead of the combustion front, H_2 and N_2 and/or CO (m/e = 28) are released from the sample. As far as the front moves, boron oxide lines appear. Further changes of these lines can be attributed to the well known spatial zones of the combustion wave.

The region t_0 - t_1 can be identified as a pre-heating zone. The temperature here increases within the range 900-1200 K (*19*) and, though it is considerably lower than the maximum T_f = 1800 K (*20*), it is already high enough for evaporation of boron oxides (for B_2O_3 T_m = 723±2 K (*21*)). The intensity of the $B_2O_2^+$ ion current within this region is significantly higher than that of $B_2O_3^+$. In Figure 7, the calculated (*26*) equilibrium ratios of oxide partial pressures over the system B(s)-B_2O_3 (l) show that B_2O_2 is the dominating component of equilibrium vapor in the range 1200-1800 K. A kinetic analysis of the $B_2O_2^+$ ion current dependence on time was made, neglecting the effects of the combustion wave moving a short distance of the peak half-width (<0.3 mm), which shows (*19*) that B_2O_2 forms in the first order heterogeneous endothermic reaction

$$B(s) + B_2O_3(g) \rightarrow B_2O_2(g) + 147\ kJ{\cdot}mol^{-1} \quad (1)$$

with a rate coefficient k_1 = 14 s^{-1} and activation energy E = 140±15 $kJ{\cdot}mol^{-1}$.

The region t_1 - t_2 is distinguished by a sharp decrease in the B_2O_2 partial pressure with the increased temperature of the combustion front, judging by the ion current of Zn^+, which increases to a maximum. This unexpected effect can occur only due to B_2O_2 consumption in the reaction for molybdenum boride formation:

$$B_2O_2(g) + Mo(s) \rightarrow Mo_xB(s) + \ldots \qquad x = 1,2 \quad (2)$$

This process proceeds with the effective rate coefficient $k_2 \sim 10^{-8}$ s^{-1} and activation energy $E \cong 0$ (*19*). The latter indicates that the rate of molybdenum borides formation is limited by either formation of B_2O_2(g) or its diffusion to the metal surface but not by reactive diffusion in the solid.

In the region t_2 - t_3, the ion currents of $B_2O_2^+$ and $B_2O_3^+$ increase again. The partial pressure of B_2O_3 becomes higher than that of B_2O_2, which corresponds to the equilibrium vapor over B_2O_3(l) (see Figure 7). In other words, there is no elemental boron in this region and evaporation of residual B_2O_3(l) takes place. So, the region t_1 - t_2 is the zone of chemical reaction and the region t_2 - t_3 is the post-combustion zone. The decrease of the boron oxide partial pressures at $t > t_3$ is due to the sample cooling.

Gas Phase in the System Ta-C. Only carbon oxides are detected in the gas phase of the combustion wave in the system Ta-C (u = 1.7 mm/s). The dynamics of the $^{12}C^{18}O^+$ and CO_2^+ ion current profiles in the combustion wave are shown in Figure 8. A thermodynamic evaluation (Figure 9) shows that, for the C(s)-O_2 system at 1500-3000 K, the partial pressure of CO is approximately 5-7 orders of magnitude higher than that for CO_2. Hence, one can assume that tantalum carbides are formed in the reaction

$$CO(g) + Ta(s) \rightarrow Ta_xC(s) + \ldots \qquad x = 1,2 \quad (3)$$

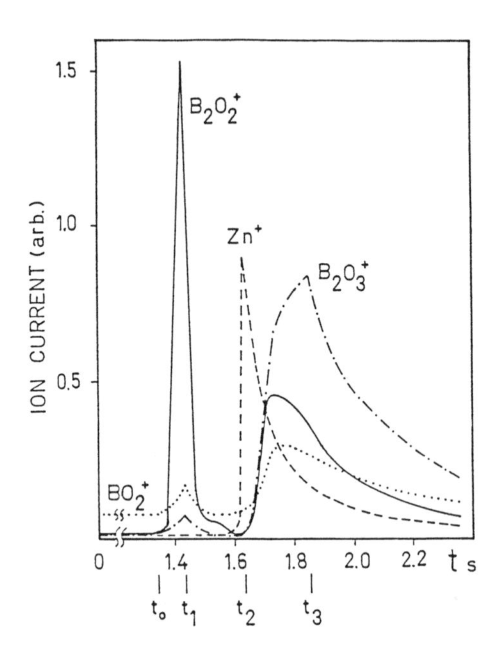

Figure 6. Boron oxides in the combustion wave of the Mo-B system ($u = 6.0$ mm/s).

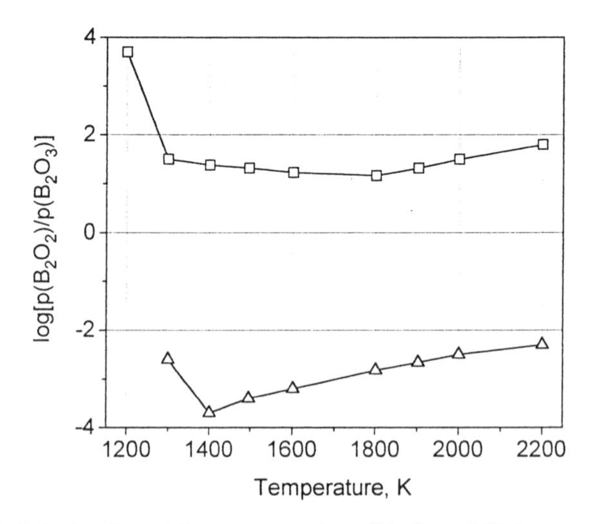

Figure 7. Calculated partial pressure ratios of B_2O_2 and B_2O_3 at equilibrium for the subsystem $B(s)$-$B_2O_3(l)$ (□) and over $B_2O_3(l)$ (Δ).

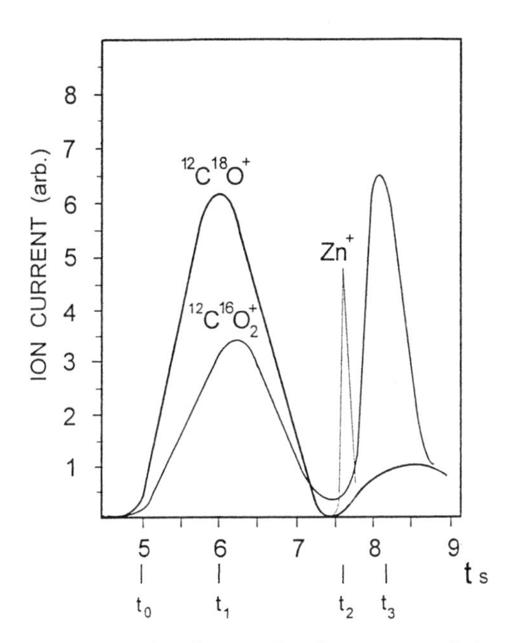

Figure 8. Carbon oxides in the combustion wave of the Ta-C system (u = 1.7 mm/s).

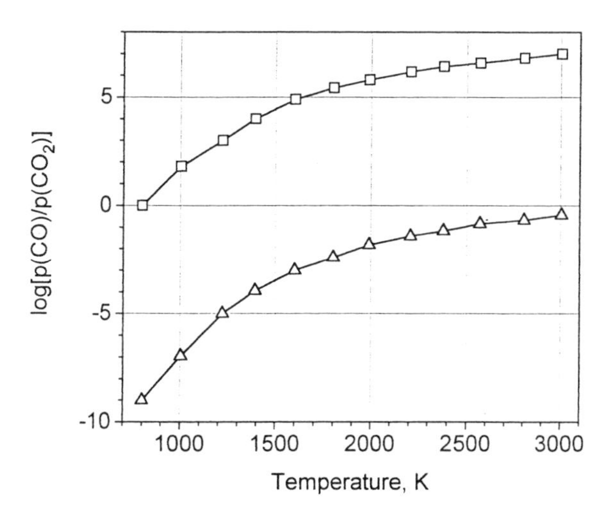

Figure 9. The calculated partial pressure ratios of CO and CO_2 in the subsystem C(s)-O_2 (□) and at the thermal dissociation of CO_2 (Δ).

When the data of Figure 8 are compared with those of Figure 6, it is apparent that the character of the concentration changes for the oxides in the combustion front is the same for both systems. For the Ta-C system, the region where the CO concentration falls almost to zero is also observed. This region corresponds to the maximum of temperature in the combustion front, as for the Mo-B system. At the same time, the combustion wave in the Ta-C system is much wider and the dimensions of the pre-heating (t_0 - t_1), reaction (t_1 - t_2) and post-combustion (t_2 - t_3) zones are different than those in the Mo-B system. It should be also noted that, in the Ta-C system, unlike the Mo-B system, the maximum of the ion current for the highest oxide (CO_2) is shifted in relation to maximum for the sub-oxide (CO) towards the zone of chemical reaction. This could be due to the higher concentration of CO in comparison with B_2O_2 (see Figures 7 and 9) in the pre-heating zone. The considerable increase in the width of the pre-heating (3 times) and reaction (2.5 times) zones, combined with the significant temperature gradients, makes it impossible to estimate kinetic parameters for the reactions in the combustion wave of the Ta-C system.

Width of the Reaction Zone. The dimensions of the zones in the combustion waves of the Mo-B and Ta-C systems defined from the plots in Figures 6 and 8 are presented in Table I. The values of the reaction zone width in the system Mo-B are in good agreement with the data obtained earlier (*20*).

Table I - Dimensions of Combustion Wave Zones (in mm)

System	*Wave width*	*Pre-heating zone*	*Reaction zone*	*Post-combustion zone*
Mo-B	3.1	0.6	1.2	1.3
Ta-C	5.3	1.7	2.8	0.8

A comparison of the calculated relationships between the partial pressures of the highest oxides (B_2O_3, CO_2) and suboxides (B_2O_2 , CO) in the presence and absence of elemental boron and carbon (Figures 7, 9) with the distribution of boron and carbon oxides along the combustion wave (Figures 6, 8) indicates that a zone of minor post-combustion probably exists for the Mo-B system, but post-combustion is entirely absent in Ta-C system. It is also obvious from this comparison that, by the time t_2, i.e., by the end of the reaction zone, the interactions in the systems are completed and boron in Mo-B system as well as carbon in Ta-C system are completely reacted. Thus, the model of wide zones (*14,15*) which assumes the interaction of the overwhelming bulk of the reactants far behind the front in a reactive diffusion mode, is not realized in the combustion of the systems under study.

Thermochemistry of the Combustion Wave. In order to clarify the mechanism of the gas-phase reactants transport in the Mo-B and Ta-C systems and to obtain the data for estimation of its efficiency, the calculations of condensed and gas phase equilibrium compositions were performed depending on the amount of transferred reactants. The adiabatic combustion temperatures (T_{ad}) and compositions of the products were calculated using the package of programs THERMO created by Dr. A. A. Shiryaev at the Institute of Structural Macrokinetics, RAS (*28*). The following assumptions were made in the model construction:

1) oxygen was considered as the third initial component of the systems and its amount is assumed to be corresponding to the admixed oxygen content;
2) the reacting system was considered as consisting of *metal-oxygen* and *reactant-oxygen* subsystems, the gas phases of which are spatially separated and occupy half of the free volume in the system;
3) it was assumed that reactants are transported between the subsystems only through diffusion of gases and the τ_D exceeds the time for establishment of any equilibrium;
4) the system was assumed to be isothermal at (T_{ad}).

All possible equilibria in the systems were considered. The simulations were made for a constant pressure equal to 101 kPa that corresponds to a system with open pores, and/or for the case of constant volume that corresponds to closed pores. Argon was chosen as the bath gas. The free volume of the initial system was calculated for an average particle size and a relative density of 0.6.

System Mo-B. Calculations show that at T_{ad} = 1800 K, the changes of the equilibrium composition of the Mo-B system with boron transport are the same for P = const and V = const. The transformation of condensed phases in the system is shown in Figure 10. The composition of the equilibrium gas phase (Figure 11) is uniquely determined by temperature according to the Gibbs phase rule. Using the results of Figure 10, the process of MoB(s) formation can be divided into four stages.

At stage 1 (up to 2 at.% of transported B), some insignificant changes of condensed phases occur mainly due to Mo_2O(s) disappearance and the beginning of B_2O_3(l) formation in the Mo-subsystem

At stage 2 (up to 5 at.% of B), formation of Mo_2B(s) starts in the molybdenum subsystem according to the reaction

$$Mo(s) + B_2O_2(g) \rightarrow Mo_2B(s) + B_2O_3(g) \quad (4)$$

and accumulation of B_2O_3(l) is accomplished. The composition of the gas phase indicated for stage 2 does not change until the end of stage 3. The removal of B_2O_3(l) from the surface is accomplished in the B-subsystem.

At stage 3 (up to 55 at.% of B), Mo(s) is completely transformed into Mo_2B(s) according to the reaction (4). The disappearance of the B_2O_3(l) in the B-subsystem in the previous stage leads to a decrease of the B_2O_3(g) partial pressure and opens the way to its transport from the Mo-subsystem and B_2O_2(g) regeneration by the reaction

$$B(s) + B_2O_3(g) \rightarrow B_2O_2(g) \quad (5)$$

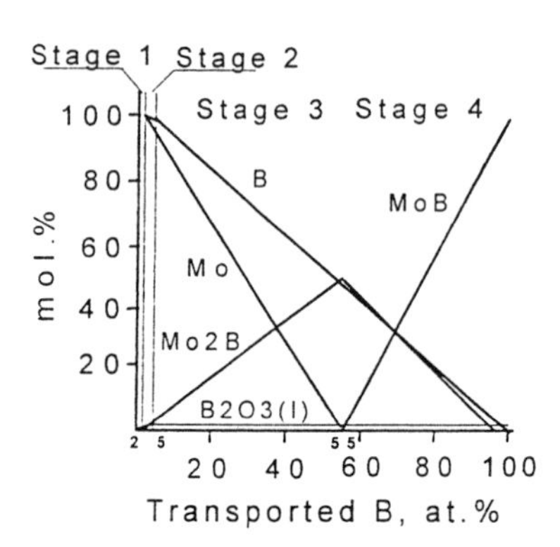

Figure 10. Calculated composition of the condensed phases of the Mo-B system as a dependence on the amount of transported boron at $P(V)$ = const.

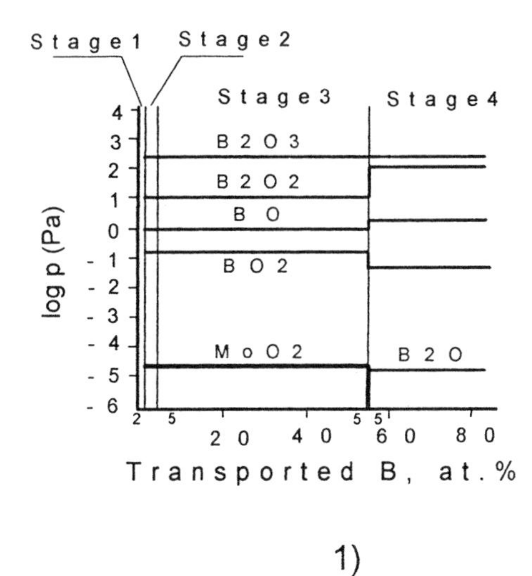

1)

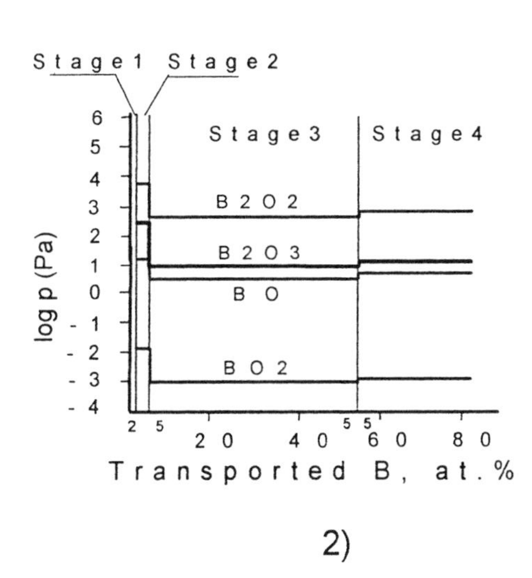

2)

Figure 11. Calculated composition of the gas phase in subsystems of the Mo-B system as a dependence on the amount of transported boron at $P(V)$ = const: 1) Mo-O subsystem, 2) B-O subsystem.

The partial pressures reached in the B-subsystem in the beginning of stage 3 stay practically constant until the end of the process.

At stage 4, the formation of MoB(s) is accomplished according to the reaction

$$Mo_2B(s) + B_2O_2(g) \rightarrow MoB(s) + B_2O_3(g) \quad (6)$$

One can see from the results of Figure 11 that the relationships between the partial pressures in the subsystems at stages 3 and 4 are consistent the diffusion flows of B_2O_2 from boron to molybdenum, with B_2O_3 going in the opposite direction. The key reactions which form the closed cycle of gas-phase transport in the Mo-B system are presented in Figure 12.

System Ta-C. The dependence of the process mechanism at $T_{ad} = 2700$ K on external conditions is a significant peculiarity of the Ta-C system. The changes of the condensed and gas phases compositions as a function of the transported carbon amount are shown in Figures 13 and 14, respectively, for P = const and V = const.

At stage 1, started after about 7×10^{-4} at.% of carbon as CO is transported from the C-subsystem, the oxidation of CO to CO_2 and the formation of $Ta_2C(s)$ and $Ta_2O_5(l)$ takes place on tantalum. The transfer of CO_2 as well as TaO and TaO_2 into the C-subsystem starts. In the systems with open pores (P = const) the filtration of excess CO occurs at this stage.

The main process at stage 2, started after the attainment of balance for CO and CO_2 counterflows, is $Ta_2C(s)$ formation. The composition of the gas phase is not changed therewith. In the C-subsystem, CO_2 is reduced to CO and TaC(s) is formed due to the tantalum oxides transfer. Stage 2 persists up to full consumption of Ta(s). Simultaneously, the growth of TaC(s) on carbon stops.

Subsequently, in the systems with closed pores (V = const), the CO and CO_2 partial pressures in the Ta- and C-subsystems increase sharply and the transition $Ta_2C(s) \rightarrow TaC(s)$ (stage 3) starts. In the systems with open pores (P = const), the growth of the carbon oxides pressure at the end of stage 2 is substantially lower because of gas expansion (filtration). In such systems, the beginning of stage 3 is preceded by removal of $Ta_2O_5(l)$ from the tantalum surface and completion of $Ta_2C(s)$ accumulation. This transition stage ends at P=const with the complete $Ta_2O_5(l)$ disappearance. As this takes place, the partial pressures of CO in the Ta- and C-subsystems, as well as CO_2 in the C-subsystem, remain constant up to the end of the process. The partial pressure of CO_2 in the Ta-subsystem falls to the previous level.

The closing stage 3 occurs mainly in the same manner at P = const and V = const. In both cases, the formation of TaC(s) from $Ta_2C(s)$ dominates at this stage. The lack of $Ta_2O_5(l)$ is a special feature of the closing stage at P = const. In a kinetic sense, this can mean that, in the systems with open pores, stage 3 proceeds faster than in the systems with the closed pores because of better accessibility of the $Ta_2C(s)$ surface. The combustion process in the Ta-C system, as well as in the Mo-B system, proceeds in the closed cycle mode, the key reactions of which are shown in Figure 15.

In conclusion, it has to be said that the abundance of admixed oxygen in realistic green mixtures of Mo+B and Ta+C is not essential to establish the equilibria forming the

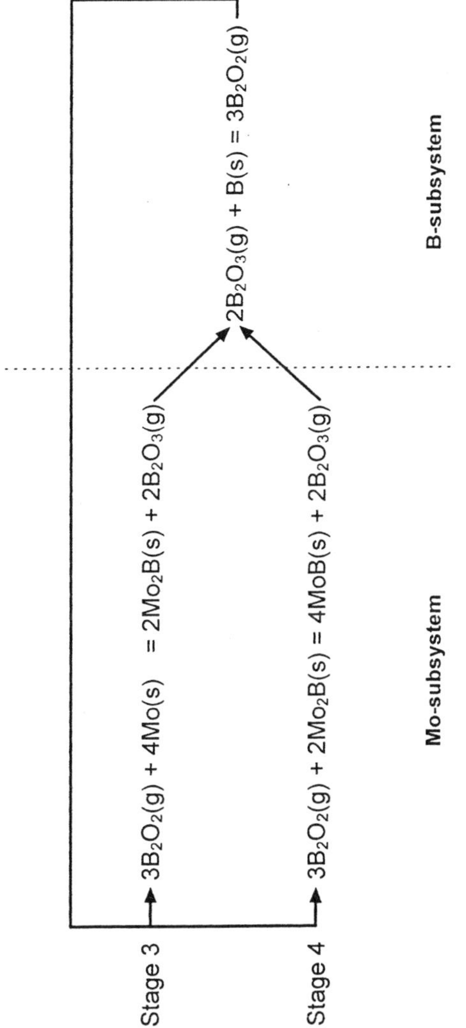

Figure 12. Reaction cycle of boron transport in the system Mo-B.

a)

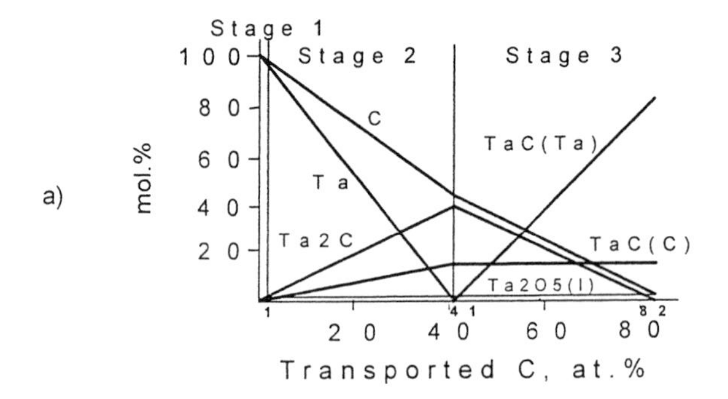

b)

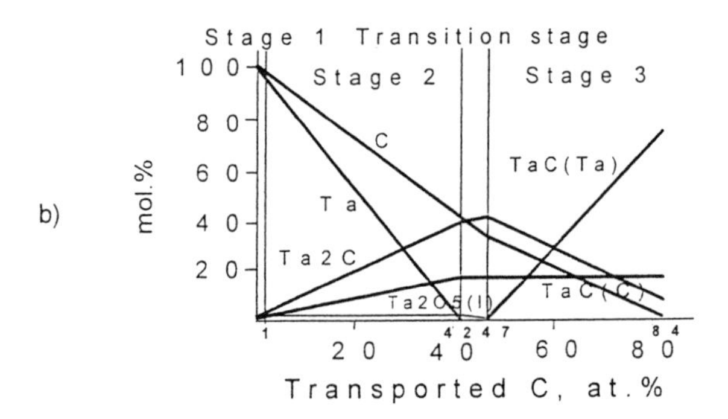

Figure 13. Calculated composition of the condensed phases of the Ta-C system as a dependence on the amount of transported carbon: a) $V = \text{const}$, b) $P = \text{const}$.

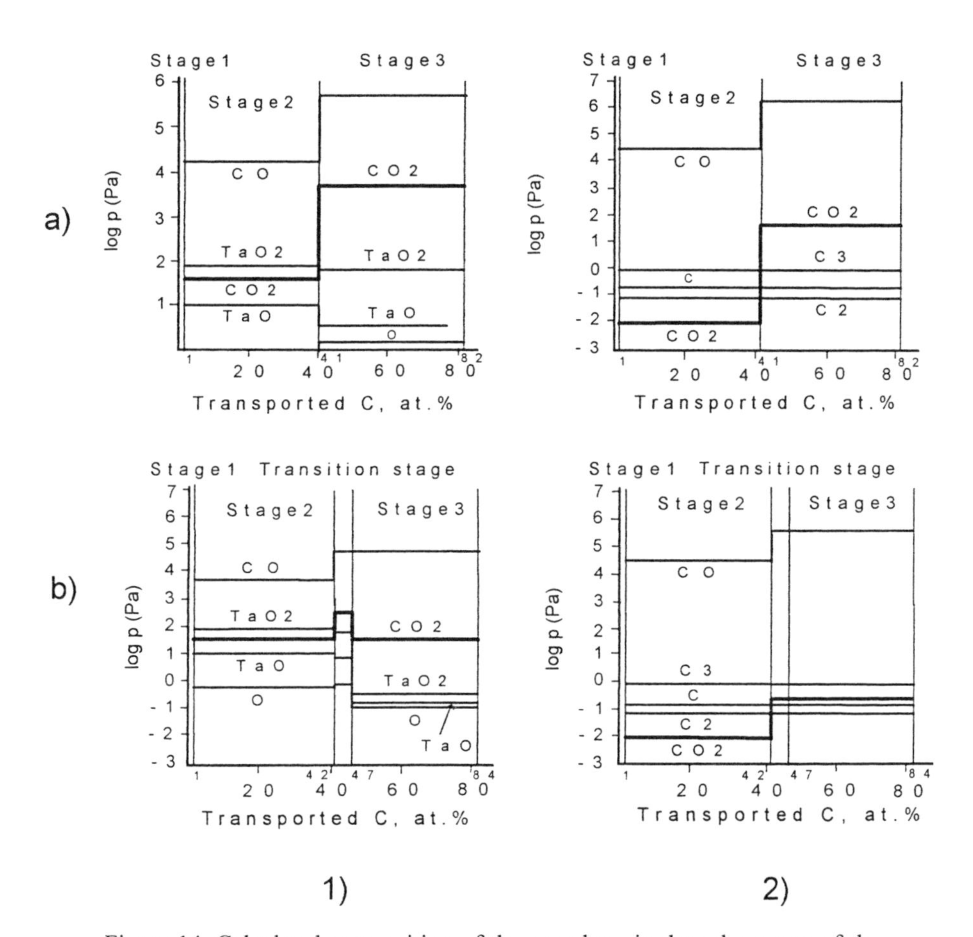

Figure 14. Calculated composition of the gas phase in the subsystems of the Ta-C system as a dependence on the amount of transported carbon: a) V = const, b) P = const; 1) Ta-O subsystem, 2) C-O subsystem.

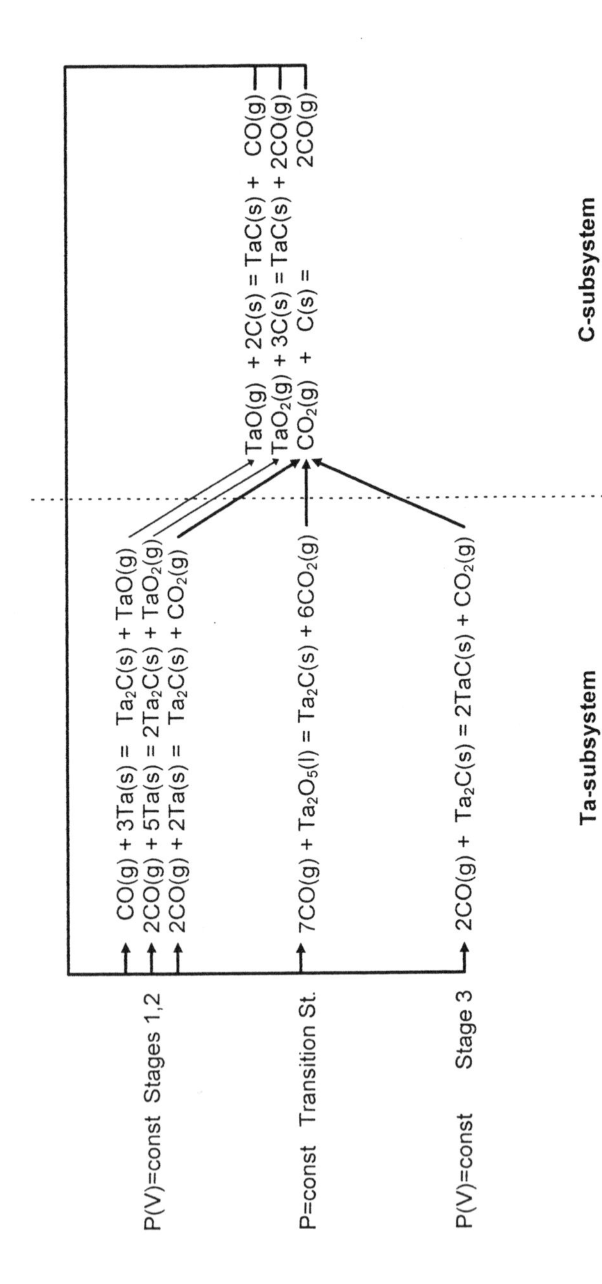

Figure 15. Reaction cycle of carbon transport in the system Ta-C.

closed reaction cycles with regeneration of gas carriers. A partial pressure of oxygen $< 1.4\times10^{-8}$ Pa at $T_{ad} = 1800$ K is optimum for Mo-B and < 0.61 Pa at $T_{ad} = 2700$ K for the Ta-C system (*29*). An excess of oxygen with respect to these limiting values shifts many equilibria in the Mo-B system and all equilibria in Ta-C system towards the oxidation of the final products. One can suppose that this effect occurs in other systems as well. In this case, the conversion and product purity for combustion synthesis has to depend on the degree of "self-purification" (*24*) of the admixed oxygen excess that realistic systems reach at the initial stages of the process. Therefore, an acceleration of combustion by oxygen containing additives (*4,30*) (if possible) is hardly justified with respect to the purity of the final product.

Efficiency of Gas-Phase Transport. The quantitative evaluations of gas-phase transport efficiency were made according to the model described elsewhere (*31,32*). The gas-phase transport of the reactants occurs mainly at stages 3 and 4 in the Mo-B system and at stages 2 and 3 in the Ta-C system, when counterflows of oxide carriers between the subsystems are balanced by oxygen. The values of the partial pressure gradients for the basic gas-phase carriers at these stages (see Figures 11 and 14) were used for estimation of the diffusion flows. It is obvious that these values are substantially lower than the gradients at the beginning of the process. Since the temperature profiles of the combustion waves are unknown for the systems under consideration, it was assumed that the temperature changes linearly on the front width L from the ignition temperature. This assumption also leads to underestimation of the conversion, as it is qualitatively known (*20*) that the temperature in any point of a combustion wave is higher than that corresponding to a linear law. Consequently, the results obtained must be taken as a lower limit of the gas-phase transport efficiency.

The estimations were made using the equation

$$\frac{d\eta}{dz} = \frac{LS_k}{Rulp} \cdot \sqrt{T} \cdot \sum n_i \beta_i c_i \cdot \exp(a_i - b_i/T) \qquad (7)$$

where η is the fractional conversion, z is the dimensionless coordinate along the combustion front width, S_k is the specific surface area of the transfer, R is the gas constant, l is the average distance of the transfer, ρ is the initial concentration of the reactant, n_i is the stoichiometric coefficient, β_i is the combination of temperature independent terms in the common equation for the diffusion coefficient (*33*), c_i is the carrier concentration in the subsystem where the transfer occurs, a_i and b_i are tabulated coefficients of the $p(T)$ equation.

The data obtained show that, in the Mo-B system, not less than 87% of the initial substances convert to the final product MoB(s) due to one-way gas-phase transport of boron (as B_2O_2) to molybdenum. As mentioned above, the assumptions made in the model construction cause the value of η to decrease. Hence, one can assume that gas-phase transport results in $\eta \cong 1$ for this system. It was found that $\eta \leq 1$ for the Ta-C system with two-way transport: carbon (as CO) onto tantalum and tantalum (as TaO and TaO_2) onto carbon. These results do not cast any doubt on the sufficiency of gas-phase transport and can mean that metal transport does not occur for the reasons which are not yet clear. Such an assumption fits with the absence of TaO and TaO_2 in the gas phase of the combustion wave in the Ta-C system (see Figure 8).

Conclusions

The study of the detailed chemical mechanisms of combustion synthesis in the systems considered as classic examples of gasless solid-state combustion or the so called "solid flame" shows that, in fact, these processes are neither gasless, nor solid-state.

It can be assumed that thermal ignition of *solid+solid* systems leads to formation of an active gas phase with sufficient partial pressures of the carriers. Oxides and suboxides usually dominate in the vapors due to admixed and ambient oxygen. At high temperatures, the fast diffusion of these species at small distances between fine particles combined with the high rate of heterogeneous reaction(s) and the high exothermicity of solid products formation results in the strong local auto-heating.

If the process is initiated by heating of the sample as a whole, the required partial pressure of gas carriers is reached simultaneously in the entire volume. A multitude of hot points appear and, in fact, the process occurs on the scale of one particle of the powder. Due to the low gradients of temperature and short reaction times for fine particles (~1-10 μm) the differences in the rates of heat transfer and diffusion have no time to manifest themselves and auto-accelerated fast reactions take place in the whole volume. The processes of this type are well known in combustion and are described by the theory of thermal explosions.

In the case of the point or surface heating of *solid+solid* systems, the gas carriers are formed only in the heating zone and the scale of the process changes. As a result of the marked temperature gradient and high heat conductivity of solid particles, the heat transfer proceeds faster than diffusion and regeneration of gas carriers. In that case, the zone of chemical reaction propagates in the frontal regime if known critical conditions are met. As this takes place, the local characteristics are averaged over the scale of front and the process is described by the thermal theory of combustion. The pulsating combustion of some *solid+solid* systems can be also explained in the kinetic terms of gas-phase carriers formation and consumption.

Unfortunately, in the few decades which have elapsed after publication of the first results, approaches to control the combustion synthesis in the condensed systems have not been found. There are reasons to think that the true preparative and technological potentialities of these processes still largely remain *terra incognita* because of that. The gas-phase transport concept formulated in the present work is being expanded to other *solid+solid* systems. At the same time, the role of the gas phase in combustion of other systems, for example *liquid+solid*, is far from evident. One would like to hope that the study of the real chemistry of these interesting and important high-temperature processes leads to effective control of the combustion synthesis.

Acknowledgments

It is necessary to note that this work would be impossible without the active participation of Dr. I. A. Yuranov and Dr. A. A. Fomin who are the co-authors of all publications on this problem. The experiments on contactless interaction in the Ta-C system were made by Dr. S. G. Vadchenko. The software for DMS was developed with the participation of Dr. V. G. Abramov. The technical service was provided by Mr. A. V. Zharenov and Mr. N. A.Somsikov. The translation from Russian was made by Mrs. A. G. Moukasyan. The author considers it as pleasant duty to express them, his friends and colleagues, very sincere and informal gratitude.

In 1995-1997, the work was supported by the Russian Foundation for Basic Research under the grant No. 095-03-09187.

References

1. Merzhanov, A. G.; Borovinskaya, I. P. *Repts USSR Acad. Sci.* **1972**, *204*, 336.
2. Hlavacek, V. *Ceram. Bull.* **1991**, *70*, 240.
3. Merzhanov, A. G. *Ceramics Engineering and Science Proceedings;* AmCerSoc Publ., 1995; Vol.16.
4. Merzhanov, A. G. *Comb. Sci. Techn.* **1994**, *98*, 307.
5. Brezinsky, K. *26th Symposium (International) on Combustion;* The Combustion Institute: Pittsburgh, 1997 (in press).
6. Calcote, H. F.; Felder, W.; Keil, D. G.; Olson, D. B. *23rd Symposium International) on Combustion;* The Combustion Institute: Pittsburgh, 1990; pp.1739-1744.
7. Gerhold, B. W.; Inkrott, K. E. *Combust. Flame,* **1995**, *100*, 144.
8. Glassman, I.; Davis, K. A.; Brezinsky, K. *24th Symposium (International) on Combustion;* The Combustion Institute: Pittsburgh, 1992; pp.1877-1882.
9. Pityulin, A. N.; Scherbakov, V. A.; Borovinskaya, I. P.; Merzhanov, A. G. *Combust., Explos., Shock Waves,* **1979**, *15*, 432.
10. Mukasyan, A. S.; Vadchenko, S. G.; Khomenko, I. *Combust. Flame* (in press).
11. Shkiro, V. M.; Borovinskaya, I.P. *Combust., Explos., Shock Waves,* **1976,** *12*, 945.
12. Merzhanov, A. G. *Proc. IV Symposium on Chemical Problems Connected with the Stability of Explosives;* Molle, Sweden, May 31-June 2, 1976; pp.381-401.
13. Aldushin, A. P.; Khaikin, B. I. *Combust., Explos., Shock Waves,* **1974**, *15*, 273.
14. Merzhanov, A. G.; Khaikin, B. I. *Prog. Energy Combust. Sci.* **1988**, *14*, 1.
15. Merzhanov, A. G., *Combustion and Plasma Synthesis of High-Temperature Materials;* VCH: New York, 1990; pp.1-53.
16. Merzhanov, A. G. *Arch. Procesow Spalania,* **1974**, *5*, 17.
17. Maksimov, E. I.; Merzhanov, A. G.; Shkiro, V. M. *Combust., Explos., Shock Waves,* **1965**, *1*, 15.
18. Merzhanov, A. G.; Borovinskaya, I. P.; Shkiro, V.M. *Vestnik AN SSSR* **1984**, *10*, 141 (in Russian).
19. Kashireninov, O. E.; Yuranov, I. A., *25th Symposium (International) on Combustion*; The Combustion Institute: Pittsburgh, 1994; pp.1669-1675.
20. Zenin, A. A.; Merzhanov, A. G.; Nersisyan, G.A. *Combust., Explos., Shock Waves,* **1981**, *17*, 63.
21. *Handbook of Chemistry and Physics;* 73rd Edition, Linde, D. R., Ed.; CRC Press: Ann Arbor, 1992-1993.
22. Bloshenko, V. N.; Bokii, V. A.; Merzhanov, A.G. *Combust., Explos., Shock Waves,* **1988**, *24*, 218.

23. Merzhanov, A. G.; Rogachev, A. S.; Mukasyan, A. S.; Khusid, B. M.; Borovinskaya, I.P.; Khina, B. B. *J. Eng. Phys.* **1990**, *59*, 809.
24. Merzhanov, A. G. *Combust. Sci. Techn.* **1995**, *105*, 295.
25. Zvuloni, R.; Gomez, A.; Rosner, D. E. *J. Propulsion and Power,* **1991**, *7*, 9.
26. *Thermodynamic Properties of Individual Substances. Tables;* Glushko, V. P.; Gurvich, L. V. Eds.; Science: Moscow, 1978-1982, Vol. 1-4.
27. Inghram, M. G.; Porter, F.; Chupka, W. A. *J. Chem. Phys.* **1956**, *25*, 498.
28. Shiryaev, A. A. *Intern. J. SHS* **1995**, *4*, 351.
29. Kashireninov, O. E.; Yuranov, I. A.; Fomin, A. A. *High Temp. Sci.* **1991**, *32*, 79.
30. Novikov, N. P.; Borovinskaya, I. P.; Boldyrev, V. V. *Combust., Explos., Shock Waves,* **1977**, *13*, 280.
31. Yuranov, I. A.; Fomin, A. A.; Shiryaev, A. A.; Kashireninov, O. E. *J. Mater. Synth. and Proc.* **1994**, *2*, 239.
32. Fomin, A. A.; Yuranov, I. A.; Shiryaev, A. A.; Kashireninov, O. E. *Chem. Phys. Reports,* **1996**, *15(7)*, 1091.
33. Ried, R. C.; Sherwood, T. K. *The Properties of Gases and Liquids;* 2nd Ed.; McGraw-Hill: New York, 1966.

Chapter 18

Comparison of the Syntheses of Vanadium, Niobium, and Molybdenum Carbides and Nitrides by Temperature-Programmed Reaction

R. Kapoor and S. T. Oyama[1]

Department of Chemical Engineering, Virginia Polytechnic Institute and State University, Blacksburg, VA 24061–0211

The carbides and nitrides of the early transition metals, vanadium, niobium, and molybdenum, are known to possess good catalytic properties. The compounds are synthesized by a temperature programmed reaction (TPR) method where a reactive gas is reacted with a precursor oxide as the temperature is uniformly increased. Results under similar reaction conditions are presented to compare the progress of the reaction, the formation of intermediate phases, and the development of surface areas. The increase in surface area is influenced by the phenomena of pseudomorphism and topotaxy. It is believed that pseudomorphism, found in all of the above syntheses, is associated with the development of internal pores, while topotaxy, found in some of the nitrides, maximizes this process to yield high surface area products.

Studies in the past few decades have shown that carbides and nitrides of early transition metals, such as V, Ti, Mo, W, Nb, etc., possess useful electrical, magnetic, mechanical, and optical properties (*1*). Since then, newer applications other than their traditional use in structural materials and cutting tools have been investigated (*2*). One of these applications has been catalysis where numerous studies have shown activity resembling those exhibited by noble group elements (*3*). The catalytic activity of these compounds is believed to be due to modification of the crystalline and electronic structure of the parent metal by the presence of the non-metal component (*2,4,5*).

Catalytic materials need high exposures or specific surface areas to be used economically. This has led to development of novel preparative methods, such as temperature programmed reaction (TPR) (*6*), laser pyrolysis (*7*), etc., to produce finely divided transition metal carbides and nitrides. Among these, TPR remains a practical method for making large batches with moderate to high surface area products.

The TPR process involves: *(i)* a suitable solid precursor, such as an oxide of the transition metal, *(ii)* a gaseous carburizing or nitriding medium, such as a hydrocarbon or ammonia, flowing over the precursor, and *(iii)* a heating program for raising the

[1]Corresponding author

temperature of the sample to levels where reaction occurs. This last component gives rise to the name TPR.

In these syntheses the surface area increases with the progress of the reaction *via* development of pores as the density of the precursor increases without much change in exterior volume. This phenomenon, where the overall size and shape of a material remains constant over the course of a solid state transformation, is called pseudomorphism (*8*). Another related phenomena that affects the development of surface area is topotaxy. Here the transformation proceeds in a manner such that the crystallographic planes of the final product bear a relationship with the planes of starting material throughout the bulk (*8*).

This chapter presents a review of the synthesis of vanadium, niobium, and molybdenum carbides and nitrides by the TPR method. Factors responsible for development of surface area during transformation are discussed.

Experimental

A typical preparation of V, Nb, and Mo carbide or nitride by the TPR method consists of loading 0.4 g of their respective higher oxides, V_2O_5, Nb_2O_5, and MoO_3, in a microreactor and passing 0.68 μmol s^{-1} (1000 cm^3 min^{-1}) of a carburizing mixture of 20 % CH_4/H_2 or a nitriding mixture of 20 % NH_3/He over the samples while heating the reactor contents at a uniform heating rate. In this study a constant heating rate of 0.083 K s^{-1} is used for all three syntheses. In the case of Mo_2N and NbC (*9*) the samples are held isothermal at final temperatures of 1000 K for 0.5 h, and 1373 K for 0.7 h, respectively. The system pressure is approximately one bar.

A schematic of the synthesis unit is shown in Figure 1. The reactor is placed in a furnace controlled by a programmable temperature controller (Omega CN 9000) with its own dedicated thermocouple (chromel-alumel) placed at the furnace wall. Another thermocouple is placed in a well extending into the reactor bed to measure sample temperature. The effluent gas is analyzed by a mass spectrometer (Ametek MA 100, quadrupole gas analyzer) through a variable leak valve (Granville Phillips, model 203). The signals from the mass spectrometer and the reactor bed thermometer (Omega, Model 650) are monitored in real time by computer through an interface. The trace of the constituent components and their fragments in the effluent allows study of the progress of reaction.

In nitride synthesis, the ammonia reduces the oxidic precursor to sub-oxides or oxynitrides before final nitridation. The process is manifested by features appearing on the MS traces of H_2, N_2 , and H_2O. No features are observed for the NH_3 signal, tracked as mass 15, due to the high signal level of the parent mass 17.

In carbide synthesis, the 20 % CH_4/H_2 mixture similarly reduces and carburizes the oxides, and the H_2O, CO and CO_2 signals provide the information on the progress of the reaction. Again, no features are observed in the CH_4 and H_2 signals due to their presence in large excess. In all cases the reaction intermediates are quenched at intermediate points to identify the solid phases by X-ray diffraction (XRD).

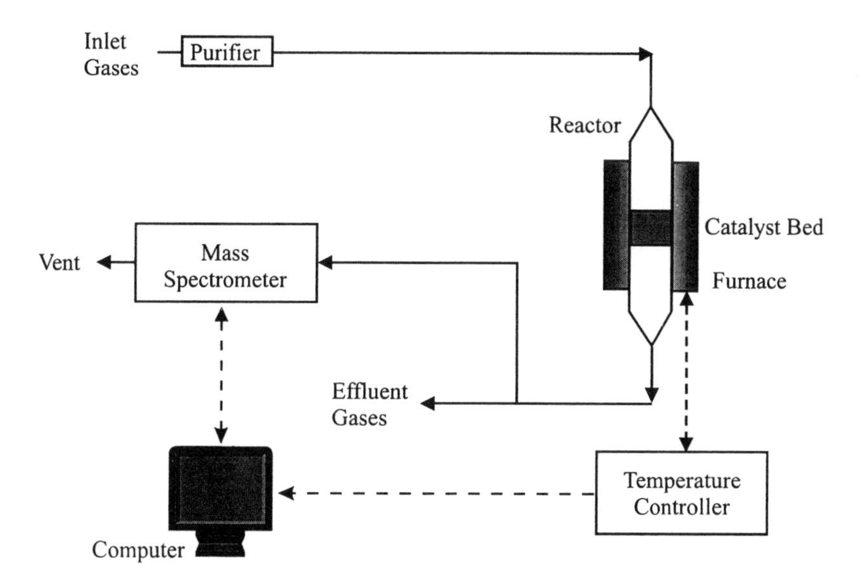

Fig. 1: Schematic of the synthesis unit.

Results and Discussion

In the synthesis of carbides or nitrides, the reduction and the associated carburization or nitridation follow the same general pattern. As expected, different metal oxides form different solid intermediate phases at different transformation temperatures.

In this section the salient features of carbide and nitride synthesis will be presented first. The role of ammonia and methane/hydrogen as reductant in the development of surface area will be discussed next, with references to the significance of pseudomorphic and topotactic phenomena in these transformations.

Nitrides. Figure 2 shows H_2, H_2O, and N_2 mass spectroscopic signals as a function of temperature during nitridation of V_2O_5, Nb_2O_5, and MoO_3. In all cases at higher temperature the H_2 and the N_2 background signals rise due to ammonia decomposition. These occur at 770 K for VN and Mo_2N, and 960 K for NbN.

The H_2O peak is the most revealing in all the three cases, tracking the reductive processes and the temperatures at which they take place during the reaction. In the case of VN four distinct peaks are observed at 625, 666, 742, and 864 K. The first three are reported to correspond to the following sequence $V_2O_5 \rightarrow V_6O_{13} \rightarrow V_2O_4/VO_2 \rightarrow V_2O_3$, while the last one to $V_2O_3 \rightarrow VO \rightarrow VN$ (*10*). The N_2 features for VN show the same three peaks before extensive ammonia decomposition takes place, and is due to the release of the excess nitrogen from ammonia during hydrogen consumption in the reductive process. Conversely, the peak in H_2 at 880 K is due to the release of excess hydrogen during the final nitridation step $VO \rightarrow VN$.

Of the three metal nitrides, Mo_2N synthesis has been studied the most extensively (*11-18*). The reaction is reported to occur *via* two parallel reactions: $MoO_3 \rightarrow MoN_xO_{1-x} \rightarrow Mo_2N$ and $MoO_3 \rightarrow MoO_2 \rightarrow Mo_2N$, with the latter occuring when the presence of water is significant (*18*). Prolonged exposures to ammonia at elevated temperatures can lead to the irreversible reduction of nitride to Mo-metal. In our synthesis, this is avoided by nitriding isothermally at 1000 K for 0.5 h, enough time to just allow the water signal to return to baseline. The H_2O trace shows two distinct peaks at 760 and 1000 K with an exotherm at 774 K. The exotherm is due to the high heat of reaction and the use of a relatively high heating rate.

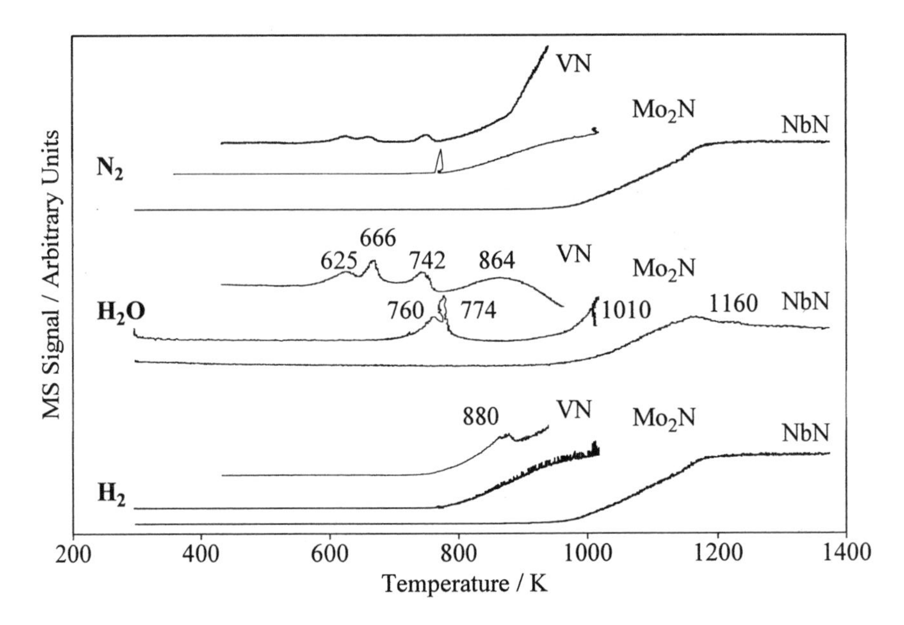

Fig. 2: Mass spectrometer trace of H_2, H_2O, and N_2 during TPR synthesis of nitrides.

The NbN synthesis occurs at relatively high temperature with no intermediate features (*19*). The H_2O peak at 1160 K represents a single step transformation of $Nb_2O_5 \rightarrow NbN$ *via* an NbN_xO_{1-x} intermediate (*20*). No NbO_2 phase has been detected. This is further confirmed by the lack of reaction between ammonia and NbO_2 in a separate experiment (*19,20*).

The rate of reduction to sub-oxides and final nitridation is known to be limited by bulk diffusion of oxygen (*14*). At the surface, the ammonia decomposition reaction is fast, irreversible, and non-limiting. In case of VO $\rightarrow$ VN the activation energy, E_a, progressively increases from 203 to 240 kJ mol^{-1} during the transformation (*10*). This range is same as that found for diffusion of oxygen in oxides (*14*). The progressive increase in E_a is attributed to an increase in strength of metal-oxygen bonds due to increase in bond order with reduction. The oxygen diffusivities, calculated by modeling the TPR process as an activated diffusion process, also fall in the range expected for oxygen in oxides (*21*).

Carbides. Figure 3. shows the MS traces of H_2O and CO for carbides. As a general trend, the synthesis process involves reduction of the starting oxide to suboxides at low temperatures by H_2 (strong H_2O peaks) and carburization at high temperatures by CH_4 (strong CO peak).

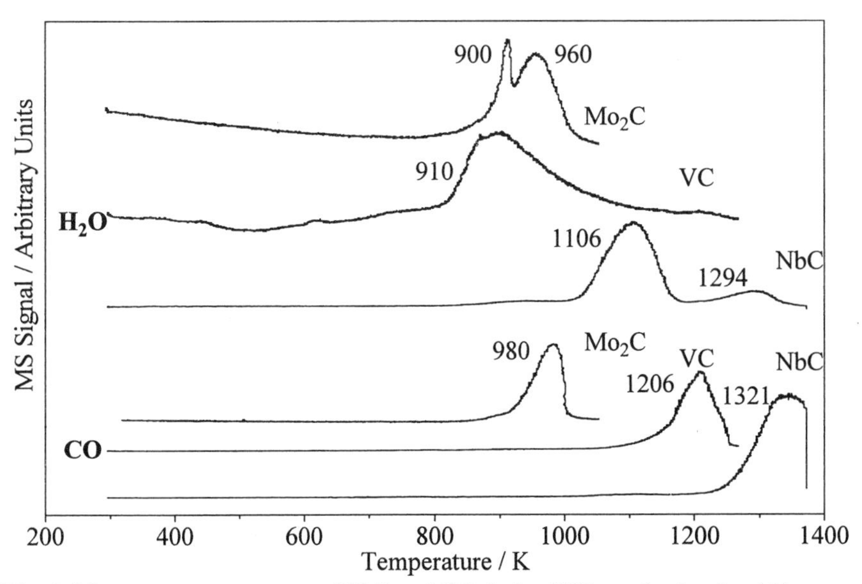

Fig. 3: Mass spectrometer trace of H_2O and CO during TPR synthesis of carbides.

The VC synthesis involves a two step process, $V_2O_5 \rightarrow V_2O_3$ indicated by an H_2O peak at 900 K, and $V_2O_3 \rightarrow$ VC indicated by a CO peak at 1206 K (*22*). The role of H_2 in the CH_4/H_2 mixture in the first reduction step has been established by reducing V_2O_5 by H_2 under the same reaction conditions (*22*). The TPR (not shown) in this case establishes that the reduction to V_2O_3 occurs at the same temperature.

The Mo_2C synthesis is reported (*23*) to proceed as $MoO_3 \rightarrow MoO_2 \rightarrow Mo_2C$ with the first step brought about by H_2 and the second by CH_4. Fig. 3 shows two H_2O peaks at 910 and 960 K corresponding to the two reductive stages, and a single CO peak at 980 indicative of the carburization process.

The NbC synthesis occurs at high temperatures as $Nb_2O_5 \rightarrow NbO_2 \rightarrow$ NbC (*9*). The last stage is believed to proceed *via* an oxycarbide phase, NbO_xC_y (*9*). Fig. 3 shows a water peak at 1106 K signaling the first stage, and a broad CO peak at 1321 K for the last stage. This last peak falls in the isothermal regime beyond 1373 K.

Development of Surface Area

The syntheses of VC and VN illustrate the typical differences in reaction of oxides with ammonia and methane-hydrogen mixtures. Because the ammonia decomposition reaction occurs readily on these surfaces, and ammonia is a strong reducing agent both kinetically and thermodynamically (*24*) it reacts at lower temperatures than hydrogen or methane, and starts the reductive process early. This early reduction is moderate and the vanadium pentoxide undergoes transformation to numerous sub-oxides before final nitridation. In sharp contrast, hydrogen reacts at higher temperatures and produces very few sub-oxides. For example, the transformation $V_2O_5 \rightarrow V_2O_3$ is a three step reduction process under ammonia starting at 600 K, but a single step process under methane-hydrogen beginning at 800 K. The surface area developed during $V_2O_5 \rightarrow V_2O_3$ under ammonia and methane-hydrogen is about 40 and 10 m^2g^{-1}, respectively (*10,22*). Considering that this difference of 30 m^2g^{-1} is the same as the final difference in surface area of the products VN (90 m^2g^{-1}) and VC (60 m^2g^{-1}), it may be argued that the early reduction by ammonia is responsible for the generation of excess surface area.

This observation may be related to topotactic transformations that commonly occur in nitride synthesis, but are absent in carbide synthesis. For example, in VN synthesis, when a sample of V_2O_5 crystals with preferential orientation along [001] planes is nitrided, the product conserves the oriented nature, with a preferential orientation along [200] planes (*8*). However, when the same V_2O_5 crystals are carburized, the preferential orientation is lost, and the powder x-ray diffraction shows a randomly oriented sample (*22*).

These topotactic reactions are believed to be structurally-constrained reactions. In case of transition metal oxides the diffusing anions leave anion vacancies that may coalesce along certain crystallographic directions to form plane defects. These plane defects may then collapse to compress the structure and form shear planes that can interact and reorder at high temperatures to an oriented crystal (*25*).

The topotactic reactions are non-equilibrated and irreversible processes and can produce routes to the synthesis of metastable compounds (*8*). Thus in this case, the reducing strength of ammonia provides access to suboxide phases that maximizes development of pores, and hence, surface area. In the case of reduction by hydrogen, lack of reaction at temperatures where the sub-oxides are thermodynamically stable, forces bypassing of the extra stages of reduction. This probably causes inefficient development of pores.

The development of surface area is also influenced by pseudomorphism, a phenomenon where the exterior morphology and dimensions do not vary appreciably despite a density change during transformation (*8*). The difference in volume manifests itself as internal cracks and pores. Scanning electron microscopy (SEM) studies on the exterior morphology of transition metal oxides and its carbides and nitrides establish the transformations to be pseudomorphic. It is to be noted that while the structurally constrained nature of topotaxy favors pseudomorphism, the former is not a necessary requirement for the latter. Indeed, carbide formation is pseudomorphic but not topotactic. Together, it may be postulated that while pseudomorphism is responsible for increases in surface area during carburization and nitridation of transition metal oxides, topotaxy maximizes the surface area development whenever it occurs.

The same is observed for Mo_2N and Mo_2C synthesis from MoO_3 (*16,26*), where the former is produced with a higher surface area. In fact, an interesting route for making high surface area molybdenum carbide is by carburizing a nitride. This reaction is topotactic and involves substitution of nitrogen in the lattice with carbon (*26*).

Determination of topotaxy in NbN synthesis has been limited by lack of availability of preferentially oriented crystals. However, its high synthesis temperature suggests a non-topotactic reaction. The NbC is expected to follow the trend of other carbides of not being topotactic. Under similar TPR conditions, both NbC and NbN are produced with similar surface areas (~ 21 m^2g^{-1}).

Table I: Summary of intermediate phases, temperatures and surface areas in carbide and nitride syntheses

Transformation with Peak Temperatures	Final Temperature, K	Surface Area m^2g^{-1} (*Ref.*)
Carbides		
$MoO_3 \rightarrow (MoO_2, MoO_xC_y) \rightarrow Mo_2C$ 910 960	1050	90 (*23*)
$V_2O_5 \rightarrow V_2O_3 \rightarrow VC$ 900 1206	1265	60 (*22*)
$Nb_2O_5 \rightarrow NbO_2 \rightarrow NbO_xC_y \rightarrow NbC$ 1106 1321 1373*	1373*	21 (*9*)
Nitrides		
$MoO_3 \rightarrow MoO_2, MoO_xN_y \rightarrow Mo_2N$ 760 973*	973*	116 (*11*)
$V_2O_5 \rightarrow V_6O_{13} \rightarrow V_2O_4/VO_2 \rightarrow V_2O_3 \rightarrow VO \rightarrow VN$ 625 666 742 864 880	940	90 (*10*)
$Nb_2O_5 \rightarrow NbO_xN_y \rightarrow NbN$ 1160 1160*	1160*	23 (*19*)

*These samples were held isothermally at the peak temperature for 0.5 - 0.7 h.

Conclusions

The temperature programmed reaction synthesis of carbides and nitrides of V, Mo, and Nb, is presented under same reaction conditions and allows direct comparison of the temperatures at which various intermediates and products are formed. In general, the nitrides are formed at lower temperatures than the carbides for the same metal, because of the greater reducing-nitriding strength of ammonia over hydrogen-methane mixtures. The transformations of oxides to carbides or nitrides under TPR synthesis conditions are pseudomorphic, and in some cases of nitride synthesis are topotactic.

Pseudomorphism is responsible for the substantial increase in surface area of the starting material, while topotacticity is believed to maximize this process.

Acknowledgments

The authors thank Dr. V. L. S. Teixeira da Silva and Ms. V. Schwartz for providing the data on NbC and NbN synthesis, respectively, and financial support from the Exxon Education Foundation and the U.S. Department of Energy under the Advanced Coal Research at U.S. Universities Program, Grant # DE-P522-95PC95207.

Literature Cited

1. Toth, L. E. , *Transition Metal Carbides and Nitrides;* Academic Press: New York, 1971.
2. *The Chemistry of Transition Metal Carbides and Nitrides;* Oyama, S. T., Ed.; Blackie Academic & Professional: London, 1996.
3. Levy, R.; Boudart, M. *Science* **1973**, *181*, 547.
4. Oyama, S. T.; Haller, G. L. In *Catalysis, Specialist Periodical Reports;* Bond, G. C.; Webb, G., Eds.; The Chemical Society, London, 1981; Vol. 5; pp 333.
5. Oyama, S. T. *Catalysis Today* **1992**, *15*, 179.
6. Oyama, S. T., Ph.D. Dissertation, Stanford University, 1981.
7. Ochoa, R.; Bi, X. X.; Rao, A. M.; Eklund, P. C. In *The Chemistry of Transition Metal Carbides and Nitrides;* Oyama, S. T., Ed.; Blackie Academic & Professional, London, 1996.
8. Volpe, L.; Boudart, M. *Catal. Rev. -Sci. Eng.* **1985**, *27*, 515.
9. Teixeira da Silva, V. L. S.; Schmal, M; Oyama S. T. *J. Solid State Chem.* **1996**, *123*, 169.
10. Kapoor, R.; Oyama S. T. *J. Soid State Chem.* **1992**, *99*, 30.
11. Choi, J-G.; Curl, R. L.; Thompson, L. T. *J. Catal.* **1994**, *146*, 218.
12. Wise, R. S.; Markel, E. J. *J. Catal.* **1994**, *145*, 335.
13. Wise, R. S.; Markel, E. J. *J. Catal.* **1994**, *145*, 344.
14. Schlatter, J. C.; Oyama, S. T.; Metcalf, J. E., III; Lambert, J. M., Jr., *Ind. Eng. Chem. Res.* **1988**, *27*, 1648.
15. Ranhotra, G. S.; Bell, A. T.; Reimer, J. A. *J. Catal.* **1987**, *108*, 40.
16. Volpe, L.; Boudart, M. *J. Solid State Chem.* **1985,** *59,* 332.
17. Clark, P. A.; M.S. Thesis, Ohio State University, 1995.
18. Jaggers, C. H.; Michaels, J. N.; Stacy, A. M. *Chem. Mater.* **1990**, *2*, 150.
19. Schwartz, V.; Oyama, S. T., in preparation.
20. Kim, H. S.; Shin, C. H.; Bugli, G.; Bureau-Tardy, M.; Djéga-Mariadassou, G. *Appl. Catal. A: General* **1994**, *119*, 223.
21. Kapoor, R; Ph.D. Dissertation, Clarkson University, 1994.
22. Kapoor R.; Oyama, S. T. *J. Solid State Chem.* **1995**, *120*, 320.
23. Lee, J. S.; Oyama, S. T.; Boudart, M. *J. Catal.* **1987**, *106*, 125.
24. Oyama, S. T. *J. Catal* **1992**, *133*, 35.
25. Bertrand, O.; Dufour, L. C. *Phys. Stat. Sol. A* **1980**, *60*, 507.
26. Volpe L.; Boudart, M. *J. Solid State Chem.***1985**, *59*, 348.

Polymerization Synthesis and Characterization

Chapter 19

Frontal Polymerization: Self-Propagating High-Temperature Synthesis of Polymeric Materials

John A. Pojman, Dionne Fortenberry, Akhtar Khan, and Victor Ilyashenko

Department of Chemistry and Biochemistry, University of Southern Mississippi, Hattiesburg, MS 39406

Frontal polymerization is a mode of converting monomer into polymer via a localized reaction zone that propagates. Such fronts can exist with free-radical polymerization or epoxy curing. The conditions for the existence of the free-radical frontal polymerization regime are considered. The factors affecting velocity, conversion, and molecular weight are considered. Special attention is paid to fronts with solid monomers and the effects of particle size and green density on the front velocity compared to SHS. The future directions of research with frontal polymerization are considered, especially regarding applications to materials synthesis.

In 1967 at the Institute of Chemical Physics at Chernogolovka (Russia) Alexander G. Merzhanov and his colleagues discovered the process of Self-Propagating High-temperature Synthesis (SHS) to prepare technologically useful ceramics and intermetallic compounds (*1-3*). A compressed pellet of reactants was ignited at one end that resulted in a self-propagating combustion wave. The method had the advantages that the initial stimulus was the only energy input required and that superior materials were produced. In 1972 Chechilo and Enikolopyan applied the same approach to the free-radical polymerization of vinyl monomers. Using a steel reactor under high pressure (> 3000 atm) they studied descending fronts of methyl methacrylate with peroxide initiators (*4, 5)* In 1991 Pojman rediscovered this phenomenon using methacrylic acid at ambient pressure in standard test tubes (*6*).

Although the two processes are similar because they both are thermal waves (Figure 1), they differ in the "High Temperature" aspect. In SHS the combustion waves propagate with velocities ranging from 0.1 to 1500 cm/min with front temperatures as high as 2000 °C (*7*). However, propagating polymerization fronts have velocities on the order of 1 cm/min with front temperatures of 200 °C. Nonetheless, they share many similarities because of their common mechanism of propagation. In this chapter we will consider basic aspects of frontal polymerization, types of systems that can be studied, interferences with stable propagation, the nature of the product produced and prospects for materials synthesis. We will first provide an overview of frontal polymerization of liquid monomers and examine in detail acrylamide polymerization fronts and compare them to results for SHS.

Basic Phenomena

An experiment can be performed in a glass tube filled with reactants. For free-radical polymerization systems the reactants are a monomer and a few percent of a peroxide or nitrile initiator. Epoxy curing uses the resin and a stoichiometric amount of an amine curing agent and/or 20 - 40% of BCl_3-amine complex. An external heat source, when applied at the top of the tube, starts a descending front that appears as a slowly moving (~1cm min^{-1}) region of polymer formation. **Although most of the systems we will describe can be performed in a standard test tube, significant pressure can build up that can lead to explosions. Therefore, all experiments should always be performed behind a safety shield.** Monomers whose boiling point is significantly lower than the front temperature must be polymerized under pressure. Figure 2 shows n-butyl acrylate polymerization (the reason for the silica gel will be explained later). In the absence of any kind of thermal or convective instabilities the front moves with a constant velocity determined by a plot of the front position as a function of time (Figure 3). As we will see discuss in a later section, there are a number of instabilities that can occur, especially with liquid systems, that prevent stable propagation.

Frontal polymerization works with a wide variety of systems and has been demonstrated with neat liquid monomers such as methacrylic acid, *(6, 8-10)* n-butyl acrylate (*10*), styrene, methyl methacrylate, and triethylene glycol dimethacrylate (*11*). Fronts can also be performed with solid monomers such as acrylamide (with initiator) *(12)* or transition metal nitrate acrylamide complexes (without initiator) (*13*). Pojman et al. demonstrated frontal polymerization of acrylamide, methacrylic acid and acrylic acid, each in dimethyl sulfoxide or dimethyl formamide (*14*). Frontal curing of epoxy resins has been demonstrated with aliphatic curing agents (*15*).

Unfortunately, we do not know the necessary and sufficient conditions for self-sustaining fronts. However, we do know several factors that favor the frontal mode. The monomer must have a boiling point below the front temperature to prevent heat loss from vaporization and bubbles that can obstruct the front. (The boiling point can be raised by applying pressure.) The front temperature should be large. Thus, highly exothermic reactions are the most likely candidates for frontal polymerization because the heat production must exceed the heat losses, especially if a reinforced or filled composite is sought. The geometry of the system also plays a role. If the surface area to volume ratio is too large, even a reactive system will be quenched. For example, at room temperature the only system we have found that can propagate in a 3 mm glass tube is acrylamide -- no liquid monomer is sufficiently reactive and exothermic.

The reaction rate at the initial temperature must be vanishingly small but rapid at the front temperature. The front temperature is determined by the enthalpy of the reaction, heat capacity of the product and the amount of heat loss. Free-radical polymerization is ideal because for most peroxide and nitrile initiators the rate of polymerization at ambient temperature is low but high at elevated temperatures. Amine-cured epoxies suffer from the problem of short pot life but cationic cured systems are very similar to free-radical systems (*10*).

Temperature Profiles

A polymerization front has a sharp temperature profile, (*11*) and profile measurements can provide much useful information. From Figure 4 it may appear that the chemical reaction is occurring in a zone about 0.5 cm in width. This is incorrect. If the chemical reaction has an infinitely high energy of activation, the chemical reaction will occur in an infinitely narrow region. In actuality, the 0.5 cm represents a pre-heat zone. The temperature below that at which significant chemical reaction occurs follows an exponential profile, which can be described with the following relationship in terms of

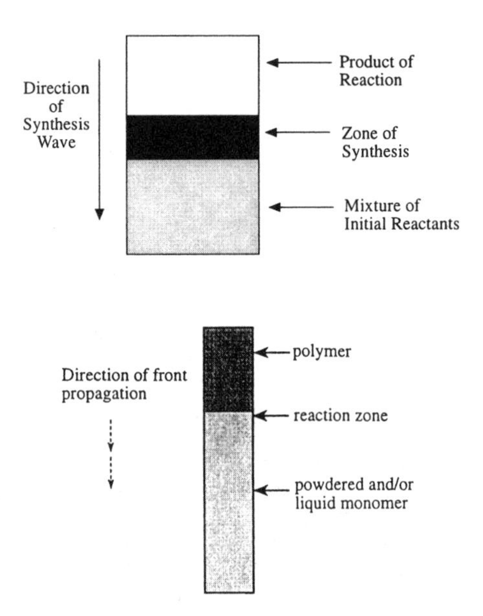

Figure 1. A schematic diagram of the SHS process (top) and frontal polymerization (bottom).

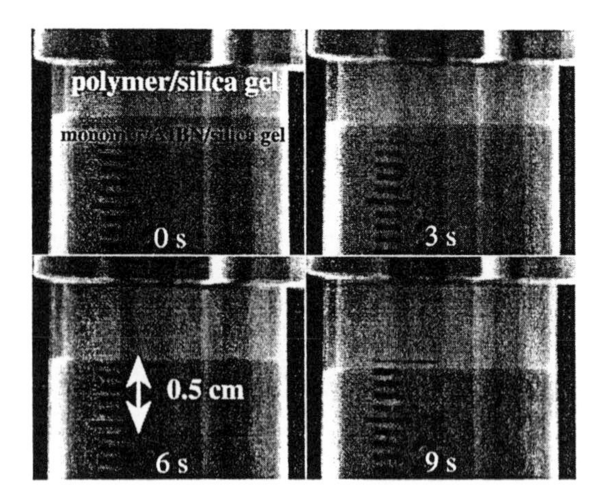

Figure 2. A montage of an n-butyl acrylate front propagating under 50 atm pressure. [AIBN] = 4% w/v.

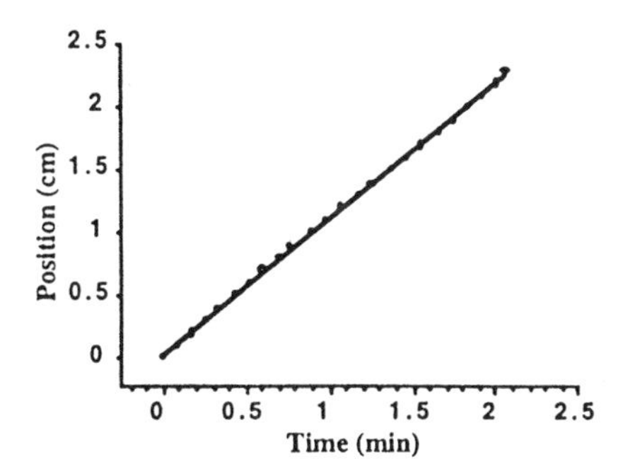

Figure 3. The front velocity (1.09 cm/min) is determined from the slope of a plot of front position versus time for n-butyl acrylate polymerization under pressure. $[AIBN]_0$ = 0.00457 molal; initial temperature = 18.9 °C

the front velocity (c), thermal diffusivity (κ) and temperature difference (ΔT): *(16)*

$$T(x) = T_0 + \Delta T \exp\left(\frac{cx}{\kappa}\right) \quad (1)$$

Because heat is a catalyst in these reactions, conversion is directly proportional to the difference between the maximum and initial temperatures. A higher maximum temperature indicates higher conversion. More stable initiators give higher conversion, as can be seen in Figure 4 with benzoyl peroxide (BPO) and t-butyl peroxide (tBPO). The methacrylic acid front with tBPO was significantly slower in spite of having the highest reaction temperature. This means that the effective activation energy of a polymerization front is directly correlated to the activation energy of the initiator decomposition.

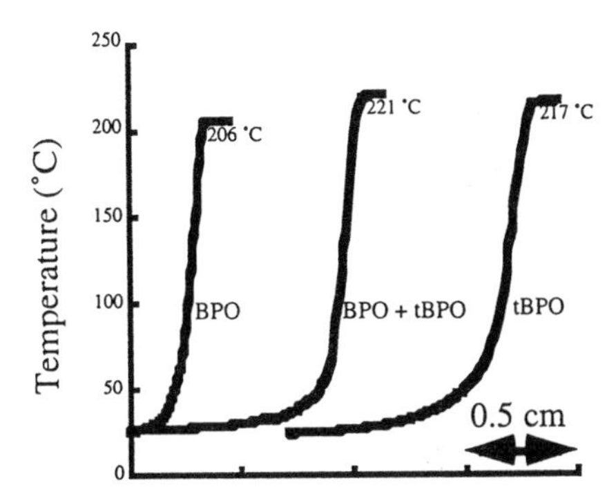

Figure 4. The temperature profiles of methacrylic acid polymerization fronts with BPO alone, with benzoyl peroxide and t-butyl peroxide (tBPO) and with t-butyl peroxide alone. [BPO] = [tBPO] = 0.0825 mol/kg. Reactions were performed in 1.5 cm (i.d.) tubes. The temperatures indicated correspond to the maximum temperatures reached for each initiator. While the temperature at the monomer/polymer interface was not directly measured, it is believed to be *ca.* 100 °C

Effect of Pressure

We would not normally expect that moderate pressure would affect the rates of chemical reactions. Nonetheless, Pojman et al. found that the front velocity was a function of the applied pressure, even at low values of less than 30 atm (*10*). As the pressure was increased, the velocity decreased, exactly opposite the behavior observed by Chechilio and Enikolopyan at high pressures (*5*)! At the low pressures employed in our experiments, rate constants were not affected but rather the size of bubbles.

Bubbles can increase the velocity of fronts in standard closed test tubes initially at ambient pressure by as much as 30% compared to fronts free of bubbles under high pressure. The expansion of bubbles is part of the velocity by forcing unreacted monomer up and around the cooling polymer plug that is contracting; poly(methacrylic acid) is *ca.* 25% more dense than its monomer. This means that the pressure increases during the reaction because the tube is sealed, except for leakage around the initial polymer plug (*10*).

There are three sources of bubbles. All the thermal initiators investigated (except sodium persulfate, which is insoluble in most monomers) produce volatile by-products, such as CO_2, methane and acetone. It is an inherent problem with all commercially available peroxide or nitrile initiators.

Another source of bubbles is dissolved gas and water in the monomer. Gases can be removed under vacuum but water is extremely difficult to remove from methacrylic acid and TGDMA. The only certain solution to all three sources is to perform reactions under pressure. We did so using a custom built reactor, shown in Figure 5 that allowed temperature and pressure control.

To test pressure control in the reactor, we set the reactor pressure to 400 psi and measured pressure every 0.1 minutes. The pressure, on average, remained 400 psi as shown in Figure 6a. When the temperature of the reactor was set to 26 °C, it took less than two minutes for the system to equilibrate to the set temperature as shown in Figure 6b.

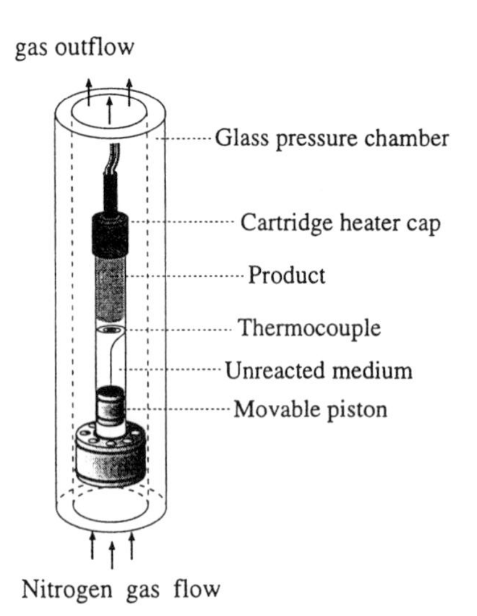

Figure 5. Reactor for the study of frontal polymerization under controlled pressure and temperature.

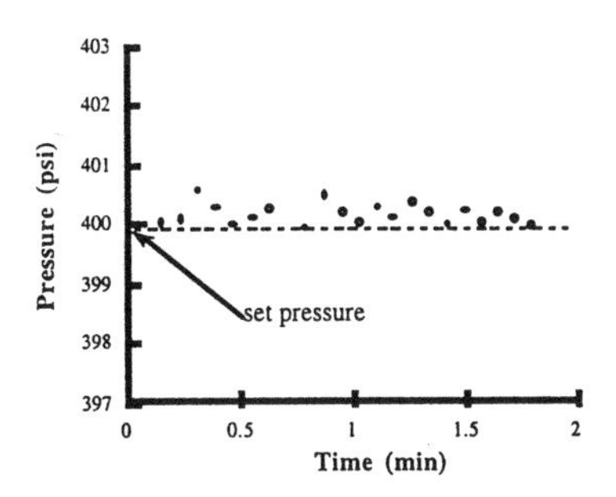

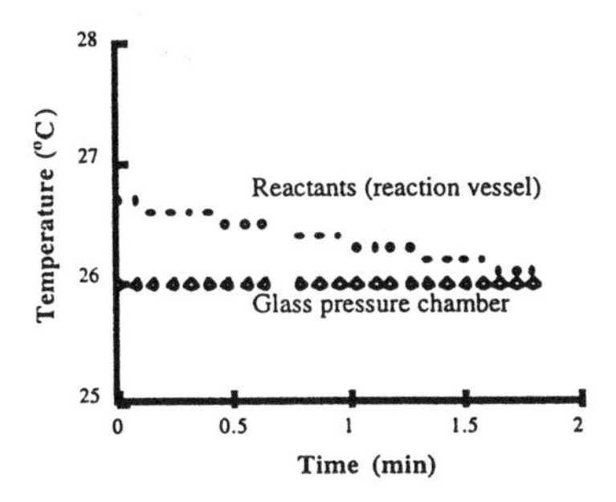

Figure 6a. (Top) Graph depicting pressure control in the reactor. Figure 6b. (Bottom) Graph depicting temperature control in the reactor.

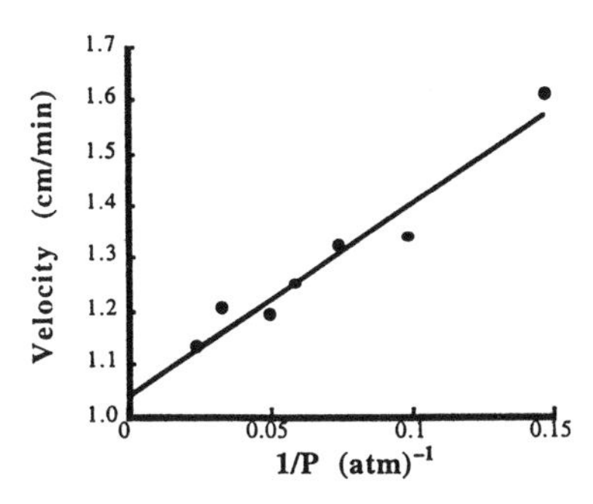

Figure 7. The front velocity of butyl acrylate polymerization as a function of applied pressure. AIBN initiator (1.7 % w/w); Cabosil (5.7 % w/w); T_0 = 24 °C.

Figure 7 shows the front velocity as a function of the inverse of the applied pressure. Since bubbles increase velocity and pressure suppresses bubbles, pressure decreases velocity. As the pressure is increased, the velocity decreases because the volume of the bubbles is decreased, following Boyle's law. Pressure is not affecting the rate of polymerization but the rate of front propagation because of the expansion of the medium. A similar phenomenon was observed by Merzhanov with SHS materials that created a porous product (*17*).

We can write the velocity as:

$$vel(p) = vel_0 + \frac{const}{p} \tag{2}$$

where the constant will be a function of the number of moles of gas produced in the front. Therefore, the higher the initiator concentration, the higher is the applied pressure necessary to obtain the true front velocity, which is equal to the y-intercept in a plot of velocity versus 1/p.

Front Velocity Dependence on Initiator Concentration

Chechilo et al. studied frontal polymerization of methyl methacrylate with benzoyl peroxide as the initiator. By placing several thermocouples along the length of the metal reaction tube, they could infer the front velocity and found a 0.36 power dependence for the velocity on the benzoyl peroxide concentration (*18*). More detailed studies for several initiators showed 0.223 for t-butyl peroxide, 0.324 for BPO and 0.339 for cyclohexylperoxide carbonate (*4*). Figure 8 shows the bubble-free velocities for n-butyl acrylate fronts as a function of initiator concentration, with power function dependencies similar to those of Chechilo et al.

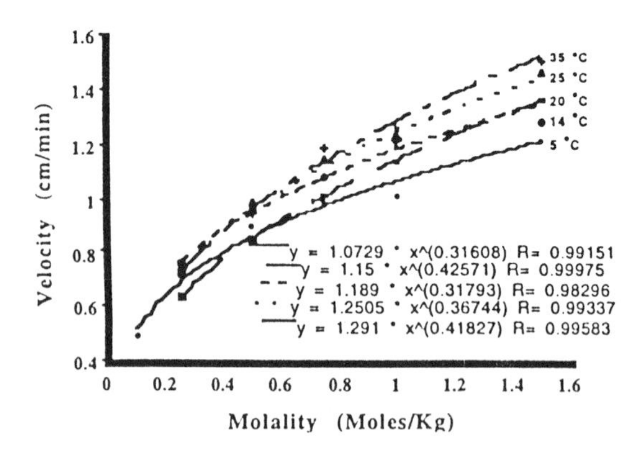

Figure 8. The n-butyl acrylate front velocity as a function of AIBN concentration for several initial temperatures. Pressure was maintained at 50 atm.

Pojman et al. studied binary systems with two non-interfering polymerization mechanisms (*19*). With a cationic/amine cured epoxy and a free-radical cured diacrylate, they observed there was a minimum in the front velocity as a function of the relative concentration of each component. No comparable study has been made with copolymerization fronts.

Properties of Polymers

Pojman et al. investigated the molecular weight distributions for poly(n-butyl acrylate) produced frontally and found the trend that one would expect from the classical steady-state theory of free-radical polymerization (*9*). Poly(n-butyl acrylate) produced in fronts had unimodal molecular weight distributions with $M_w < 10^5$ ($M_w/M_n = 1.7$ - 2.0). The average molecular weight decreased with increasing initiator concentrations. Poly(methacrylic acid) produced frontally had bimodal distributions because of crosslinking via anhydride formation.

Pojman et al. studied the conversion of methacrylic acid in fronts and found it could be as low as 70% with BPO as initiator and greater than 90% with tBPO (*11*). n-Butyl acrylate fronts can yield conversion greater than 96% (*20*).

Solid Monomers

Studies had been done in SHS to observe the effect of various parameters on wave-front velocity and product morphology. Among the parameters studied were the green (unreacted) density and particle sizes of the reactants. All the systems so far considered use liquid monomers or solid monomers in solution. Relatively little work has been done with solid monomers. Pojman et al. demonstrated frontal acrylamide polymerization with a variety of free-radical initiators (*12*). Savostyanov et al. studied transition metal complexes of acrylamide without initiator (*13*). No studies were performed on the effect of particle size and/or green density. We therefore investigated those two factors to compare to work done in intermetallic SHS systems.

It was thought that the particle size trends in frontal free-radical polymerization would be similar to particle size trends in SHS. In Ti + C and Ti + B systems in SHS, as particle size increases, velocity decreases (*7*). This trend is seen because in many SHS systems, one component must melt so that the other can diffuse into the melt and begin the reaction. The same trend was expected in frontal polymerization as seen in Figure 9.

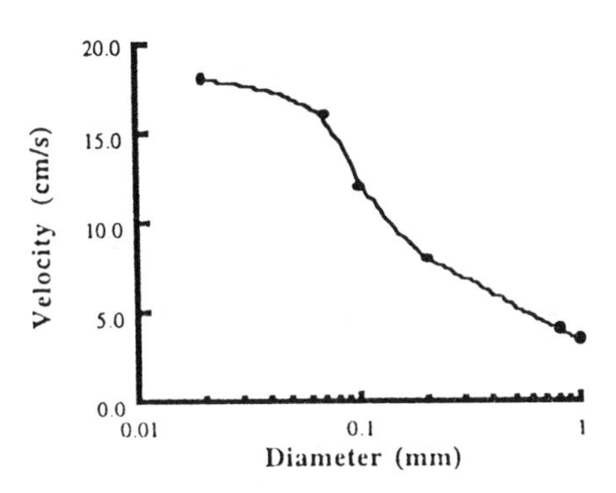

Figure 9. Dependence of velocity on particle size of Ti + B system as a function of titanium particle size. Adapted from reference (*7*)

Solid monomer-solid initiator systems were tried with a variety of monomers and initiators. Two types of monomer systems were used: 1) monomers whose melting points were lower than the adiabatic reaction temperature. Acrylamide worked well but n-octadecyl acrylate did not because it was not sufficiently reactive ; 2) High melting point monomers with a reactive diluent in which the monomer could dissolve at high temperature. Zinc dimethacrylate/acrylamide or zinc dimethacrylate/methacrylic acid both supported frontal propagation with BPO initiator.

To achieve reproducible front velocities a mold was made, and a four-ton hydraulic press was used to pack the mixture. The resulting pellet was then sanded down to fit into a 16 x 150 mm test tube. The sanding of the pellets had no effect on the green density which was to be determined later. Unfortunately, zinc dimethacrylate adhered to the mold. Acrylamide was used then as the sole monomer.

Another change made was the switch from benzoyl peroxide to potassium persulfate as the initiator. Fronts run with potassium persulfate did not release the large amounts of gas that fronts run with benzoyl peroxide and AIBN did. The reason was that when potassium persulfate decomposed, the resulting product (bisulfate) was not volatile as was the case with benzoyl peroxide (CO_2) and AIBN (N_2) (*21*).

For the study of particle size trends in acrylamide fronts, particle size ranges were 710-500, 500-408, 408-297, 297-210, 210-149 μm. The initiator was kept in the 210-149 μm range. The procedure was as follows: acrylamide and 4% (w/w) potassium persulfate were mixed in a tumbler for thirty minutes. The mixture was then placed in the mold and sufficient force applied until the density of the pellet reached the desired value. The pellet was then fit into a 16 x 150 mm test tube. The front was ignited with a soldering iron. To obtain reproducible front velocities, a constant green density had to be maintained. The green density was kept to 1.14 to 1.15 g/cm^3, and the particle size was varied. There was no dependence on particle size as can be seen in Figure 10.

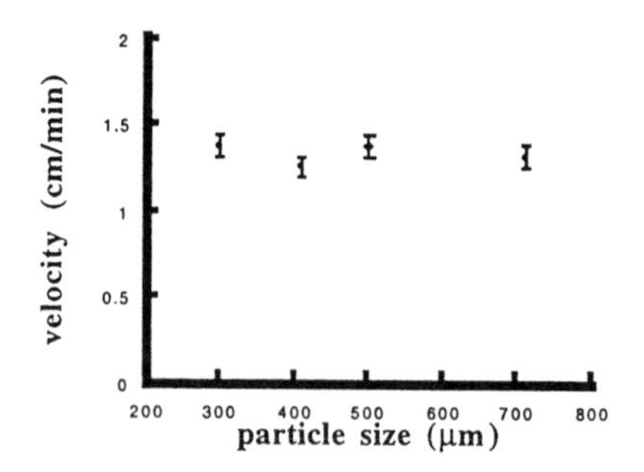

Figure 10. Graph depicting the dependence of velocity on particle size at constant green density.

The velocity dependence on the particle size was not the same as observed in SHS. This result is due to a basic difference in frontal polymerization and SHS reactions. SHS reactions are stoichiometric. In order for reaction to occur, one molecule of a species has to react with another molecule. In frontal polymerization, as in all addition polymerization, once a molecule of initiator is decomposed, it can begin the polymerization for hundreds of monomers. Acrylamide melts at 84 °C. Once melting has occurred, the initiator particles are surrounded by a sea of monomer and reaction can proceed; particle size is unimportant.

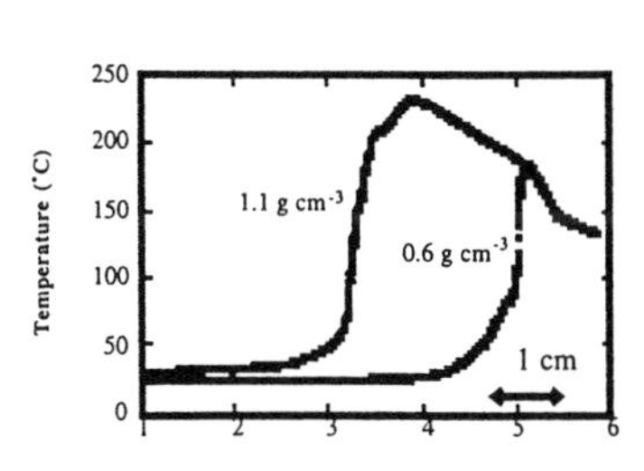

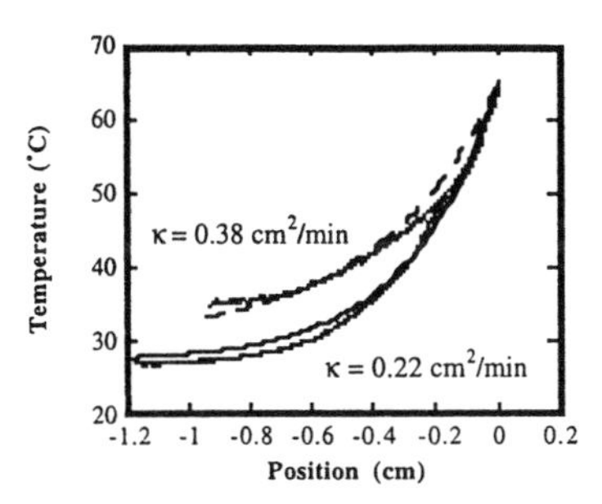

Figure 11. (Left) Temperature profiles of acrylamide polymerization with potassium persulfate initiator at two green densities. (Right) Fits of equation 1 to preheat zones. Thicker lines are curve fits to data points (thinner lines).

We studied the effect of green density on the front temperature profile to determine the effective thermal diffusivity. Figure 11 compares the profiles at two different densities. It was thought that as green density was increased, velocity would also increase because the sample is packed closer, and heat diffusion is facilitated. Temperature profiles of acrylamide fronts packed to different green densities show that heat diffusion in samples of greater green densities is greater than heat diffusion in samples of lesser green densities. The front temperature in a sample of lower green density was only 185 °C. Whereas the front temperature in a sample of greater green density was 235 °C. Trapped air in the sample of lesser green density insulated the sample and obstructed heat diffusion. However, why the temperature maxima differ is not clear. Conversion may be affected by porosity but we are not able to propose a mechanism. Also, a slower front may allow more heat loss to the surroundings.

A temperature profile of acrylamide polymerized in liquid nitrogen had been done as well (Figure 12). It showed a 400 °C degree temperature increase. We were puzzled as to why our acrylamide system only showed a temperature increase of only about 200 °C. Upon doing a TGA of our sample, we found that at 400 °C, only 20% of the sample remains (Figure 12). The rest is decomposed which is why our systems do not have a 400 °C temperature increase.

Increased green density should lead to increased heat diffusion, which should lead to increased velocity. This result was observed (Figure 13) when the velocity was corrected for the greater length present in less dense samples. (A lower density means a longer pellet, so that a front "covers more ground" but only because the reactants are spread over a longer distance.)

The velocity is a linear function of persulfate concentration (Figure 13). This differs from previous results of Pojman et al. who found a logarithmic velocity dependence with AIBN initiator and a power function dependence with persulfate (*12*). Those experiments were done without careful control of the green density, which we now see is a crucial parameter.

The polymer produced is filled with large pores, on the order of 100 μm (Figure 14). Conversion is around 50%, which we estimate from the ΔT of the front. It is half the value found when the front is run with the tube immersed in liquid nitrogen (*10*).

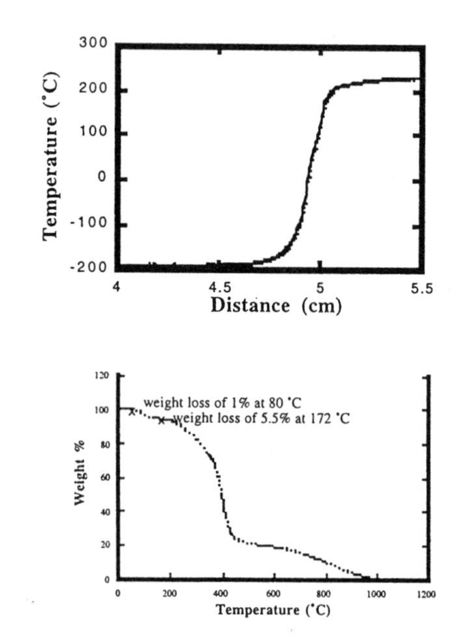

Figure 12. (Top) Temperature profile of acrylamide with lauroyl peroxide done in liquid nitrogen showing a 400 °C temperature increase. (Bottom) TGA of acrylamide with potassium persulfate showing decomposition at 400 °C.

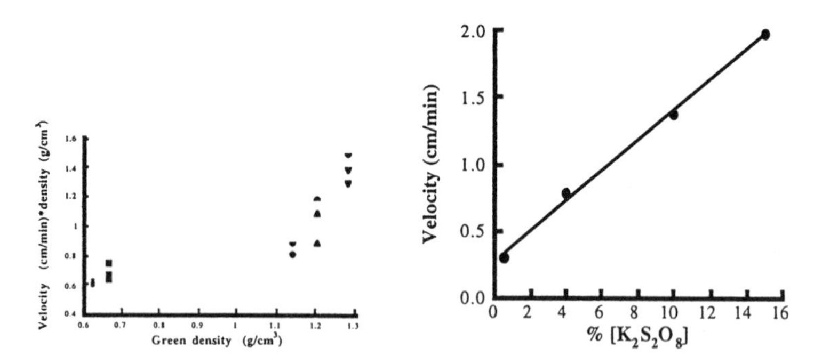

Figure 13. (Left) Graph depicting velocity dependence on green density. Multiple experiments are plotted to indicate the range of reproducibility. (Right) Velocity dependence on initiator concentration at fixed green density of 1.1 g cm^{-3}

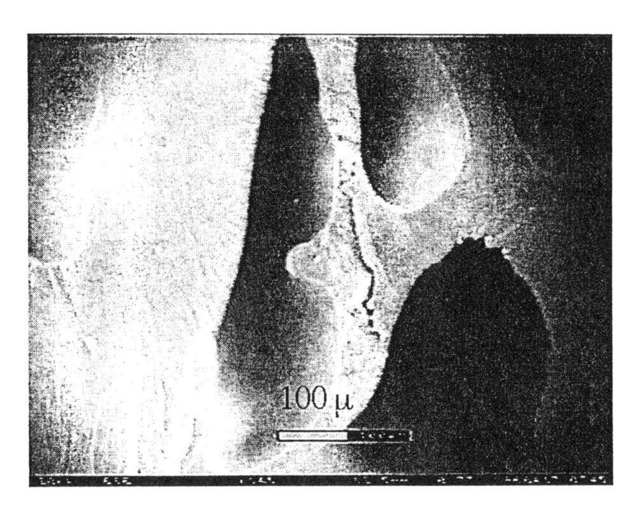

Figure 14. The morphology of poly(acrylamide) produced frontally, as seen by SEM. Green density 1.14 g cm^{-3} 4% $K_2S_2O_8$

Unstable Propagation

Systems with liquid monomers are susceptible to buoyancy-driven convection. If a front were to propagate upward then the hot polymer-monomer solution in the reaction zone could rise because of buoyancy, removing enough heat at the polymer monomer interface to quench the front. Whether an ascending front is stable depends on the initial viscosity, front temperature and velocity *(22)*. The greater the front velocity, the lower the initial viscosity that will allow convection-free propagation.

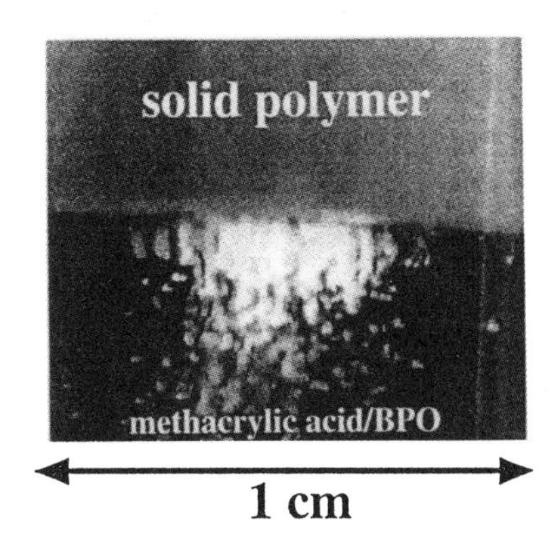

Figure 15. Descending "fingers" from a front of poly(methacrylic acid) polymerization.

Descending fronts with monomers such as methacrylic acid that produce solid but uncrosslinked product can exhibit "fingering" in which partially polymerized material sinks from the front (Figure 15). Nagy and Pojman showed that this instability could be suppressed by rotating the tube around the axis of propagation *(23)*. The rotation increased the front velocity and caused a curved front.

Figure 16. Rayleigh-Taylor instability in a descending front of n-butyl acrylate polymerization.

The most pernicious convective instability occurs with monomers that produce a molten polymer at the front, such as n-butyl acrylate, styrene and methyl methacrylate. A Rayleigh-Taylor instability (*24, 25*), which also appears as "fingers" as the more dense molten polymer streams down from the reaction zone and destroys the front (Figure 16). The only currently available methods to study frontal polymerization with thermoplastics are to add a crosslinking monomer to produce a thermoset or to increase the viscosity with a viscosifier such as ultrafine silica gel (CAB-O-SIL). To prepare pure poly(n-butyl acrylate) frontally, Pojman et al resorted to performing the reaction under weightless conditions of a sounding rocket (*26*).

Systems with solid monomers are immune to those convective instabilities, but other instabilities can arise that are generic to any thermally propagating front. For frontal polymerization of methacrylic acid or the frontal curing of cationically-cured epoxies, lowering the initial temperature or increasing the amount of heat loss can cause the onset of periodic "spinning" modes of propagation (*10, 27*). (In general, amine cured epoxies are not susceptible because of their low energies of activation.) Figure 17 shows infrared images of a front of methacrylic acid polymerization in which the monomer solution was initially at 0 °C. A "hot spot" or spin mode can be seen propagating around the surface of the descending front. The product has clearly visible spiral patterns.

Maintaining the initial system above the critical temperature for the onset of the instability and reducing heat losses can prevent such instabilities, although it is usually not possible to know the critical temperature a priori. We do know three factors that favor periodic modes: high front temperature, high energy of activation and low initial temperature. These factors are summarized by the Zeldovich number (*16*).

$$Z = \frac{T_m - T_o}{T_m}\frac{E_{eff}}{RT_m} \quad (3)$$

For adiabatic systems with one step reaction kinetics, for all values of Z less than 8.4, planar fronts occur, but Solovyov et al. have found that models with realistic polymerization kinetics are more stable than predicted by equation 3 (*28*).

SHS systems exhibit similar behavior (*29, 30*). Because of the extremely high front temperatures, diluting the initial reactant mixture tends to lead to the instability. (The Zeldovich number does not increase monotonically with T_m.)

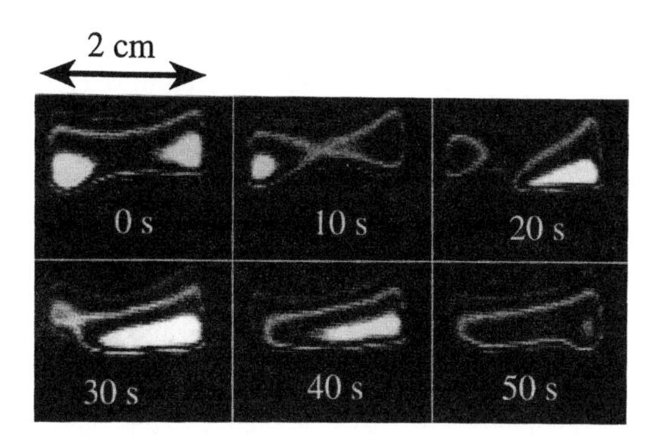

Figure 17. A montage of infrared images of methacrylic acid frontal polymerization under unstable conditions.

Material Synthesis

We expect that this approach will ultimately have three benefits over traditional methods of polymer synthesis: 1) reduced energy costs, 2) reduced waste production, and 3) unique morphologies. A desirable feature of frontal polymerization is the rapid and uniform conversion of monomer to polymer. Performing an adiabatic polymerization of a neat monomer is difficult at best and dangerous at worst, but it can be advantageous because the heat of the reaction is used to increase the rate of reaction. Also, the absence of solvent eliminates the need to separate the polymer from the solvent and residual monomer, which requires energy and can have environmental ramifications.

Thermosets and Composites

The curing of thick sections of large composites with the existing methods such as autoclaving, pultrusion, etc., is difficult because the internal temperature lags behind the surface temperature. If the surface cures faster than the interior regions, resin can be trapped in the interior, which leads to non-uniform composition and voids. Another problem is thermal spiking, during which the heat produced in the interior can not escape and builds up. The autocatalytic nature of the curing process can lead to a thermal runaway and non-uniform composition. Very slow heating rates can solve this problem but at the cost of increased processing time.

The continuous production of composites is a promising technological application of frontal polymerization. The approach reduces imperfections and production costs for a high value-added products. White developed two continuous curing processes for producing graphite fiber composites (*31, 32*). The first is used for large rectangular shaped objects. Layers of prepreg (graphite fiber material impregnated with the epoxy thermoset resin) are placed into a mold that is heated from below. Pressure (25 psi) is applied until the front reaches near the top of the mold at which point the pressure is released and another layer of prepreg is inserted; the process is repeated until the desired thickness. White has shown that the process produces product with fewer voids in less time.

The second approach is for the production of cylindrical objects, such as energy-storing fly wheels. A filament winding procedure has been developed in which resin-impregnated fibers are wound onto a heated mandrel at a rate that matches that of the expanding cure front.

Thermal curing encourages high conversion by maintaining low viscosity at the elevated front temperature. Unfortunately, the low viscosity causes sedimentation of the filler particles that can result in nonuniform composition. Nagy and Pojman demonstrated that frontal polymerization could be used to overcome this problem. They prepared a novel material, a thermochromic composite that gradually changes color from 80 °C to 140 °C (*33, 34*). Batch reactions produced a product that could not be machined and was nonuniform in composition because of phase separation and sedimentation during polymerization. A composite produced by a front was uniform because the very fast reaction in the front locked the components together before sedimentation could occur. This was the first example of material produced frontally that is superior to that produced in a traditional batch reaction.

We envision a different approach to the manufacture of large composite parts. At room temperature, the resin would be forced into the fiber/particle matrix. A cure front would be ignited that would spread throughout the lay-up. The composites have to be limited to simple geometrical shapes with thick sections such that the heat losses are minimized to sustain self-propagation. A great deal of work must be done on the relationship between the mechanical properties of the composite and the conditions of frontal curing. Specifically, high energy of activation curing agents with long pot lives need to be developed.

Conclusions

Frontal polymerization shares many of the characteristics of SHS for ceramic production -- rapid conversion to unique products without the continual input or removal of heat. Some of the same problems occur such as periodic modes of propagation but with liquid monomers frontal polymerization is also greatly affected by buoyancy-driven convection.

The major difference between SHS and free-radical frontal polymerization of a solid monomer is that the latter is not a stoichiometric process. As a result, acrylamide frontal velocities are not affected by the monomer particle size. The most important factor affecting the front velocity and the front temperature is the green density.

Acknowledgments

This work was supported by the National Science Foundation (CTS-9319175 and the Mississippi EPSCoR program), the Air Force Office of Scientific Research and NASA's Microgravity Materials Science Program (NAG8-973). Special acknowledgment is made to Joe Whitehead for preparing SEM image.

References

(1) Merzhanov, A. G.; Borovinskaya, I. P. *Dokl. Nauk SSSR* **1972**, *204,* 336-339.
(2) No format in JACS style for this reference type: please edit the JACS style.
(3) Merzhanov, A. G.; Khaikin, B. I. *Prog. Energy. Combust. Sci.* **1988**, *14,* 1-98.
(4) Chechilo, N. M.; Enikolopyan, N. S. *Dokl. Phys. Chem.* **1975**, *221,* 392-394.
(5) Chechilo, N. M.; Enikolopyan, N. S. *Dokl. Phys. Chem.* **1976**, *230,* 840-843.
(6) Pojman, J. A. *J. Am. Chem. Soc.* **1991**, *113,* 6284-6286.
(7) Varma, A.; Lebrat, J.-P. *Chem. Eng. Sci.* **1992**, *47,* 2179-2194.
(8) Pojman, J. A.; Craven, R.; Khan, A.; West, W. *J. Phys. Chem.* **1992**, *96,* 7466-7472.
(9) Pojman, J. A.; Willis, J. R.; Khan , A. M.; West, W. W. *J. Polym. Sci. Part A: Polym Chem.* **1996**, *34,* 991-995.

(10) Pojman, J. A.; Ilyashenko, V. M.; Khan, A. M. *J. Chem. Soc. Faraday Trans.* **1996**, *92,* 2825-2837.
(11) Pojman, J. A.; Willis, J.; Fortenberry, D.; Ilyashenko, V.; Khan, A.*J. Polym. Sci. Part A: Polym Chem.* **1995**, *33,* 643- 652.
(12) Pojman, J. A.; Nagy, I. P.; Salter, C. *J. Am. Chem. Soc.* **1993**, *115,* 11044-11045.
(13) Savostyanov, V. S.; Kritskaya, D. A.; Ponomarev, A. N.; Pomogailo, A. D. *J. Poly. Sci. Part A: Poly. Chem.* **1994**, *32,* 1201-1212.
(14) Pojman, J. A.; Curtis, G.; Ilyashenko, V. M. *J. Am. Chem. Soc.* **1996**,*115,* 3783-3784.
(15) Davtyan, S. P.; Arutyunyan, K. A.; Shkadinskii, K. G.; Rozenberg, B. A.; Yenikolopyan, N. S. *Polymer Science U.S.S.R.* **1978**, *19,* 3149-3154.
(16) Zeldovich, Y. B.; Barenblatt, G. I.; Librovich, V. B.; Makhviladze, G. M. *The Mathematical Theory of Combustion and Explosions*; Consultants Bureau: New York, 1985.
(17) Merzhanov, A. G. *Usp. Khim.* **1976**, *45,* 827-848.
(18) Chechilo, N. M.; Khvilivitskii, R. J.; Enikolopyan, N. S. *Dokl. Akad. Nauk SSSR* **1972**, *204,* 1180-1181.
(19) Pojman, J. A.; Elcan, W.; Khan, A. M.; Mathias, L. *J. Polym. Sci. Part A: Polym Chem.* **1997**, *35,* 227-230.
(20) Khan, A. M.; Pojman, J. A. *Trends Polym. Sci. (Cambridge, U.K.)* **1996**, *4,* 253-257.
(21) Thomson, R. A. M. in *Chemistry and Technology of Water Soluble Polymers*; Finch, C. A.,Ed., Plenum: New York, 1983; pp 31-69.
(22) Bowden, G.; Garbey, M.; Ilyashenko, V. M.; Pojman, J. A.; Solovyov, S.; Taik, A.; Volpert, V. *J. Phys. Chem. B* **1997**,*101,* 678-686.
(23) Nagy, I. P.; Pojman, J. A. *J. Phys. Chem.* **1996**, *100,* 3299-3304.
(24) Taylor, G. *Proc. Roy. Soc. (London)* **1950**, *Ser. A. 202,* 192-196.
(25) Rayleigh, L. *Scientific Papers, ii*; Cambridge University Press: Cambridge, 1900.
(26) Pojman, J. A.; Khan, A. M.; Mathias, L. J. *Submitted to Microg. sci. tech.*
(27) Pojman, J. A.; Ilyashenko, V. M.; Khan, A. M. *Physica D* **1995**,*84,* 260-268.
(28) Solovyov, S. E.; Ilyashenko, V. M.; Pojman, J. A. *Chaos in press*
(29) Maksimov, Y. M.; Merzhanov, A. G.; Pak, A. T.; Kuchkin, M. N. *Combust. Explos. Shock Waves* **1981**, *17,* 393-400.
(30) Maksimov, Y. M.; Pak, A. T.; Lavrenchuk, G. V.; Naiborodenko, Y. S.; Merzhanov, A. G. *Comb. Expl. Shock Waves* **1979**, *15,* 415-418.
(31) White, S. R.; Kim, C. *J. Reinforced Plastics and Comp.* **1993**,*12,* 520-535.
(32) Kim, C.; Teng, H.; Tucker, C. L.; White, S. R. *J. Comp. Mater.* **1995**, *29,* 1222-1253.
(33) Nagy, I. P.; Sike, L.; Pojman, J. A. *J. Am. Chem. Soc.* **1995**, *117,* 3611-3612.
(34) Nagy, I. P.; Sike, L.; Pojman, J. A. *Adv. Mat.* **1995**, *7,* 1038-1040.

Chapter 20

Synthesis and Characterization of Linear Tetraphenyl–Tetramethyldisiloxane Diacetylene Copolymers

Eric J. Houser and Teddy M. Keller[1]

Chemistry Division, Materials Chemistry Branch, Naval Research Laboratory, Code 6120, 4555 Overlook Avenue, SW, Washington, DC 20375–5320

New phenyl substituted siloxyl diacetylene polymers have been prepared in high yields. The syntheses are one-pot, two step procedures involving the reaction of dilithiobutadiyne with various dichlorodisiloxanes. Spectroscopic and microanalytical data support the proposed structures. The polymers are thermally crosslinked to give hard, void-free thermosets which can be further heated to give ceramic products with high weight retention.

Materials possessing high thermal and oxidative stabilities are in high demand for applications in the aerospace and related industries. Polymeric fiber-reinforced carbon-carbon composites are frequently used in many of these applications due to their high thermal stability and strength to weight ratio, but these materials suffer from a relatively low oxidative stability (*1, 2*). Research has shown that the oxidative stability of carbon-based materials can be greatly increased by the introduction of inorganic elements such as silicon or boron (*2*).

Preceramic polymers have been used successfully in the synthesis of a variety of useful ceramics such as SiC, Si_3N_4 and BN (*3-5*). The use of polymers in the formation of ceramic materials has potential processing advantages such as in forming complex shapes and giving homogeneous final ceramic products. However, obtaining pure materials, such as SiC, from polymer precursors can prove difficult in that small amounts of carbon and silicon oxide are frequently present in the final product (*3*). The pyrolysis of several silylene or siloxyl diacetylene polymers to SiC-containing ceramics has been reported (*6*).

In the field of composite materials, inorganic-organic hybrid polymers offer great promise as precursors to ceramic matrix materials. In these applications, high purity ceramics are often not necessary and preceramic polymers allow the introduction of inorganic elements such as silicon and boron in quantities which can be directed by polymer structure and stoichiometry.

Our research efforts have focused on the synthesis of hybrid inorganic-organic diacetylenic polymers as precursors to materials with high thermal and oxidative

[1]Corresponding author

stabilities (*7,8*). The diacetylene units act as crosslinking agents which is necessary for high weight retention upon pyrolysis (*9*). Heat treatment of these polymers to 300 °C gives hard, void-free thermosets. The thermosets can be further pyrolyzed up to 1500 °C affording ceramics which exhibit excellent thermal and oxidative stabilities. We recently reported on the properties of the diacetylene polymer poly(tetramethyldisiloxyldiacetylene) (*8*). We were interested in the effects of phenyl substitution on the thermal and oxidative stability in this type of polymer. To this end we have synthesized several new polymers and copolymers via the reaction of 1,3-dichloro-1,3-dimethyl-1,3-diphenyldisiloxane, **1**, or 1,3-dichlorotetraphenyldisiloxane, **2**, and 1,3-dichlorotetramethyldisiloxane, **3**, with 1,4-dilithiobutadiyne, **4**. The molar ratios of **2** and **3** were varied in order to tailor the structure of the copolymers. Related copolymers, formed from **4** and 1,7-bis(chlorotetramethyldisiloxyl)-m-carborane, **6**, were also prepared. We hereby present preliminary work describing the syntheses, characterization, and thermal evaluation of these new hybrid polymers.

Experimental

All reactions were carried out under an inert atmosphere using standard Schlenk techniques unless otherwise noted. Tetrahydrofuran (THF) was distilled from sodium/benzophenone under inert atmosphere immediately prior to use. Hexachlorobutadiene was obtained from Aldrich Chemical Co. and purified by distillation. n-Butyllithium (2.5M in hexanes) was obtained from Aldrich Chemical Company and used as received. 1,3-Dichloro-1,3-dimethyl-1,3-diphenyldisiloxane was purchased from Hüls Chemical Company and purified by distillation under N_2 from Mg chips. 1,3-Dichlorotetramethyldisiloxane and 1,3-dichlorotetraphenyldisiloxane were obtained from Gelest, Inc. and were purified by distillation under N_2 from Mg chips. 1,7-Bis(chlorotetramethyldisiloxyl)-m-carborane was obtained from Dexsil Corp. and used as received. 1,4-Dilithiobutadiyne was prepared according to literature procedures (*6a*). **Caution**: The synthesis of 1,4-dilithiobutadiyne as described involves the use of n-BuLi which is pyrophoric and should thus be used only with the necessary precautions. Thermogravimetric analyses (TGA) were performed on a TA Instruments SDT 2960 DTA-TGA thermogravimetric analyzer. Aging studies were carried out on a DuPont 951 thermogravimetric analyzer in air (60 mL/min.). Differential scanning calorimetry analyses (DSC) were performed on a DuPont 910 instrument. All thermal measurements were carried out at a heating rate of 10 °C/min and a nitrogen flow rate of 60 mL/min. unless otherwise noted. Gel permeation chromatography (GPC) data were collected using a Hewlett-Packard Series 1050 pump and two Altex μ-spherogel columns (size 10^3 and 10^4 Å, respectively) connected in series. GPC values were referenced to polystyrene standards. Infrared spectra were obtained using a Nicolet Magna 750 FTIR spectrometer. ^1H and ^{13}C NMR spectra were recorded on a Bruker AC-300 NMR spectrometer in $CDCl_3$. Elemental analyses were performed by E and R Microanalytical Labs. Corona, N.Y.

Preparation of poly(1,3-dimethyl-1,3-diphenyldisiloxyldiacetylene), 7. A mixture of 1, 4-dilithiobutadiyne (6.3 mmol) in THF(10 mL)/hexane (10 mL) was cooled in an ice bath. To this slurry, was added a solution of 1,3-chloro-1,3-dimethyl-1,3-

diphenyldisiloxane (2.0 g, 6.3 mmol) in THF (5 mL). The ice bath was removed and the resulting solution was stirred for 90 minutes at room temperature at which time the reaction was judged to be complete by the disappearance of a band at 2140 cm^{-1} in the infrared spectrum (*10*). The solution was then diluted with 30 mL of diethyl ether and 50 mL of ice water. The aqueous layer was separated and extracted twice with diethyl ether (20 mL). The organic extracts were combined, dried over $MgSO_4$, and filtered through a Celite pad. Removal of volatiles under reduced pressure gave a viscous brown material. Further heating of the polymeric mixture to 75 °C at 0.1 Torr for six hours afforded a brown solid (1.67 g, 87 %). Elem. Anal. ($C_{18}H_{16}Si_2O$) Calcd: C, 71.00; H, 5.30; Si, 18.45. Found: C, 71.03; H, 5.37; Si, 18.57.

Preparation of poly(tetraphenyldisiloxyldiacetylene), 5a. The procedure that was used in the preparation of **7** was adapted for the synthesis of **5a.** The reaction of 1,4-dilithiobutadiyne (6.3 mmol) with 1,3-dichlorotetraphenyldisiloxane (2.88 g, 6.3 mmol) gave a brown solution of crude polymer. Work-up gave a solid, brown product (2.31g, 85 %). Elem. Anal. ($C_{28}H_{20}Si_2O$) Calcd: C, 78.46; H, 4.70; Si, 13.10. Found: C, 76.71; H, 5.21; Si, 12.40.

Preparation of tetraphenyl/tetramethyldisiloxyldiacetylene copolymers, 5b-d. A similar procedure to that used in the preparation of **7** was used to prepare copolymers **5b-d** by reacting 1,4-dilithiobutadiyne with appropriate amounts of both 1,3-dichlorotetraphenyldisiloxane and 1,3-dichlorotetramethyldisiloxane.

75/25 tetraphenyldisiloxyl/tetramethyldisiloxyl units, 5b. The reaction of 1,4-dilithiobutadiyne (6.38 mmol) with 1,3-dichlorotetraphenyldisiloxane (2.16 g, 4.8 mmol) and 1,3-dichlorotetramethyldisiloxane (0.33 g, 1.6 mmol) gave a brown solution. Work-up gave a solid, brown product (1.83 g, 78 %). Elem. Anal. ($C_{23}H_{18}Si_2O$) Calcd: C, 75.36; H, 4.95; Si, 15.32. Found: C, 75.13; H, 5.14; Si, 15.08.

50/50 tetraphenyl/tetramethyldisiloxyl, 5c. The reaction of 1,4-dilithiobutadiyne (6.38 mmol) with 1,3-dichlorotetraphenyldisiloxane (1.44 g, 3.2 mmol) and 1,3-dichlorotetramethyldisiloxane (0.65 g, 3.2 mmol) gave a brown solution. Work-up gave a solid, brown product (1.58 g, 81 %). Elem. Anal. ($C_{18}H_{16}Si_2O$) Calcd: C, 71.00; H, 5.30; Si, 18.45. Found: C, 71.15; H, 5.46; Si, 18.74.

25/75 tetraphenyl/tetramethyldisiloxyl, 5d. The reaction of 1,4-dilithiobutadiyne (6.38 mmol) with 1,3-dichlorotetraphenyldisiloxane (0.72 g, 1.6 mmol) and 1,3-dichlorotetramethyldisiloxane (0.97 g, 4.8 mmol) gave a brown solution. Work-up gave a solid, brown product (1.31 g, 84 %). Elem. Anal. ($C_{13}H_{14}Si_2O$) Calcd: C, 64.41; H, 5.82; Si, 23.17. Found: C, 64.15; H, 6.23; Si, 23.27.

Preparation of tetraphenyldisiloxyl/1,7-bis(tetramethyldisiloxyl)-m-carborane copolymers, 5e-f.

90/10 tetraphenyldisiloxyl/1,7-bis(tetramethyldisiloxyl)-m-carborane, 5e. A similar procedure to that used in the preparation of **7** was used in the reaction of 1,4-

dilithiobutadiyne (6.38 mmol) with 1,3-dichlorotetraphenyldisiloxane (2.59 g, 5.74 mmol) and 1,7-bis(tetramethyldisiloxyl)-m-carborane (0.30 g, 0.63 mmol). Work-up gave a solid, brown product (2.34 g, 85 %). Elem. Anal. ($C_{26.6}H_{21.4}Si_{2.2}O_{1.1}B$) Calcd: C, 74.08; H, 5.00; Si, 14.33; B, 2.51. Found: C, 73.71; H, 5.26; Si, 14.68; B, 2.33.

60/40 tetraphenyldisiloxyl/1,7-bis(tetramethyldisiloxyl)-m-carborane, 5f. A similar procedure to that used in the preparation of **7** was used in the reaction of 1,4-dilithiobutadiyne (6.38 mmol) with 1,3-dichlorotetraphenyldisiloxane (1.73 g, 3.83 mmol) and 1,7-bis(tetramethyldisiloxyl)-m-carborane (1.22 g, 2.55 mmol). Work-up gave a solid, brown product (2.35 g, 84 %). Elem. Anal. ($C_{22.4}H_{25.6}Si_{2.8}O_{1.4}B_4$) Calcd: C, 61.27; H, 5.88; Si, 17.91; B, 9.85. Found: C, 60.98; H, 6.13; Si, 17.94; B, 9.87.

Results and Discussion

Synthesis. The synthesis of the tetraphenyl/tetramethyldisiloxyldiacetylene polymers and copolymers is shown in Scheme 1. The polymeric materials **5a-f** were isolated as brown solids with yields in the 78-85 % range. These materials were prepared having molar ratios of **4** to **3** incorporated into the main chain as follows: 100:0, **5a**; 75:25, **5b**; 50:50, **5c**; and 25:75, **5d**. Copolymers **5e** and **5f** were prepared having **2** and **6** incorporated into the main chain in molar ratios of 90:10 and 60:40, respectively. The synthetic strategy used is a one-pot, two step reaction sequence adapted from previously published procedures (*7,8*). The reaction of **4** with appropriate ratios of **2** and either **3** or **6** resulted in the formation of **5a-f** with concomitant formation of LiCl. Poly(1,3-dimethyl-1,3-diphenyldisiloxyl diacetylene), **7**, was prepared similarly from the reaction of **4** with 1,3-dichloro-1,3-dimethyl-1,3-diphenyldisiloxane.

The polymers and copolymers **5a-f** were characterized by FTIR spectroscopy. A strong absorption around 2068 cm^{-1} confirms the presence of the internal butadiyne groups. The structure of polymer **5a** was supported by absorptions at 3048 cm^{-1} (Ar-H), 1123 cm^{-1} (Si-Ph), and 1053 cm^{-1} (Si-O-Si). Copolymers **5b-d** showed additional stretches at 1261 and 850-790 cm^{-1} ($Si\text{-}CH_3$) consistent with the proposed copolymer composition. Copolymers **5e-f** showed an additional absorption at 2595 cm^{-1} (B-H) verifying the presence of the carborane moiety.

^1H NMR spectroscopy supports the proposed structures of **5a-f** with resonances in the 7.7-7.2 ppm region for the phenyl groups and 0.2-0.6 ppm for the silicon methyl protons with intensities consistent with the expected copolymer composition. $^{13}C\{^1H\}$ NMR showed two resonances in the 83-90 ppm region for the diacetylenic carbons. The resonances for the phenyl and methyl carbons occurred in the 132-138 and 0.3-2.0 ppm regions, respectively. For **5e-f** the carborane cage carbons appeared at 68 ppm.

Poly(1,3-dimethyl-1,3-diphenyldisiloxyldiacetylene), **7**, was also characterized using gel permeation chromatography (GPC). GPC analysis showed a broad molecular weight distribution with a peak maximum at approximately 10,000 (relative to polystyrene) molecular weight. This polymer was chosen as a representative material and the GPC results are consistent with previous observations (*11*).

Thermal curing of **7** was studied by DSC. When heated to 450 °C, polymer **7** showed a prominent exotherm with a peak maximum at 314 °C. Upon cooling and rerunning the sample, an essentially featureless DSC trace was observed (Figure 1). This

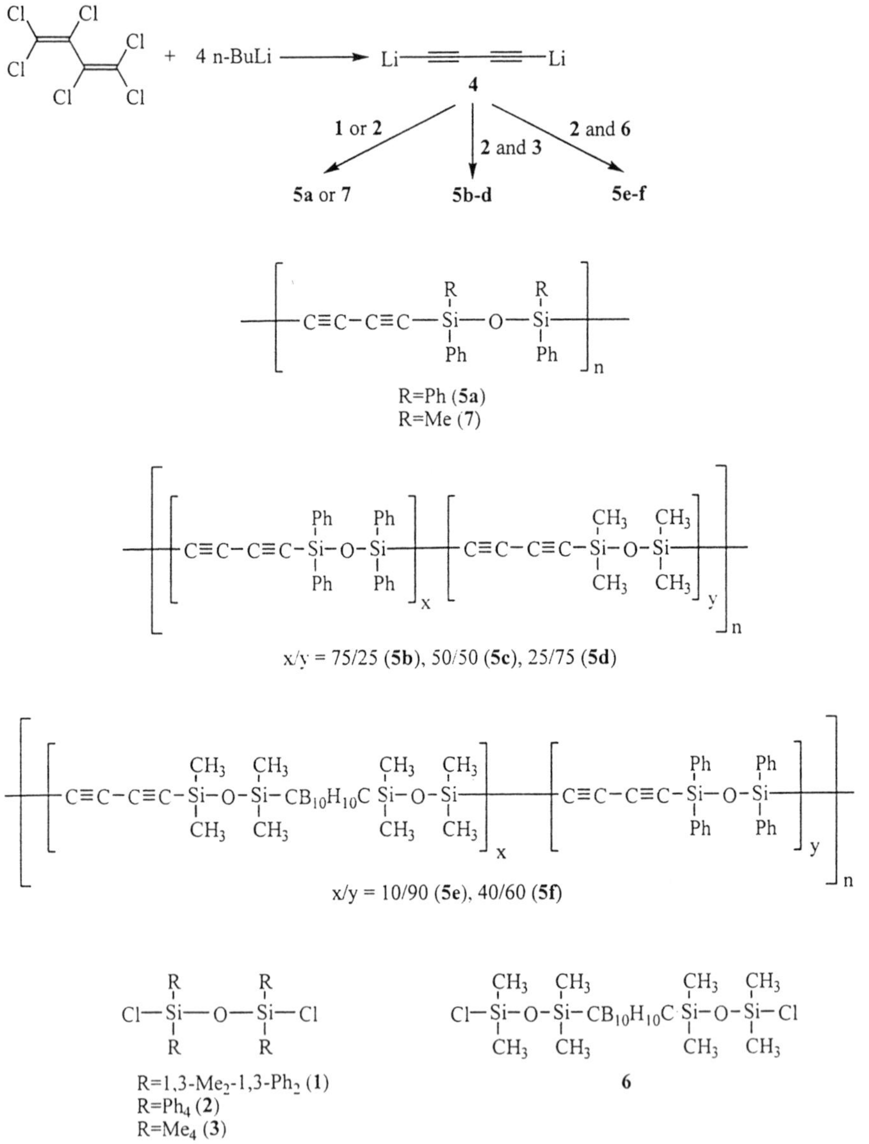

$Cl_2C{=}CCl{-}CCl{=}CCl_2$ + 4 n-BuLi ⟶ $Li{-}C{\equiv}C{-}C{\equiv}C{-}Li$ (**4**)

4 ⟶ (1 or 2) **5a** or **7**; (2 and 3) **5b-d**; (2 and 6) **5e-f**

$$\left[-C{\equiv}C{-}C{\equiv}C{-}Si(R)(Ph){-}O{-}Si(R)(Ph)-\right]_n$$

R=Ph (**5a**)
R=Me (**7**)

$$\left[\left[-C{\equiv}C{-}C{\equiv}C{-}Si(Ph)_2{-}O{-}Si(Ph)_2-\right]_x\left[-C{\equiv}C{-}C{\equiv}C{-}Si(CH_3)_2{-}O{-}Si(CH_3)_2-\right]_y\right]_n$$

x/y = 75/25 (**5b**), 50/50 (**5c**), 25/75 (**5d**)

$$\left[\left[-C{\equiv}C{-}C{\equiv}C{-}Si(CH_3)_2{-}O{-}Si(CH_3)_2{-}CB_{10}H_{10}C{-}Si(CH_3)_2{-}O{-}Si(CH_3)_2-\right]_x\left[-C{\equiv}C{-}C{\equiv}C{-}Si(Ph)_2{-}O{-}Si(Ph)_2-\right]_y\right]_n$$

x/y = 10/90 (**5e**), 40/60 (**5f**)

$$Cl{-}Si(R)_2{-}O{-}Si(R)_2{-}Cl$$

R=1,3-Me_2-1,3-Ph_2 (**1**)
R=Ph_4 (**2**)
R=Me_4 (**3**)

$$Cl{-}Si(CH_3)_2{-}O{-}Si(CH_3)_2{-}CB_{10}H_{10}C{-}Si(CH_3)_2{-}O{-}Si(CH_3)_2{-}Cl$$

6

Scheme 1. Preparation of Linear Inorganic-Organic Hybrid Diacetylene Polymers.

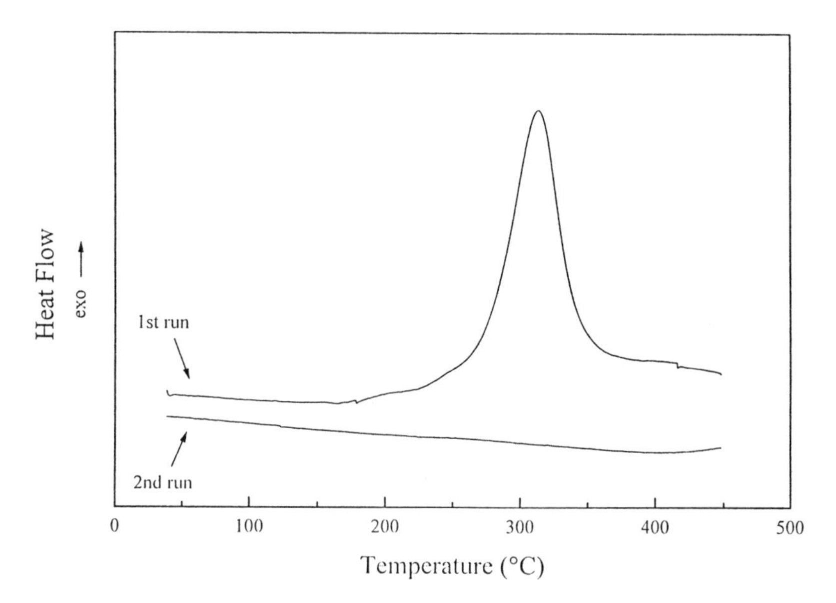

Figure 1. DSC Trace of **7**.

exotherm is attributed to the thermally induced crosslinking of the internal diacetylene groups. This assignment is supported by infrared studies which showed the disappearance of the butadiyne stretch (2160 cm^{-1}) upon heating a sample of **7** to 450 °C (Figure 2). DSC studies on copolymers **5b-f** showed similar behavior with exotherm peak maxima occurring at 311 °C (**5b**), 300 °C (**5c**), 298 °C (**5d**), 315 °C (**5e**), and 311 °C (**5f**).

Thermogravimetric analysis of **5a-f** showed these materials have high thermal stabilities. When heated to 1500 °C (10 °C/min) under a nitrogen atmosphere these polymers gave char yields of 68 % (**5a**), 71 % (**5b**), 77 % (**5c**), 76 % (**5d**), 72 % (**5e**), and 74 % (**5f**). Oxidative stabilities of the chars obtained from heat treatment of the polymers to 1500 °C under N_2 were determined by cooling to the sample to room temperature and heating to 1500 °C under an air atmosphere. The chars obtained from **5a-d** showed initial weight loss beginning around 650 °C with weight retentions at 1500 °C of 38 %, **5a**; 43 %, **5b**; 63 %, **5c**; and 84 %, **5d** (Figure 3). The trend of lower oxidative stability in air with an increasing mole percent of phenyl groups in **5a-d** is consistent with the formation of a char with a higher carbon content (*12*). It should be noted that precursor materials with high carbon content do not always give high carbon ceramic chars (*13*). Oxidative stabilities of the chars obtained from heat treatment of **5e-f** to 1500 °C showed increased oxidative stability over the chars of **5a-d** with similar phenyl group content. At 1500 °C in air the weight retentions for the chars of **5e** and **5f** were 52 % and 99 %, respectively (Figure 4). This latter result clearly demonstrates the increased oxidative stability obtained from incorporating both boron and silicon into the polymeric chain (*7*). Examination of **7** by TGA showed 78 % weight retention at 1500 °C under a nitrogen atmosphere. Heating a sample of the resulting char of **7** to 1500 °C in air gave a weight retention of 57 %. Thus, the thermooxidative properties of **7** are similar to those of copolymer **5c**.

Aging studies were performed on the thermoset and ceramic derived from poly(1,3-dimethyl-1,3-diphenyldisiloxyldiacetylene), **7**. The thermoset sample was prepared by heat treatment of the polymer to 450 °C (10 °C/min) under N_2 (98 % weight retention). Aging of the thermoset was studied by heating the sample in air for 5 hours sequentially at 250, 300, 350, and 400 °C followed by 15 hours at 450 °C (Figure 5). After 5 hours at 250 °C the thermoset showed a slight weight gain with 101 % weight retention. After 5 hours at 300 °C the thermoset showed a slight weight loss returning to 100 % weight retention. More substantial weight loss was observed during aging at higher temperatures with weight retentions of 97 and 91 % after 5 hours at 350 and 400 °C, respectively. After 15 hours at 450 °C, the sample showed 44 % weight retention which had nearly reached a plateau. This behavior is similar to that of other crosslinked siloxane systems (*14*).

Aging studies were also performed on the char obtained from heat treatment of **7**. The char sample was prepared by heating the polymer to 1000 °C (10 °C/min) under N_2. Aging of the ceramic was studied by heating the sample in air sequentially at 400 °C, 500 °C, 600 °C, and 700 °C for five hours at each temperature (Figure 6). The aging at 400 °C resulted in a slight weight gain with 101 % weight retention after 5 hours. Higher temperatures resulted in significant weight loss with weight retentions of 83 and 50 % after 5 hours at 500 and 600 °C, respectively, with the weight retention stabilizing at 49 % at 700 °C.

Conclusion

Diacetylene disiloxyl polymers and copolymers containing phenyl and methyl substituents were synthesized and their structures characterized by FTIR, ^{1}H and $^{13}C\{^{1}H\}$ NMR spectroscopies. Copolymers containing tetraphenyldisiloxyl and 1,7-bis(tetramethyldisiloxyl)-m-carborane units in the polymer backbone were also studied. DSC and TGA analyses were used to evaluate the thermal and oxidative stabilities of these new materials and their conversion into thermosets and ceramics. The copolymer-derived ceramic products exhibited good thermal stabilities up to 1500 °C under a nitrogen atmosphere. The oxidative stability of the chars of phenyl-substituted copolymers was shown to decrease with increasing percentage of phenyl groups. The oxidative stability of the char of the 60/40 tetraphenyl/1,7-bis(tetramethyldisiloxyl)-m-carborane copolymer is excellent to 1500 °C, demonstrating the protecting effect of boron and silicon in these systems. Further work is currently underway to determine the thermal, oxidative, and mechanical properties of the thermosets and ceramics obtained from these copolymers.

Acknowledgements.

Support for this work was provided by The Office of Naval Research (ONR) and is gratefully acknowledged. Eric J. Houser is also grateful to the American Society for Engineering Education (ASEE) for his postdoctoral fellowship. Thanks to Dr. Tai Ho for his GPC work.

Literature Cited.

1. a) Fitzer, E. *Carbon* **1987**, *25*, 163. b) Pilato, L. A.; Michno, M. J. *Advanced Composite Materials*; Springer-Verlag: Heidelberg 1994.
2. Savage, G. *Carbon-Carbon Composites*; Chapman & Hall: New York, 1992.
3. For recent reviews on the formation of SiC and related ceramics from polymers see: a) Birot, M.; Pillot, J. - P.; Dunoguès, J. *Chem. Rev.* **1995**, *95*, 1443. b) Laine, R. M.; Babonneau, F. *Chem. Mater.* **1993**, *5*, 260. c) Wynne, K. J.; Rice, R. W. *Ann. Rev. Mater. Sci.* **1984**, *14*, 297.
4. a) Seyferth, D.; Lang, H. *Organometallics* **1991**, *10*, 551. b) Schmidt, W. R.; Interrante, L. V.; Doremus, R. H.; Trout, T. K.; Marchetti, P. S.; Maciel, G. E. *Chem. Mater.* **1991**, *3*, 257. c) Boury, B.; Corriu, R. J. P.; Leclercq, D.; Mutin, P. H.; Planeix, J.-M.; Vioux, A. *Organometallics* **1991**, *10*, 1457.
5. a) Fazen, P. J.; Remsen, E. E.; Beck, J. S.; Carroll, P. J.; McGhie, A. R.; Sneddon, L. G. *Chem. Mater.* **1995**, *7*, 1942. b) Wideman, T.; Sneddon, L. G. *Chem. Mater.* **1996**, *8*, 3-5. c) Kawaguchi, M.; Kawashima, T.; Nakajima, T. *Chem. Mater.* **1996**, *8*, 1197.
6. a) Ijadi-Maghsoodi, S.; Barton, T. J. *Macromolecules* **1990**, *23*, 4485. b) Corriu, R. J. P.; Guérin, C.; Henner, B.; Kuhlmann, T.; Jean, A. *Chem. Mater.* **1990**, *2*, 351. c) Corriu, R.; Gerbier, Ph.; Guérin, C.; Henner, B.; Fourcade, R. *J. Organomet. Chem.* **1993**, *449*, 111. d) Bréfort, J. L.; Corriu, R. J. P.; Gerbier, Ph.; Guérin, C.; Henner, B. J. L.; Jean, A.; Kuhlmann, Th. *Organometallics* **1992**, *11*

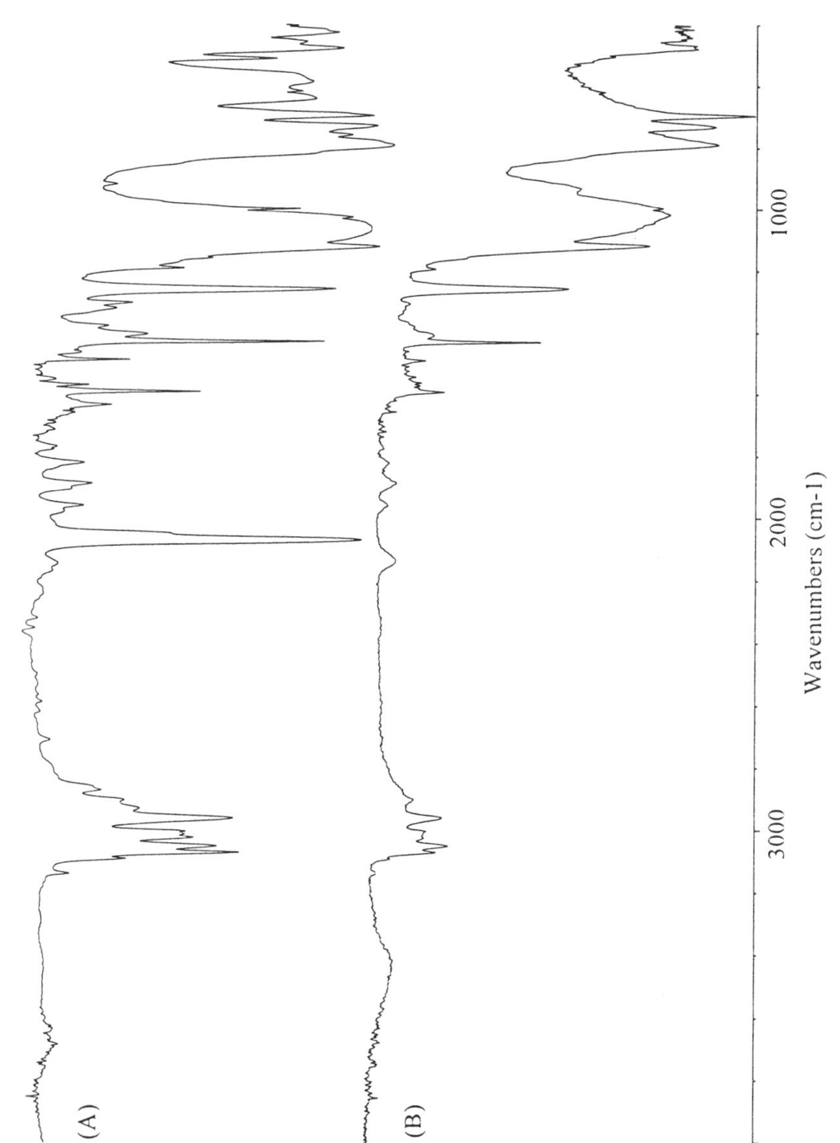

Figure 2. Infrared Spectra (KBr) of **7** (A) and Thermoset of **7** (B).

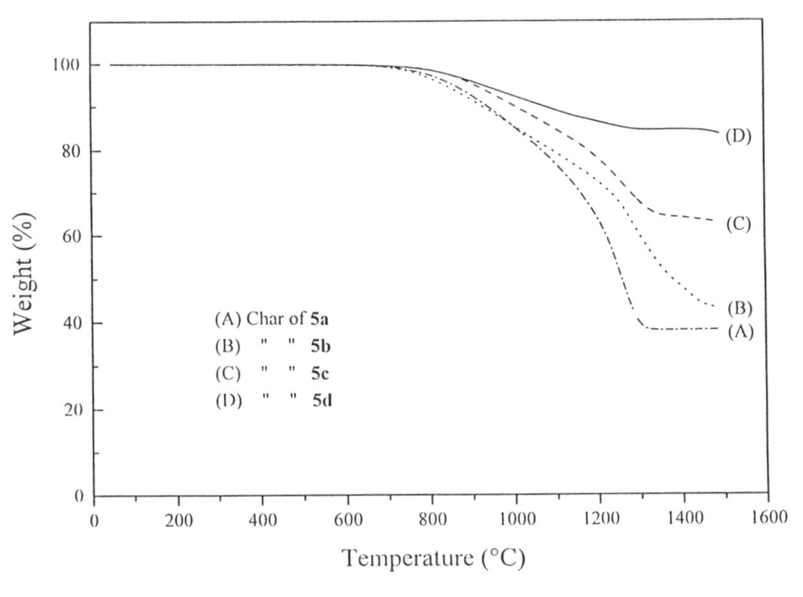

Figure 3. Oxidative Stability of Chars from Copolymers **5a-d**.

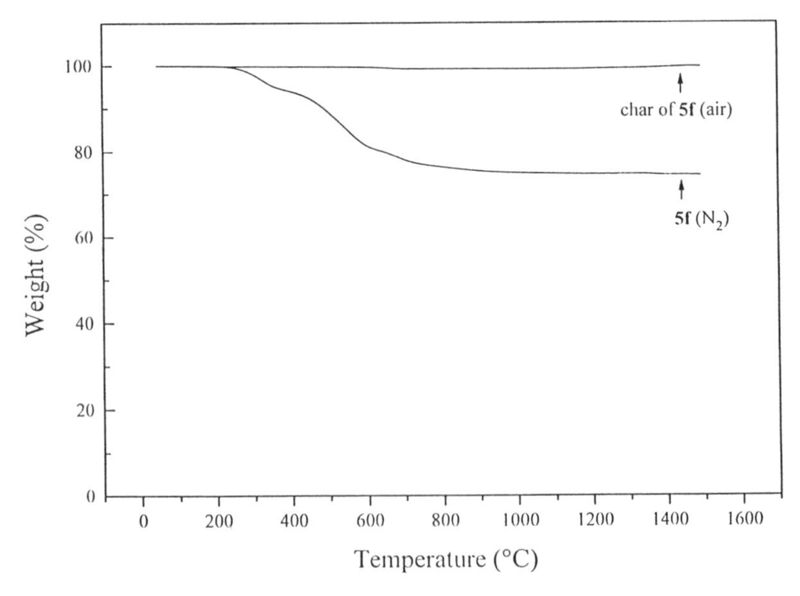

Figure 4. TGA of **5f** (N_2) and Char of **5f** (in air).

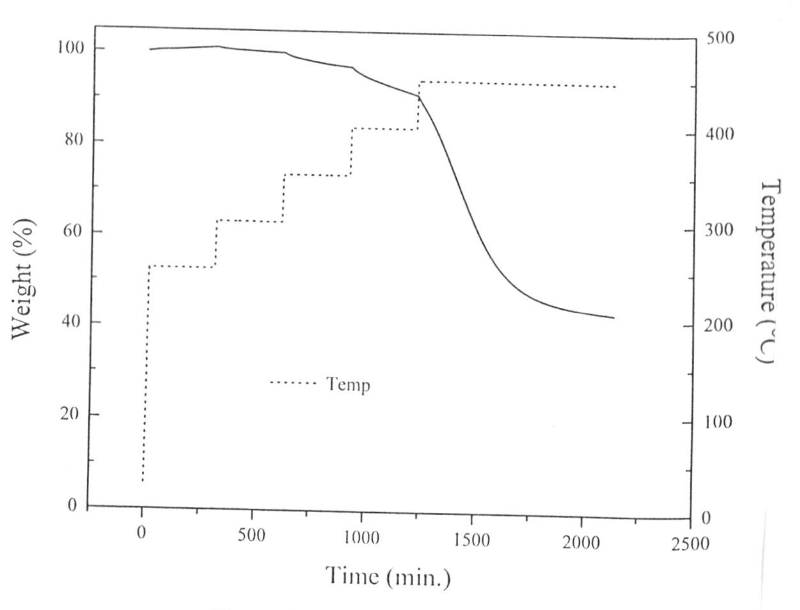

Figure 5. Aging of Thermoset of **7**.

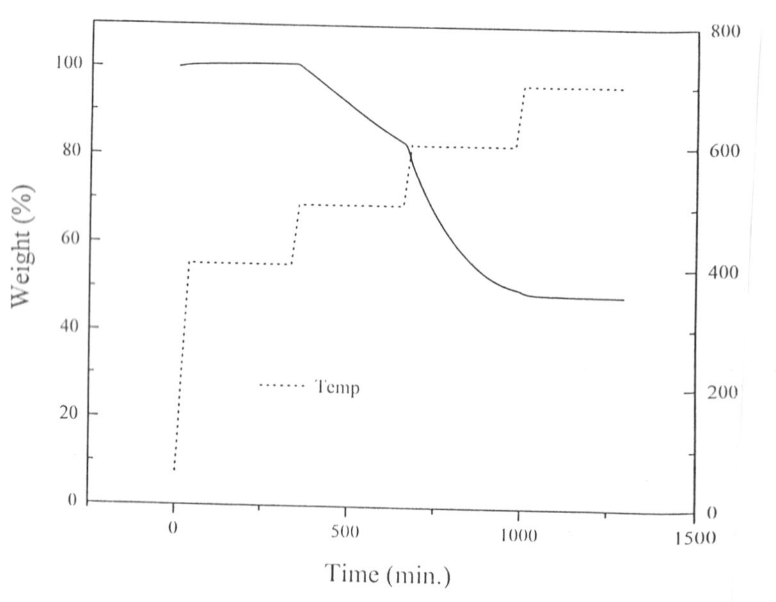

Figure 6. Aging of Char Obtained from Heat Treatment of **7** to 1000 °

2500. e) Corriu, R. J. P.; Gerbier, P.; Guérin, C.; Henner, B. J. L.; Jean, A.; Mutin, P. H. *Organometallics* **1992**, *11*, 2507.
7. Henderson, L. J.; Keller, T. M. *Macromolecules* **1994**, *27*, 1660.
8. Son, D. Y.; Keller, T. M. *Macromolecules* **1995**, *28*, 399.
9. a) Boury, B.; Corriu, R. J.; Douglas, W. E. *Chem. Mater.* **1991**, *3*, 487. b) Allcock, H. R.; McDonnell, G. S.; Riding, G. H.; Manners, I. *Chem. Mater.* **1990**, *2*, 425.
10. The 2140 cm^{-1} band in the FTIR spectrum of the crude reaction mixtures is assigned to terminal butadiyne groups which diminish in concentration as the polymer forms.
11. Son, D. Y.; Keller, T. M. In *Fire and Polymers II: Materials and Tests for Hazard Prevention* A.C.S. Symposium Ser. No. 599; Ed. by Nelson, G. L.; American Chemical Society: Washington, D.C. 1995.
12. The apparent higher carbon content in the chars with increasing percentages of phenyl substituents in the polymer was, in general, verified by elemental analysis. A low carbon content was observed for the char obtained from **5a** which is attributed to incomplete combustion of the sample. Elem. Anal. of chars from **5a**: %C, 67.40; %H, 0.0; %Si, 16.76. **5b**: %C, 70.84; %H, 0.0; %Si, 18.43. **5c**: %C, 65.72; %H, 0.0; %Si, 22.19. **5d**: %C, 62.06; %H, 0.0; %Si, 23.82.
13. See Bouillon, E.; Pailler, R.; Naslain, R.; Bacqué, E.; Pillot, J.-P.; Birot, M.; Dunoguès, J.; Huong, P. V. *Chem. Mater.* **1991**, *3*, 356.
14. Kovalenko, Y. V.; Pustyl'nik, M. L.; Bebchuk, T. S.; Muzafarova, M. N.; Zherdev, Y. V.; Nederosol, V. D.; Andrianov, K. A. *Polym. Sci. U.S.S.R.* **1978**, *20*, 1824.

Chapter 21

Blending Studies of Poly(siloxane acetylene) and Poly(carborane siloxane acetylene)

Teddy M. Keller and David Y. Son[1]

Chemistry Division, Materials Chemistry Branch, Naval Research Laboratory, Code 6120, 4555 Overlook Avenue, SW, Washington, DC 20375–5320

High temperature thermosets and ceramics have been synthesized by heat treatment of various blends of poly (siloxane-acetylene) and poly (carborane-siloxane-acetylene). The polymeric blends give high char yields on pyrolysis, and the resultant chars show excellent oxidative stability to at least 1500 °C. The thermosets and ceramic chars show similar oxidative stability to previously studied copolymers containing varying amounts of siloxane, carborane, and acetylene units within the backbone. It has been determined that only a small percentage of carborane is necessary to provide this oxidation protection. Thus, these precursor linear hybrid polymers are more cost-effective than previous polymers which contained carborane in each repeating unit.

Our current interest in inorganic-organic linear hybrid polymers as precursors to high temperature thermosets and ceramics has led us to investigate the synthesis of novel materials containing silicon, carborane, and acetylenic segments. Several poly (carborane-siloxane-acetylene)s[1,2] **1** and **2** and poly (siloxane-acetylene)s[3] **3** have been synthesized (see Scheme 1) and are being evaluated as high temperature matrix materials for composites and as precursor materials to ceramics for applications under extreme environmental conditions.

The major advantage of our approach is that the desirable features of inorganics and organics such as high thermal and oxidative stability and processability are incorporated into the same polymeric chain. The siloxane units provide thermal and chain flexibility to polymeric materials. Siloxane-acetylenic polymers have also been made but lack the thermal and oxidative stability that the carborane units possess. The chemistry involved in synthesizing poly(siloxane) and poly(carborane-siloxane) has been modified to accommodate the inclusion of an acetylenic unit in the backbone. The novel linear

[1]Current address: Department of Chemistry, Southern Methodist University, Dallas, TX 75275

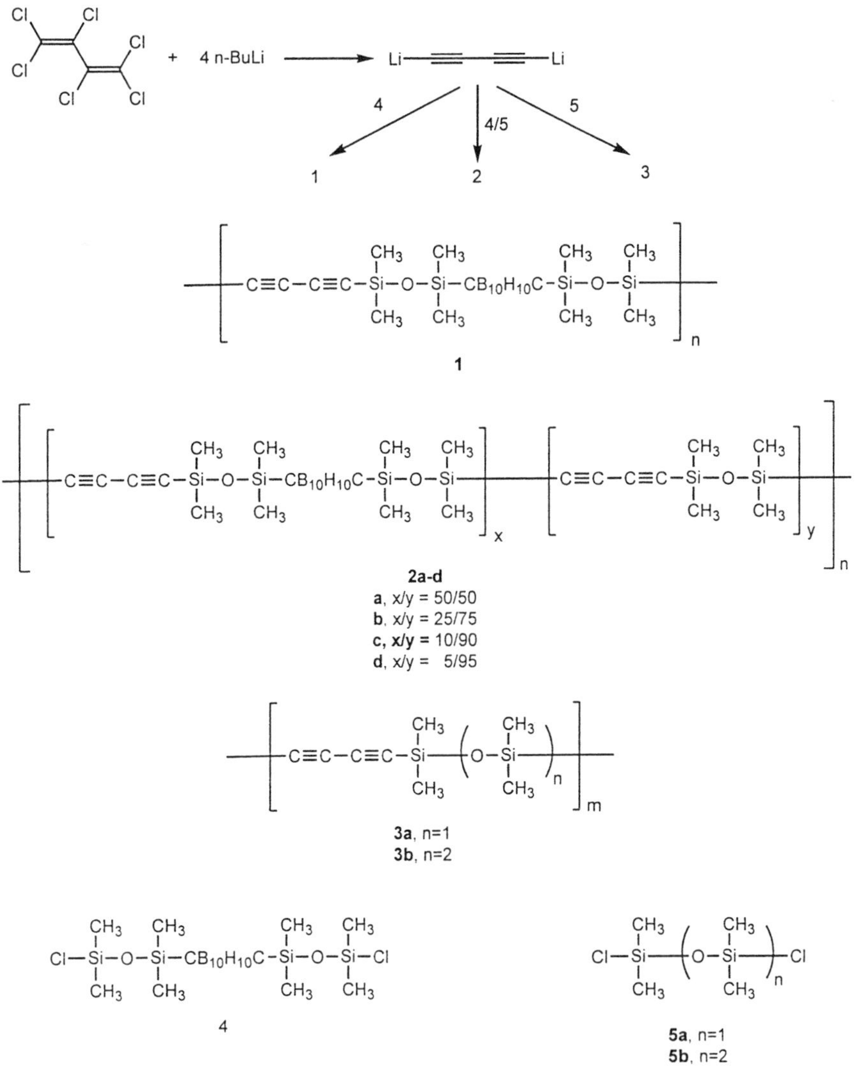

+ 4 n-BuLi → Li—≡—≡—Li

4 → 1; 4/5 → 2; 5 → 3

1

2a-d

a, x/y = 50/50
b, x/y = 25/75
c, x/y = 10/90
d, x/y = 5/95

3a, n=1
3b, n=2

4

5a, n=1
5b, n=2

Scheme 1. Preparation of linear inorganic-organic hybrid polymers

polymers have the advantage of being extremely easy to process and convert into thermosets or ceramics since they are either liquids at room temperature or low melting solids and are soluble in most organic solvents. They are designed as thermoset polymeric precursors. The cross-linked density of the thermosets is easily controlled as a function of the quantity of reactants used in the synthesis. The acetylenic functionality provides many attractive advantages relative to other cross-linking centers. The acetylene group remains inactive during processing at lower temperatures and reacts either thermally or photochemically to form conjugated polymeric cross-links without the evolution of volatiles.

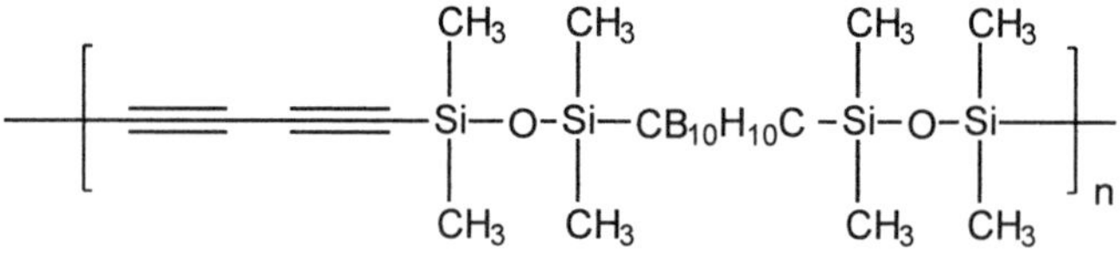

$$\left[-C \equiv C - C \equiv C - Si(CH_3)_2 - O - Si(CH_3)_2 - CB_{10}H_{10}C - Si(CH_3)_2 - O - Si(CH_3)_2 - \right]_n$$

1

$$\left[\left[-C \equiv C - C \equiv C - Si(CH_3)_2 - O - Si(CH_3)_2 CB_{10}H_{10}C Si(CH_3)_2 - O Si(CH_3)_2 - \right]_x \left[-C \equiv C - C \equiv C - Si(CH_3)_2 O Si(CH_3)_2 - \right]_y \right]_z$$

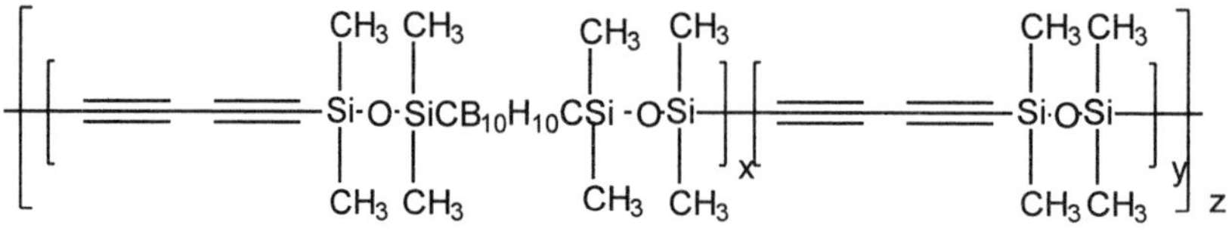

2a-d

a, x/y = 50/50
b, x/y = 25/75
c, x/y = 10/90
d, x/y = 5/95

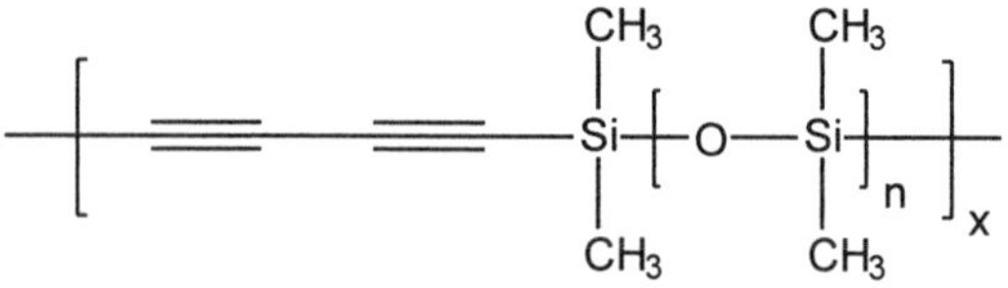

$$\left[-C \equiv C - C \equiv C - Si(CH_3)_2 \left[-O - Si(CH_3)_2 - \right]_n \right]_x$$

3a, n=1
3b, n=2

This paper is concerned with blending **1** and **3** in an attempt to arrive at similar thermoset and ceramic compositions as found for copolymer **2** upon thermal treatment. Thermal analysis studies were performed on thermosets **4** and ceramics **5** obtained from various blends of **1** and **3a.**

Experimental

The synthesis of **1**, **2**, and **3** have been reported previously.[1-3] All reactions were carried out in an inert atmosphere unless otherwise noted. Solvents were purified by established procedures. 1,3-Dichlorotetramethyldisiloxane and 1,5-dichlorohexamethyltrisiloxane were obtained from Silar Laboratories and used as received. *n*-Butyllithium (2.5 M in hexane) was obtained from Aldrich and used as received. 1,7-Bis(chlorotetramethyldisiloxyl)-*m*-carborane 1 was purchased from Dexsil Corporation. Hexachlorobutadiene was obtained from Aldrich and distilled before use. Cure and thermal analysis studies were performed on various mixtures of **1** and **3a** in milligram quantities. Thermogravimetric analyses (TGA) were performed on a DuPont SDT 2960 Simultaneous DTA-TGA analyzer. Differential scanning calorimetry analyses (DSC) were performed on a DuPont 910 instrument. Unless otherwise noted, all thermal experiments were carried out at a heating rate of 10 °C/min and a nitrogen flow rate of 50 cc/min.

Synthesis of poly (carborane-siloxane-acetylene) 1. In a typical synthesis, a 2.5M hexane solution of *n*-BuLi (34.2 ml, 85.5 mmol) in 12.0 ml of THF was cooled to -78 °C under an argon atmosphere. Hexachlorobutadiene (5.58 g, 21.4 mmol) in 2.0 ml THF was added dropwise by cannula. The reaction was allowed to warm to room temperature and stirred for 2 hours. The 1,4-dilithiobutadiyne in THF was then cooled to -78 °C. At this time, an equimolar amount of 1,7-bis(chlorotetramethyldisiloxyl)-*m*-carborane (10.22 g, 21.4 mmol) in 4.0 ml THF was added dropwise by cannula while stirring. The temperature of the reaction mixture was allowed to slowly rise to room temperature. While stirring the mixture for 1 hour, a copious amount of white solid (LiCl) was formed. The reaction mixture was poured into 100 ml of dilute hydrochloric acid resulting in dissolution of the salt and the separation of a viscous oil. The polymer **1** was extracted into ether. The ethereal layer was washed several times with water until the washing was neutral, separated, and dried over anhydrous sodium sulfate. The ether was evaporated at reduced pressure leaving a dark-brown viscous polymer **1**. A 97% yield (9.50 g) was obtained after drying in vacuo. GPC analysis indicated the presence of low molecular weight species (≈500) as well as higher average molecular weight polymers (Mw≈4900, Mn≈2400). Heating of **1** under vacuum at 150 °C removed lower molecular weight volatiles giving a 92% overall yield. Major FTIR peaks (cm^{-1}): 2963 (C-H); 2600 (B-H); 2175 (C≡C); 1260 (Si-C); and 1080 (Si-O).

Synthesis of poly (siloxane-acetylene) 3a. A mixture of 1,4-dilithiobutadiyne (6.3 mmol) in THF/hexane was cooled in a dry ice/acetone bath. To this mixture, 1,3-dichlorotetramethyldisiloxane (1.24 mL, 6.3 mmol) was added dropwise over 15 min. After addition, the cold bath was removed and the mixture was stirred at room temperature for two hours. The tan mixture was poured into 20 mL of ice-cooled saturated aqueous ammonium chloride solution with stirring. The mixture was filtered through a Celite pad and the layers were separated. The aqueous layer was extracted twice with Et_2O and the combined organic layers were washed twice with distilled water and once with saturated aqueous NaCl solution. The dark brown organic layer was dried over anhydrous magnesium sulfate and filtered. Most of the volatiles were removed at reduced pressure and the residue was heated at 75 °C for three hours at 0.1 torr to give **3a** as a thick, dark brown material (1.04 g, 92%). Polymer **3a** slowly solidifies on standing at room temperature and liquefies at approximately 70 °C. ^{1}H NMR (ppm) 0.30 (s, 12H, -Si(CH_3)); ^{13}C NMR (ppm) 1.7, 1.9 (-Si(CH_3)), 84.9 (-Si-CC-), 86.9 (-Si-CC-). Anal. Calcd. for $(C_8H_{12}OSi_2)_n$: C, 53.31; H, 6.66; Si, 31.16. Found: C, 55.81; H, 7.61; Si, 27.19.

Preparation of Homogeneous Mixtures from 1 and 3a. Molar mixtures (50/50, 25/75, and 10/90 weight amounts) of **1** and **3a** were weighed into a vial, mixed by dissolution in THF, and concentrated at reduced pressure. These compositions were used for thermal analysis studies.

Preparation of Thermoset 4. Various mixtures of **1** and **3a** were weighed into a TGA pan and cured by heating at 200, 250, 350, and 450 °C for 4 hours at each temperature under an inert atmosphere.

Conversion to Ceramic 5. Various mixtures of **1** and **3a** or **4** were weighed into a TGA pan and heated to either 1000 °C or 1500 °C under inert conditions. Upon cooling the ceramic chars were reheated to 1500 °C under a flow of air to determine the oxidative stability.

Oxidative aging studies. Various mixtures of **1** and **3a** were weighed into a TGA pan and either cured to a thermoset **4** or converted into a ceramic **5**. Thermoset **4** was then heated in sequence in a flow of air (50 cc/min) at 200, 250, 300, 350, and 400 °C for 5 hours at each temperature. Addtional heating at 450 °C was performed up to 15 hours. The ceramic compositions **5** were heated in sequence at 400, 500, and 600 °C for 5 hours. Further heat exposure at 700 °C was carried-out up to 15 hours.

Results and Discussion

Several molar mixtures (50/50, 25/75, and 10/90) of **1** and **3a** were prepared for cure and thermal analysis studies. Homogenous mixtures were obtained by dissolving the linear polymers **1** and **3a** in THF. After thorough mixing, the solvent was removed by distillation at reduced pressure. The resulting mixtures as prepared were viscous compositions. However, gummy, semicrystalline compositions formed after several days. Upon heating to 100 °C, the mixtures existed as viscous liquids.

$$\mathbf{1} + \mathbf{3a} \xrightarrow{\text{Cure}} \underset{\mathbf{4}}{\text{Thermoset}} \xrightarrow{\text{Pyrolysis}} \underset{\mathbf{5}}{\text{Ceramic}}$$

DSC Studies. DSC analyses of blends of **1** and **3a** show a homogeneous reaction initially to a thermoset. The DSC scans to 400 °C of the blends exhibit only one cure exotherm for each of the compositions studied (see Figures 1 and 2). For example, mole percent mixtures (10/90, 25/75 and 50/50) of **1** and **3a** display exotherms (polymerization

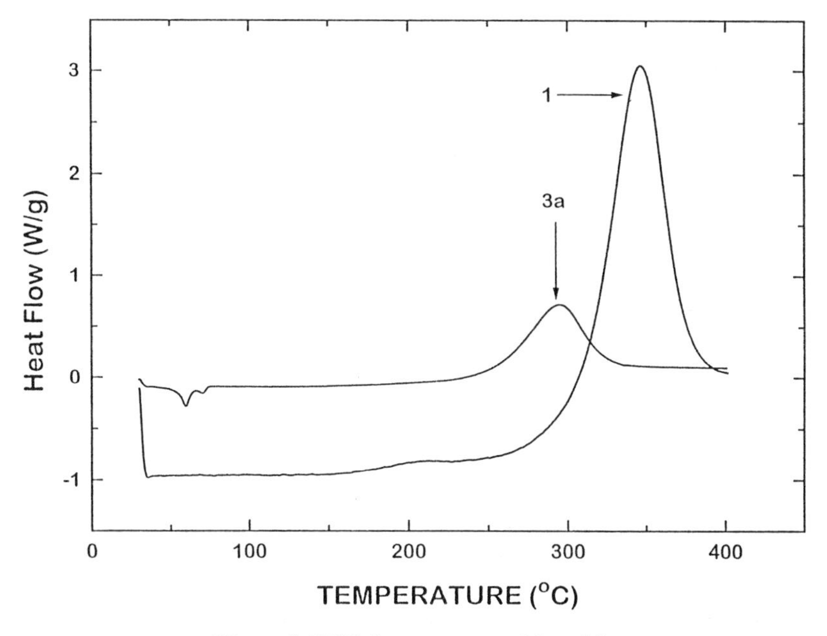

Figure 1. DSC thermograms of **1** and **3a**

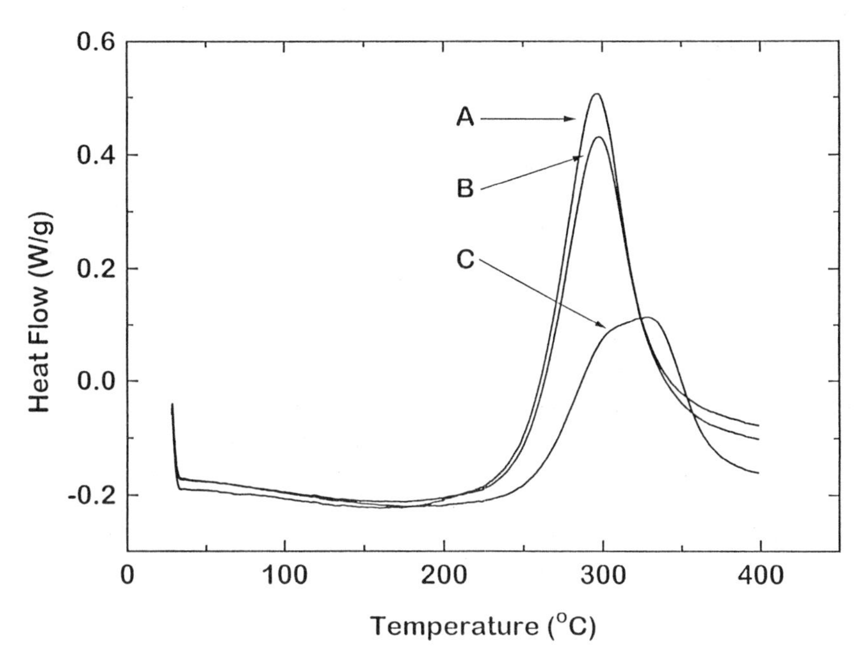

Figure 2. DSC thermograms of various mixtures of **1**/**3a**: (A) 10/90, (B) 25/75, and (C)50/50

reaction) peaking at 296, 298 and 328 °C, respectively. It is apparent from the observed cure temperature for the blends that **3a** being more reactive initially forms radicals that are not selective in the chain propagation reaction with the acetylenic units of both **1** and **3a**.[4] Samples that have been heat treated to 400 °C do not exhibit characteristic exothermic transitions. Copolymer **2c** shows a similar DSC thermogram with a strong exotherm at approximately 300 °C.[5]

Thermal and Oxidative Stability. The thermal and oxidative stability of various mixtures of **1** and **3a** was determined to 1500 °C by simultaneous TGA/DTA analysis. The scans were run at 10 °C/min at a gas flow of 50 cc/min in either nitrogen or air. When heated to 1000 °C and 1500 °C under inert conditions, the various mixtures containing **1** and **3a** afforded char yields of 79-80 and 77-78%, respectively. During the heat treatment, similar exothermic transitions (DTA) as found during the DSC scans were observed. Moreover, above 1000 °C, an exothermic transition is observed which is attributed to the formation of crystalline ceramic components such as SiC and B_4C. Upon cooling, the carbon/ceramic masses were reheated to 1500 °C in air. The oxidative stability of the charred mass was found to be a function of the amount of **1** present and the initial heat treatment. Charred samples obtained from heat treatment to 1000 °C and 1500 °C of 10/90, 25/75, and 50/50 molar weight percent of **1** to **3a** showed chars of 98, 98, 99% and 90, 97, 98% respectively, when reheated in air (see Figures 3 and 4).

The major difference observed was in the chars formed from the 10/90 mixtures. The crystalline-containing compositions (see Figure 4) lost most of their weight between 700 and 800 °C. For the amorphous compositions (see Figure 3), weight losses occurred between 600 and 700 °C and above 1000 °C. As the temperature was further increased, an acceleration in the weight loss was observed. These results indicate that the oxidative stability of the carbon/ceramic mass depends on the morphology.

Oxidative Aging Studies. Aging studies were performed on the thermosets derived from various blending compositions of **1** and **3a**. The compositions were cured by heating at 200, 250, 350, and 450 °C for 4 hours at each temperature under a nitrogen atmosphere. Aging of the thermoset was studied by heating the sample in air for 5 hours in sequence at 250, 300, 350, and 400 °C followed by 15 hours at 450 °C. Copolymer **2c** and the 10/90 composition showed similar thermo-oxidative stability upon conversion into a thermoset (see Figure 5). The stabilizing effect of the carborane unit was apparent. All of the samples gained weight during the oxidative exposure up to 350 °C. Moreover, less oxidation occurred on the surface as the amount of **1** increased. More extreme heat treatment at 400 and 450 °C showed an enhancement in oxidative stability with greater amounts of carborane (see Figure 6). The 50/50 mixture exhibited outstanding oxidative performance during the entire heat exposure. When **3a** was cured and aged under identical conditions, the sample gained almost 7% during the heat exposures from 200 to 300 °C. While at 350 and 400 °C, the sample had lost about 14% weight. Upon exposure at 450 °C for 5 hours, the sample lost another 25% weight. These results show the importance of boron at enhancing the oxidative stability of a polymeric material through the formation of a passive protective layer.

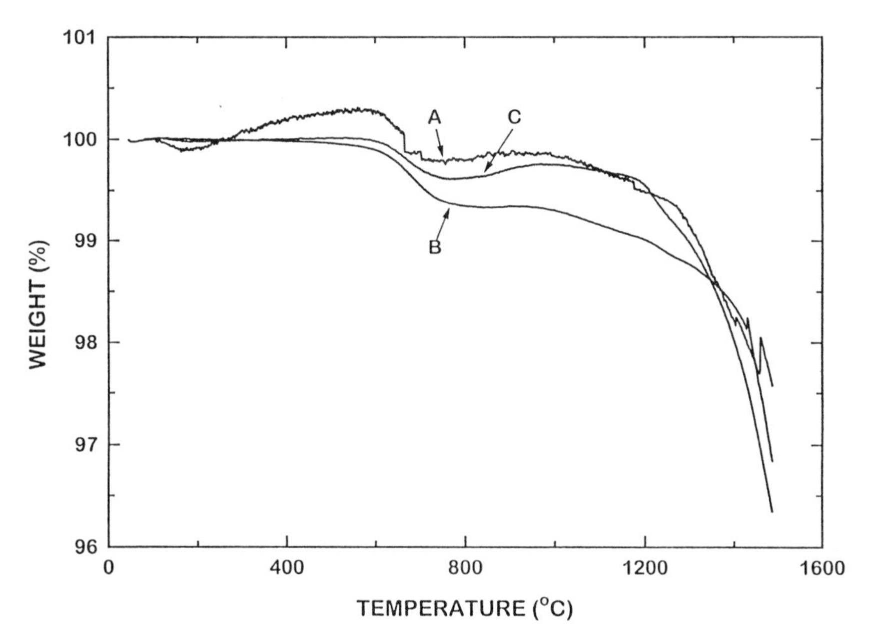

Figure 3. Oxidative stability of chars from heat treatment to 1000 °C of various mixtures of **1/3a**: (A) 10/90, (B) 25/75, and (C)50/50

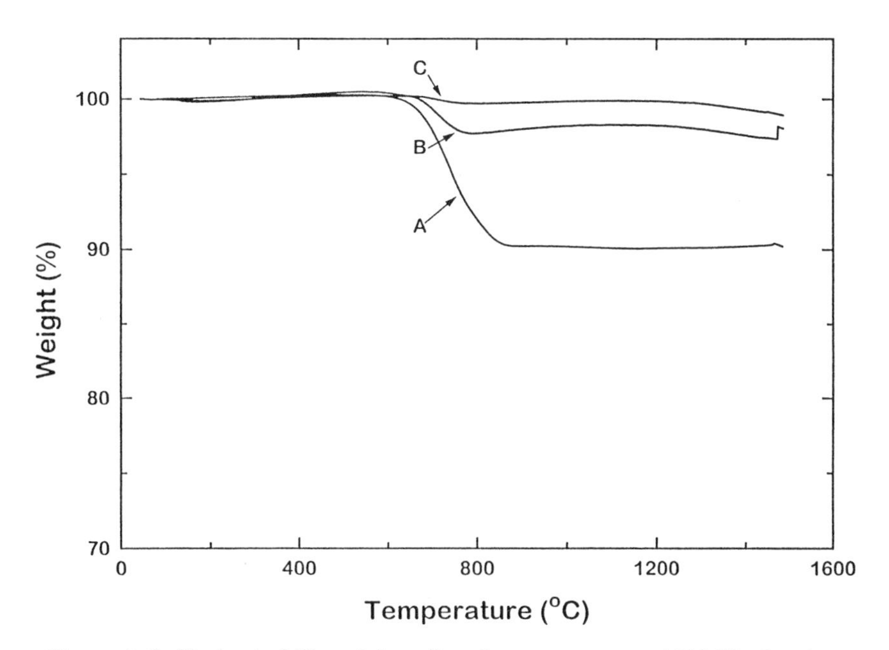

Figure 4. Oxidative stability of chars from heat treatment to 1500 °C of various mixtures of **1/3a**: (A) 10/90, (B) 25/75, and (C)50/50

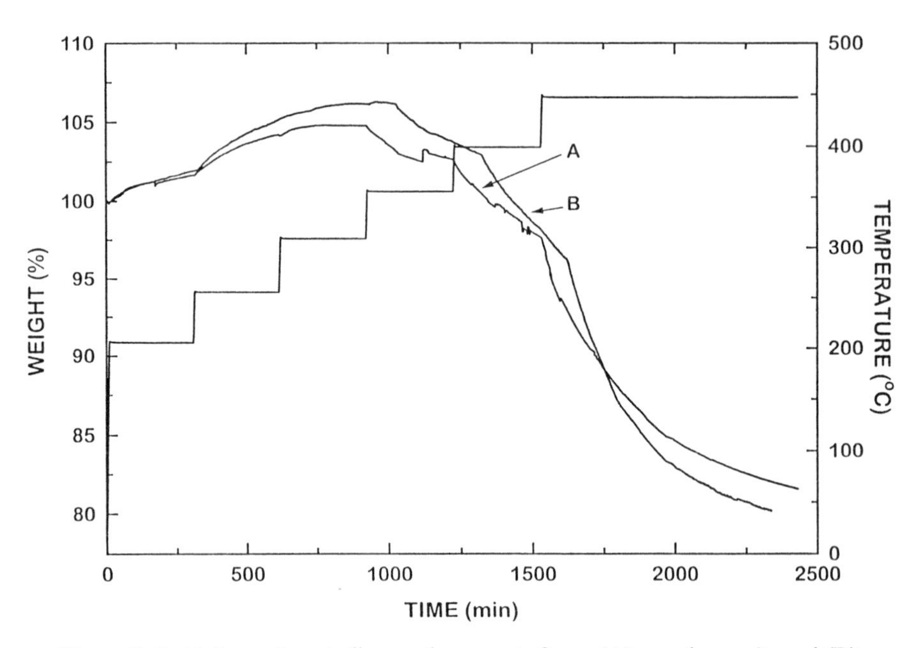

Figure 5. Oxidative aging studies on thermosets from: (A) copolymer **2c** and (B) 10/90 composition

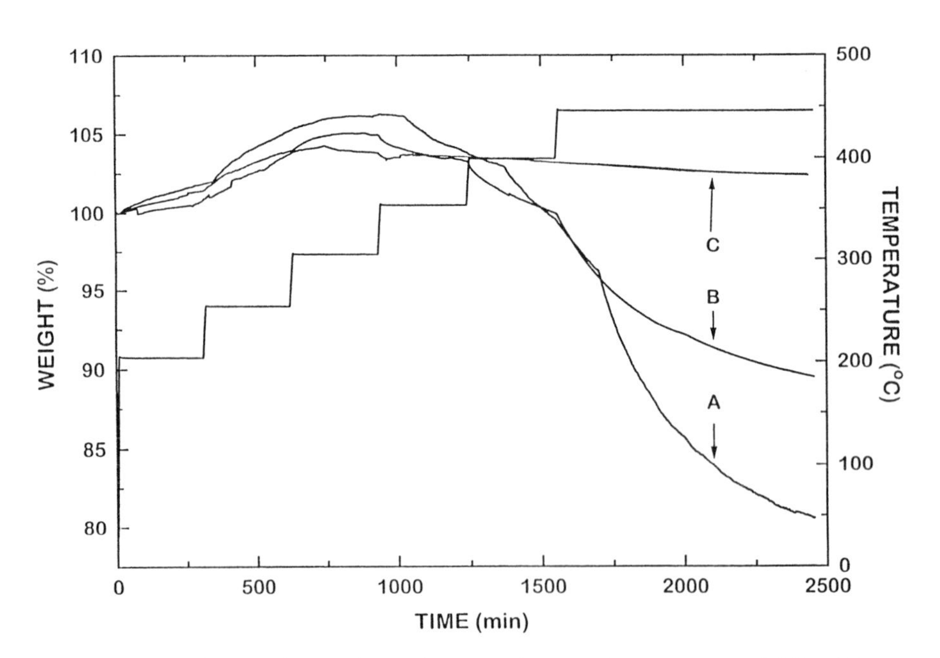

Figure 6. Oxidative aging study on thermoset from various mixtures of **1/3a**: (A) 10/90, (B) 25/75, and (C)50/50

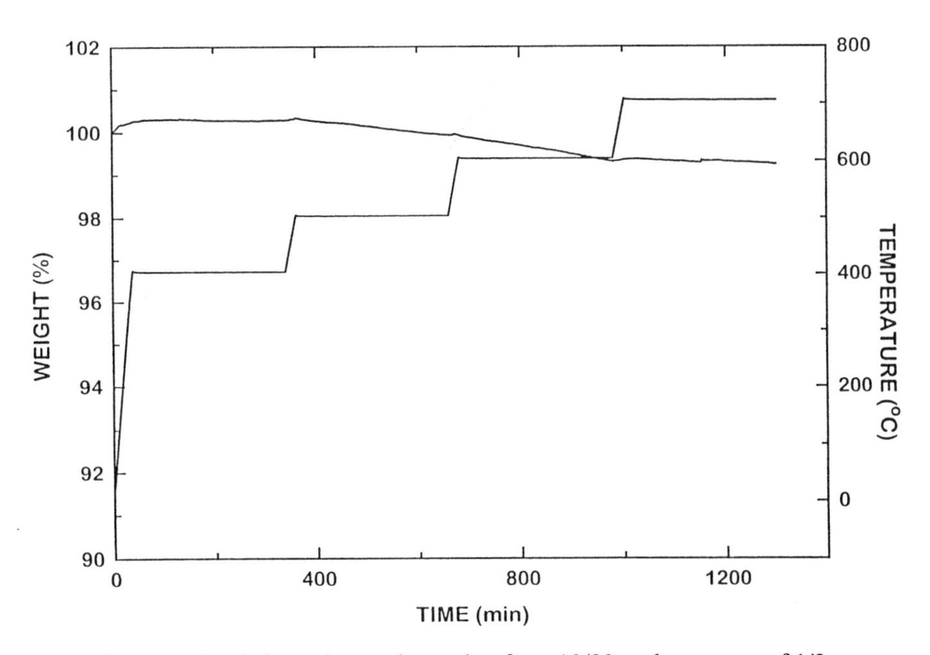

Figure 7. Oxidative aging study on char from 10/90 molar amount of **1/3a**

Long-term oxidative studies were also performed on the chars obtained from blending compositions of **1** and **3a**. A ceramic composition prepared from the 10/90 mixture was initially processed to 1000 °C under a nitrogen atmosphere. Upon cooling, the char was aged in sequence at 400, 500, 600, and 700 °C in a flow of air for 5 hours at each temperature (see Figure 7). While at 500 °C the char gained about 0.3% weight. During the entire heat treatment, the ceramic sample lost less than 1% weight. This weight loss occurred between 500 and 600 °C. At 700 °C no weight changes were observed. These observations indicate that a protective outer layer forms and insulates the interior against further oxidation. Similar results had been observed previously for the copolymer **2c**.[6] A charred sample that had been processed from **3a** in an identical manner as the blended mixture showed a 28% weight loss after 15 hours in air at 500 °C. The outer surface of the sample was coated with a white flaky residue attributed to silicon oxide. The outer surface of the chars formed from the blended mixtures upon exposure to air at elevated temperatures remained black. This observation indicates that a different outer oxidized surface forms with structural integrity when boron is present.

Conclusion

Extreme aging conditions show the importance of silicon and boron in the protection of carbon-based systems against oxidation. Thermoset and ceramic compositions formed from mixtures of **1** and **3a** show outstanding oxidative stability. Both compounds contain acetylenic units for thermal conversion to network polymers. The resistance to oxidation was a function of the amount of **1** present in the polymeric mixtures. The studies show that carbon can be protected from oxidation at various temperatures by proper incorporation of silicon and boron units into a carbon precursor material. The ceramic compositions obtained from **2** itself and mixture of **1** and **3a** show similar oxidative stabilities. Further studies are underway to evaluate and exploit the ceramic compositions as matrix materials for high temperature composites.

Acknowledgment is made to the Office of Naval Research for financial support of this work. David Y. Son wishes to acknowledge The National Research Council for an NRC Postdoctoral Fellowship.

Reference

1. Henderson, L. J.; Keller, T. M. *Macromolecules* **1994**, *27*, 1660.
2. Son, D. Y.; Keller, T. M. *J. Polym.Sci.: Part A: Polym. Chem.* **1995**, *33*, 2969.
3. Son, D. Y.; Keller, T. M. *Macromolecules* **1995**, *28*, 399.
4. Sastri, S. B.; Keller, T. M.; Jones, K. M.; Armistead, J. P. *Macromolecules*, **1993,** *26(23)*, 6171.
5. Son, D. Y.; Keller, T. M., *Fire and Polymers II*; Gordon L. Nelson; ACS Symposium Series 599, American Chemical Society, 1995 edition, Chapter 19, p. 280
6. Keller, T. M.; Son, D. Y., *Polymeric Materials Encyclopedia*; Joseph C. Salamone; CRC Press: New York 1996; Vol. 5(H-L), p. 3262

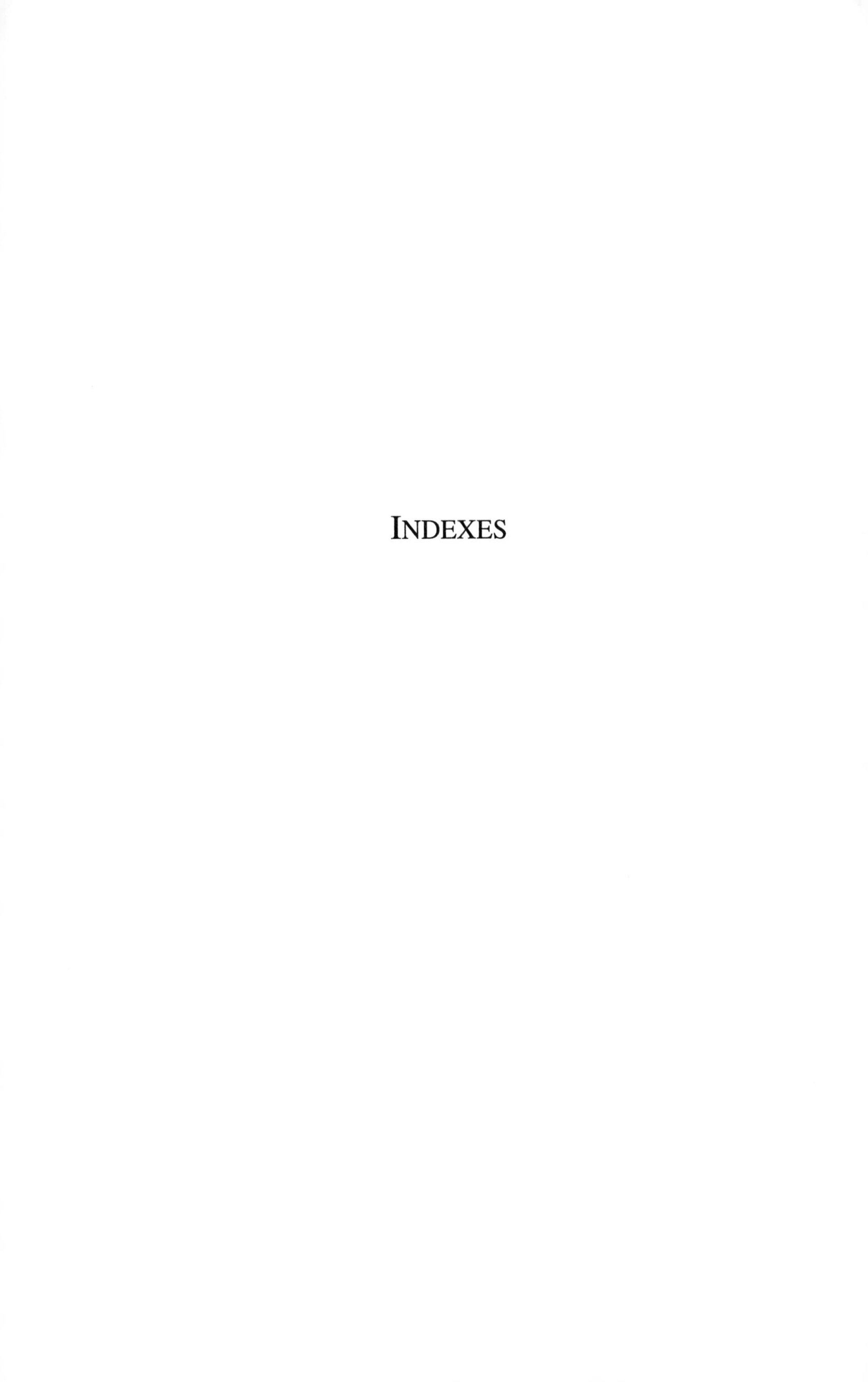

INDEXES

Author Index

Affiliation Index

Subject Index